"If you're not thinking like a Bayesian, perhaps you should be."
—John Allen Paulos, *New York Times Book Review*

"A rollicking tale of the triumph of a powerful mathematical tool."
—Andrew Robinson, *Nature*

"I recently finished reading *The Theory That Would Not Die* by Sharon McGrayne. Bayes's rule is a statistical theory that has a long and interesting history. It is important in decision making—how tightly should you hold on to your view and how much should you update your view based on the new information that's coming in. We intuitively use Bayes's rule every day."
—Alan B. Krueger, Bendheim Professor of Economics and Public Affairs at Princeton University and chair of President Obama's Council of Economic Advisers

"McGrayne . . . articulates difficult ideas in a way that the general public can understand and appreciate. . . . Readers needn't have a particular interest in statistics. . . . The book reads as if it were a love story—for an algorithm that grew up neglected, periodically taken out for a ride but mostly sitting home alone, until at long last, it finally found its rightful place of respect and appreciation in the world."
—Scott Brookhart, *IEEE Computing Now*

"An intellectual romp. . . . A masterfully researched tale of human struggle and accomplishment . . . renders perplexing mathematical debates digestible and vivid for even the most lay of audiences."
—Michael Washburn, *Boston Globe*

"[An] engrossing study. . . . A compelling and entertaining fusion of history, theory and biography."
—Ian Critchley, *Sunday Times*

"*The Theory That Would Not Die* is the first popular science book to document the rocky story of Bayes's rule. At times, her tale has everything you would expect of a modern-day thriller. . . . To have crafted a page-turner out of the history of statistics is an impressive feat. If only lectures at university had been this racy. . . . McGrayne's writing is luminous."
—David Robson, *New Scientist*

"Thorough research of the subject matter coupled with flowing prose, an impressive set of interviews with Bayesian statisticians, and an extremely engaging style in telling the personal stories of the few nonconformist heroes of the Bayesian school."
—Sam Behseta, *CHANCE*

"Carefully balanced to be accessible to a lay audience while captivating to a statistical one. Told without formulas, this eloquently written story is the history of an idea—a far-from-exhaustive, but enlightening, chronicle of the triumphs of Bayesian analysis."
—*CHANCE* interview with Sharon McGrayne

"A 'find.' . . . For the student who is being exposed to Bayesian statistics for the first time, McGrayne's book provides a wealth of illustrations to whet his or her appetite for more. It will broaden and deepen the field of reference of the more expert statistician, and the general reader will find an understandable, well-written, and fascinating account of a scientific field of great importance today."
—Andrew I. Dale, *Notices of the American Mathematical Society*

"The book is very non-technical . . . yet the reasoning behind the Rule resonates throughout. . . . A very engaging book that statisticians, probabilists, and history buffs in the mathematical sciences should enjoy."
—David Agard, *Cryptologia*

"Few ideas in the history of science have engendered so much controversy, religious fervor, revulsion, and enthusiasm as Bayes' Rule. . . . McGrayne does an admirable job of giving a voice to the scores of famous and non-famous people and data who contributed, for good or for worse. . . . The theory comes alive."
—Michele Bottone, *Significance Magazine*

"A book simply highlighting the astonishing 200 year controversy over Bayesian analysis would have been highly welcome. This book does so much more, however, uncovering the almost secret role of Bayesian analysis in a stunning series of the most important developments of the twentieth century. What a revelation and what a delightful read!"
—James Berger, Arts and Sciences Professor of Statistics, Duke University, and member, National Academy of Sciences

"I have just spent a very nice weekend reading a new book on the history of Bayesian Statistics, . . . a very compelling documented account, . . . very interesting reading."
—Jose Bernardo, *Valencia List Blog*

"It is a remarkable achievement. It taught me things, and it made me think. In fact, as a direct result of my interactions with Sharon I wrote a paper on the philosophy of statistics, which was published in *Statistical Science*. . . . This book succeeds gloriously, by never losing sight of the story, and it's a wonderful story, one that desperately deserved to be told."
—Robert E. Kass, Professor of Statistics, Carnegie Mellon University

"We now know how to think rationally about our uncertain world. This book describes in vivid prose, accessible to the lay person, the development of Bayes' rule over more than two hundred years from an idea to its widespread acceptance in practice."
—Dennis Lindley, author of *Understanding Uncertainty*

"A nice way to bring some of the rich theory—including some controversy—about statistical reasoning and Bayesian methods."
—Eric Horvitz, Microsoft Research

"This account of how a once reviled theory, Bayes' rule, came to underpin modern life is both approachable and engrossing."
—*Sunday Times*

"It does not seem an exaggeration to deem it [Bayes' rule] as significant in our times as the Darwinian theory of natural selection."
—Peter Pesic, *Times Literary Supplement*

"A lively, engaging historical account. . . . McGrayne describes actuarial, business, and military uses of the Bayesian approach, including its application to settle the disputed authorship of 12 of the Federalist Papers, and its use to connect cigarette smoking and lung cancer. . . . All of this is accomplished through compelling, fast-moving prose. . . . The reader cannot help but enjoy learning about some of the more gossipy episodes and outsized personalities."
—*Choice*

"Scientists and statisticians have fought over a deep philosophical divide about probability, which Sharon Bertsch McGrayne explores with great clarity and wit."
—Christine Evans-Pughe, *Engineering and Technology Magazine*

"A masterpiece and an invaluable contribution to the scientific literature, in general, and the actuarial literature in particular. . . . A treasure trove of citations to interesting items in the literature and biographical information on key figures in Bayesian history that we found extremely difficult to find prior to the arrival of this book."
—Tom Herzog, *Expanding-Horizons Newsletter*, Society of Actuaries

"McGrayne's book is not a textbook and does not attempt to *teach* Bayesian inferential techniques. Rather, McGrayne offers a very thorough, informative, and often entertaining (in our humble opinion) discussion of the Bayesian perspective. . . . Strongly recommended as it provides the theoretical underpinnings of the Bayesian perspective and shows how Bayesianism has been applied to *real world* inferential/statistical problems—often with great success."
—Jon Starkweather, *Go Forth and Propagate*

"McGrayne explains [it] beautifully. . . . Top holiday reading."
—*The Australian*

"Engaging. . . . Readers will be amazed at the impact that Bayes' rule has had in diverse fields, as well as by its rejection by too many statisticians. . . . I was brought up, statistically speaking, as what is called a frequentist . . . But reading McGrayne's book has made me determined to try, once again, to master the intricacies of Bayesian statistics. I am confident that other readers will feel the same."
—Michael Rawlins, M.D., *Lancet*

"A fascinating and engaging tale. . . . General readers may be quite amazed to learn how large a role mathematics and statistics have played in major events of our times. Much of the information the author has presented will be new to technical readers as well."
—Robert W. Hayden, *MAA Reviews*, Mathematical Association of America

"Recommended summer reading."
—Businessweek.com

"Fascinating. . . . I truly admire [McGrayne's] style of writing, and . . . ability to turn complex mathematical ideas into intriguing stories, centered around real people."
—Judea Pearl, Winner of the 2012 ACM Turing Award

the theory that would not die

other books by
    sharon bertsch mcgrayne

*Prometheans in the Lab: Chemistry and the Making
of the Modern World*

*Iron, Nature's Universal Element: Why People Need
Iron and Animals Make Magnets* (with Eugenie V.
Mielczarek)

*Nobel Prize Women in Science: Their Lives, Struggles,
and Momentous Discoveries*

the theory
that would
not die

how bayes' rule cracked
the enigma code,
hunted down russian
submarines, & emerged
triumphant from two
centuries of controversy

sharon bertsch mcgrayne

Yale UNIVERSITY PRESS   new haven & london

Yale University Press books may be purchased in quantity for educational, business, or promotional use. For information, please e-mail sales.press@yale.edu (U.S. office) or sales@yaleup.co .uk (U.K. office).

Designed by Lindsey Voskowsky. Set in Monotype Joanna type by Duke & Company, Devon, Pennsylvania. Printed in the United States of America.

The Library of Congress has cataloged the hardcover edition as follows:

McGrayne, Sharon Bertsch.
    The theory that would not die : how Bayes' rule cracked the enigma code, hunted down Russian submarines, and emerged triumphant from two centuries of controversy / Sharon Bertsch McGrayne.
        p. cm.
    Summary: "Bayes' rule appears to be a straightforward, one-line theorem: by updating our initial beliefs with objective new information, we get a new and improved belief. To its adherents, it is an elegant statement about learning from experience. To its opponents, it is subjectivity run amok. In the first-ever account of Bayes' rule for general readers, Sharon Bertsch McGrayne explores this controversial theorem and the human obsessions surrounding it. She traces its discovery by an amateur mathematician in the 1740s through its development into roughly its modern form by French scientist Pierre Simon Laplace. She reveals why respected statisticians rendered it professionally taboo for 150 years—at the same time that practitioners relied on it to solve crises involving great uncertainty and scanty information, even breaking Germany's Enigma code during World War II, and explains how the advent of off-the-shelf computer technology in the 1980s proved to be a game-changer. Today, Bayes' rule is used everywhere from DNA de-coding to Homeland Security. Drawing on primary source material and interviews with statisticians and other scientists, The Theory That Would Not Die is the riveting account of how a seemingly simple theorem ignited one of the greatest controversies of all time"—Provided by publisher.
    Includes bibliographical references and index.
    ISBN 978-0-300-16969-0 (hardback)
1. Bayesian statistical decision theory—History. I. Title.

QA279.5.M415 2011
519.5'42—dc22

                                2010045037

ISBN 978-0-300-18822-6 (pbk.)

A catalogue record for this book is available from the British Library.

10 9 8 7 6

When the facts change, I change my opinion. What do you do, sir?

—John Maynard Keynes

contents

preface to the
paperback edition

When I started work on this book ten years ago, Bayes' rule was almost unknown outside the ranks of specialists. I can still remember my delight when I searched the web for the word "Bayesian" in 2003 and found 100,000 websites. As of this writing, search engines produce 14 million hits. Scientific and technical interest in Bayes' rule has exploded. It's been a sea change.

But since this book's publication a little more than a year ago, Bayes has also emerged as a veritable icon of data-based decision-making and thinking.

John Allen Paulos advised readers of the *New York Times Book Review*, "If you're not thinking like a Bayesian, perhaps you should be."

Alan Krueger, chair of President Obama's Council of Economic Advisers, told an interviewer about the book: "Bayes's rule is a statistical theory that has a long and interesting history. It is important in decision making—how tightly should you hold on to your view and how much should you update your view based on the new information that's coming in. We intuitively use Bayes's rule every day."

Thus, although the word "Bayes" was too controversial to mention for decades of the twentieth century, it is no longer just for insiders-in-the-know.

For example, just as this book's hardcover edition was being released for publication in 2011, the French government formally credited Bayesian methods with locating the wreckage of Air France flight 447 after a two-year search for the plane. The remarkable story of that search is told in a new epilogue to this book.

Naturally, I am delighted with Bayes' coming of age. When I started work on the book, I saw the fabulous stories of Bayes' trials and successes as a way to tell the public about the emergence of a scientific revolution for dealing with knowledge. But almost immediately, Bayes' rule came to mean much more to me.

We seem to be living in a dogmatic age, and Bayes tells us that it's all right to start out with a half-baked idea about a situation. The Reverend Thomas Bayes himself said to begin with a "guess" and, in the absence of any other information, assign it 50–50 odds. But then Bayes' rule commits us to modifying our initial ideas every single time a new piece of information appears. And that unrelenting commitment to changing our minds in the face of new knowledge appeals to me deeply.

Critical reviews of *The Theory That Would Not Die* have been surprisingly enthusiastic. But many readers and reviewers asked for more mathematical problems where they could see Bayes at work. As a result, this edition includes an expanded appendix with several simple Bayesian math problems about hemophilia, the tragic Sally Clark miscarriage of justice, locating ships lost at sea, and the power of subjective priors in the absence of much data. Each problem illustrates a particular facet of Bayes' rule, and I would like to thank John Carlin, Daniel Gianola, Ray Hill, Albert Madansky, and Henry R. "Tony" Richardson for helping me with them.

Other readers asked for introductory reading material somewhere in difficulty between the narrative history of *The Theory That Would Not Die* and its 32-page scholarly bibliography. A list of reading suggestions has been added after the bibliography.

In closing, I would like to reiterate my thanks to all those who helped make this book possible and who encouraged it during its first year in public life.

Sharon Bertsch McGrayne
Seattle
August 2012

preface and
note to readers

In a celebrated example of science gone awry, geologists accumulated the evidence for Continental Drift in 1912 and then spent 50 years arguing that continents cannot move.

The scientific battle over Bayes' rule is less well known but lasted far longer, for 150 years. It concerned a broader and more fundamental issue: how we analyze evidence, change our minds as we get new information, and make rational decisions in the face of uncertainty. And it was not resolved until the dawn of the twenty-first century.

On its face Bayes' rule is a simple, one-line theorem: by updating our initial belief about something with objective new information, we get a new and improved belief. To its adherents, this is an elegant statement about learning from experience. Generations of converts remember experiencing an almost religious epiphany as they fell under the spell of its inner logic. Opponents, meanwhile, regarded Bayes' rule as subjectivity run amok.

Bayes' rule began life amid an inflammatory religious controversy in England in the 1740s: can we make rational conclusions about God based on evidence about the world around us? An amateur mathematician, the Reverend Thomas Bayes, discovered the rule, and we celebrate him today as the iconic father of mathematical decision making. Yet Bayes consigned his discovery to oblivion. In his time, he was a minor figure. And we know about his work today only because of his friend and editor Richard Price, an almost forgotten hero of the American Revolution.

By rights, Bayes' rule should be named for someone else: a Frenchman, Pierre Simon Laplace, one of the most powerful mathematicians and scientists in history. To deal with an unprecedented torrent of data, Laplace discovered the rule on his own in 1774. Over the next forty years he developed it into the form we use today. Applying his method, he concluded that a well-established fact—more boys are born than girls—was almost certainly the result of natural law. Only historical convention forces us to call Laplace's discovery Bayes' rule.

After Laplace's death, researchers and academics seeking precise and objective answers pronounced his method subjective, dead, and buried. Yet at the very same time practical problem solvers relied on it to deal with real-world emergencies. One spectacular success occurred during the Second World War, when Alan Turing developed Bayes to break Enigma, the German navy's secret code, and in the process helped to both save Britain and invent modern electronic computers and software. Other leading mathematical thinkers—Andrei Kolmogorov in Russia and Claude Shannon in New York—also rethought Bayes for wartime decision making.

During the years when ivory tower theorists thought they had rendered Bayes taboo, it helped start workers' compensation insurance in the United States; save the Bell Telephone system from the financial panic of 1907; deliver Alfred Dreyfus from a French prison; direct Allied artillery fire and locate German U-boats; and locate earthquake epicenters and deduce (erroneously) that Earth's core consists of molten iron.

Theoretically, Bayes' rule was *verboten*. But it could deal with all kinds of data, whether copious or sparse. During the Cold War, Bayes helped find a missing H-bomb and U.S. and Soviet submarines; investigate nuclear power plant safety; predict the shuttle *Challenger* tragedy; demonstrate that smoking causes lung cancer and that high cholesterol causes heart attacks; predict presidential winners on television's most popular news program, and much more.

How could otherwise rational scientists, mathematicians, and statisticians become so obsessed about a theorem that their argument became, as one observer called it, a massive food fight? The answer is simple. At its heart, Bayes runs counter to the deeply held conviction that modern science requires objectivity and precision. Bayes is a measure of belief. And it says that we can learn even from missing and inadequate data, from approximations, and from ignorance.

As a result of this profound philosophical disagreement, Bayes' rule is a flesh-and-blood story about a small group of beleaguered believers who struggled for legitimacy and acceptance for most of the twentieth century.

It's about how the rule's fate got entwined with the secrecy of the Second World War and the Cold War. It's about a theorem in want of a computer and a software package. And it's about a method that—refreshed by outsiders from physics, computer science, and artificial intelligence—was adopted almost overnight because suddenly it worked. In a new kind of paradigm shift for a pragmatic world, the man who had called one of the primary Bayesian methods "the crack cocaine of modern statistics. . . . seductive, addictive and destructive" began recruiting Bayesians for Google.

Today, Bayesian spam filters whisk pornographic and fraudulent e-mail to our computers' junk bins. When a ship sinks, the Coast Guard calls on Bayes and locates shipwrecked survivors who may have floated at sea for weeks. Scientists discover how genes are controlled and regulated. Bayes even wins Nobel Prizes. Online, Bayes' rule trawls the web and sells songs and films. It has penetrated computer science, artificial intelligence, machine learning, Wall Street, astronomy and physics, Homeland Security, Microsoft, and Google. It helps computers translate one language into another, tearing down the world's millennia-old Tower of Babel. It has become a metaphor for how our brains learn and function. Prominent Bayesians even advise government agencies on energy, education, and research.

But Bayes' rule is not just an obscure scientific controversy laid to rest. It affects us all. It's a logic for reasoning about the broad spectrum of life that lies in the gray areas between absolute truth and total uncertainty. We often have information about only a small part of what we wonder about. Yet we all want to predict something based on our past experiences; we change our beliefs as we acquire new information. After suffering years of passionate scorn, Bayes has provided a way of thinking rationally about the world around us.

This is the story of how that remarkable transformation took place.

Note: Observant readers may notice that I use the word "probability" a lot in this book. In common speech, most of us treat the words "probability," "likelihood," and "odds" interchangeably. In statistics, however, these terms are not synonymous; they have distinct and technical meanings. Because I've tried to use correct terminology in *The Theory That Would Not Die*, "probability" appears quite a bit.

acknowledgments

For their scientific advice and perspective and for their patience with someone who asked a multitude of questions, I am deeply indebted to Dennis V. Lindley, Robert E. Kass, and George F. Bertsch. These three also read and made perceptive comments about the entire book in one or more of its many drafts. I could not have written the book at all without my husband, George Bertsch.

For insightful guidance on various crucial threads of my narrative, I thank James O. Berger, David M. Blei, Bernard Bru, Andrew I. Dale, Arthur P. Dempster, Persi Diaconis, Bradley Efron, Stephen E. Fienberg, Stuart Geman, Roger Hahn, Peter Hoff, Tom J. Loredo, Albert Madansky, John W. Pratt, Henry R. ("Tony") Richardson, Christian P. Robert, Stephen M. Stigler, and David L. Wallace.

Many other experts and specialists spoke with me, often at length, about particular eras, problems, details, or people. They include Capt. Frank A. Andrews, Frank Anscombe, George Apostolakis, Robert A. and Shirley Bailey, Friedrich L. Bauer, Robert T. Bell, David R. Bellhouse, Julian Besag, Alan S. Blinder, George E. P. Box, David R. Brillinger, Bruce Budowle, Hans Bühlmann, Frank Carter, Herman Chernoff, Juscelino F. Colares, Jack Copeland, Ann Cornfield, Ellen Cornfield, John Piña Craven, Lorraine Daston, Philip Dawid, Joseph H. Discenza, Ralph Erskine, Michael Fortunato, Karl Friston, Chris Frith, John ("Jack") Frost, Dennis G. Fryback, Mitchell H. Gail, Alan E. Gelfand, Andrew Gelman, Edward I. George, Edgar N. Gilbert, Paul M. Goggans, I. J. "Jack" Good, Steven N. Goodman, Joel Greenhouse, Ulf Grenander, Gerald N. Grob, Thomas L. Hankins, Jeffrey E. Harris, W. Keith

Hastings, David Heckerman, Charles C. Hewitt Jr., Ray Hilborn, David C. Hoaglin, Antje Hoering, Marvin Hoffenberg, Susan P. Holmes, David Hounshell, Ronald H. Howard, David Howie, Bobby R. Hunt, Fred C. Iklé, David R. Jardini, William H. Jefferys, Douglas M. Jesseph.

Also, Michael I. Jordan, David Kahn, David H. Kaye, John G. King, Kenneth R. Koedinger, Daphne Koller, Tom Kratzke, James M. Landwehr, Bernard Lightman, Richard F. Link, Edward P. Loane, Michael C. Lovell, Thomas L. Marzetta, Scott H. Mathews, John McCullough, Robert L. Mercer, Richard F. Meyer, Glenn G. Meyers, Paul J. Miranti Jr., Deputy Commander Dewitt Moody, Rear Admiral Brad Mooney, R. Bradford Murphy, John W. Negele, Vice Admiral John "Nick" Nicholson, Peter Norvig, Stephen M. Pollock, Theodore M. Porter, Alexandre Pouget, S. James Press, Alan Rabinowitz, Adrian E. Raftery, Howard Raiffa, John J. Rehr, John T. Riedl, Douglas Rivers, Oleg Sapozhnikov, Peter Schlaifer, Arthur Schleifer Jr., Michael N. Shadlen, Edward H. ("Ted") Shortliffe, Edward H. Simpson, Harold C. Sox, David J. Spiegelhalter, Robert F. Stambaugh, Lawrence D. Stone, William J. Talbott, Judith Tanur, The Center for Defense Information, Sebastian Thrun, Oakley E. (Lee) Van Slyke, Gary G. Venter, Christopher Volinsky, Paul R. Wade, Jon Wakefield, Homer Warner, Frode Weierud, Robert B. Wilson, Wing H. Wong, Judith E. Zeh, and Arnold Zellner.

I would like to thank two outside reviewers, Jim Berger and Andrew Dale; both read the manuscript carefully and made useful comments to improve it.

Several friends and family members—Ruth Ann Bertsch, Cindy Vahey Bertsch, Fred Bertsch, Jean Colley, Genevra Gerhart, James Goodman, Carolyn Keating, Timothy W. Keller, Sharon C. Rutberg, Beverly Schaefer, and Audrey Jensen Weitkamp—made crucial comments. I owe thanks to the mathematics library staff of the University of Washington. And my agent, Susan Rabiner, and editor, William Frucht, were steadfast in their support.

Despite all this help, I am, of course, responsible for the errors in this book.

part I

enlightenment and the anti-bayesian reaction

causes in the air

Sometime during the 1740s, the Reverend Thomas Bayes made the ingenious discovery that bears his name but then mysteriously abandoned it. It was rediscovered independently by a different and far more renowned man, Pierre Simon Laplace, who gave it its modern mathematical form and scientific application—and then moved on to other methods. Although Bayes' rule drew the attention of the greatest statisticians of the twentieth century, some of them vilified both the method and its adherents, crushed it, and declared it dead. Yet at the same time, it solved practical questions that were unanswerable by any other means: the defenders of Captain Dreyfus used it to demonstrate his innocence; insurance actuaries used it to set rates; Alan Turing used it to decode the German Enigma cipher and arguably save the Allies from losing the Second World War; the U.S. Navy used it to search for a missing H-bomb and to locate Soviet subs; RAND Corporation used it to assess the likelihood of a nuclear accident; and Harvard and Chicago researchers used it to verify the authorship of the Federalist Papers. In discovering its value for science, many supporters underwent a near-religious conversion yet had to conceal their use of Bayes' rule and pretend they employed something else. It was not until the twenty-first century that the method lost its stigma and was widely and enthusiastically embraced. The story began with a simple thought experiment.

Because Bayes' gravestone says he died in 1761 at the age of 59, we know he lived during England's struggle to recover from nearly two centuries of religious strife, civil war, and regicide. As a member of the Presbyterian Church, a religious denomination persecuted for refusing to support the Church of England, he was considered a Dissenter or Non-Conformist. During his grandfather's generation, 2,000 Dissenters died in English prisons. By Bayes' time, mathematics was split along religious and political lines, and many productive mathematicians were amateurs because, as Dissenters, they were barred from English universities.[1]

Unable to earn a degree in England, Bayes studied theology and presumably mathematics at the University of Edinburgh in Presbyterian Scotland, where, happily for him, academic standards were much more rigorous. In 1711 he left for London, where his clergyman father ordained him and apparently employed him as an assistant minister.

Persecution turned many English Dissenters into feisty critics, and in his late 20s Bayes took a stand on a hot theological issue: can the presence of evil be reconciled with God's presumed benefice? In 1731 he wrote a pamphlet—a kind of blog—declaring that God gives people "the greatest happiness of which they are capable."

During his 40s, Bayes' interests in mathematics and theology began to tightly intertwine. An Irish-Anglican bishop—George Berkeley, for whom the University of California's flagship campus is named—published an inflammatory pamphlet attacking Dissenting mathematicians, calculus, abstract mathematics, the revered Isaac Newton, and all other "free-thinkers" and "infidel mathematicians" who believed that reason could illuminate any subject. Berkeley's pamphlet was the most spectacular event in British mathematics during the 1700s.

Leaping into the pamphlet wars again, Bayes published a piece defending and explaining Newton's calculus. This was his only mathematical publication during his lifetime. Shortly thereafter, in 1742, five men, including a close friend of Newton's, nominated Bayes for membership in the Royal Society. His nomination avoided any hint of controversy and described him as "a Gentleman of known merit, well skilled in Geometry and all parts of Mathematical and Philosophical Learning." The Royal Society was not the professional organization it is today; it was a private organization of dues-paying amateurs from the landed gentry. But it played a vital role because amateurs would produce some of the era's breakthroughs.

About this time Bayes joined a second group of up-to-date amateur mathematicians. He had moved to a small congregation in a fashionable resort, the cold-water spa Tunbridge Wells. As an independently wealthy bachelor—his family had made a fortune manufacturing Sheffield steel cutlery—he rented rooms, apparently from a Dissenting family. His religious duties—one Sunday sermon a week—were light. And spa etiquette permitted Dissenters, Jews, Roman Catholics, and even foreigners to mix with English society, even with wealthy earls, as they could not elsewhere.

A frequent visitor to Tunbridge Wells, Philip, the Second Earl of Stanhope, had been passionately interested in mathematics since childhood, but his guardian had banned its study as insufficiently genteel. When Stanhope was 20 and free to do as he liked, he seldom raised his eyes from Euclid. According to the bluestocking Elizabeth Montagu, Stanhope was "always making mathematical scratches in his pocket-book, so that one half the people took him for a conjuror, and the other half for a fool." Because of either his aristocratic position or his late start Stanhope never published anything of his own. Instead, he became England's foremost patron of mathematicians.

The earl and the Royal Society's energetic secretary, John Canton, operated an informal network of peer reviewers who critiqued one another's work. At some point Bayes joined the network. One day, for example, Stanhope sent Bayes a copy of a draft paper by a mathematician named Patrick Murdoch. Bayes disagreed with some of it and sent his comments back to Stanhope, who forwarded them to Murdoch, who in turn replied through Stanhope, and so on around and around. The relationship between the young earl and the older Reverend Bayes seems to have ripened into friendship, however, because Stanhope paid Bayes at least one personal visit at Tunbridge Wells, saved two bundles of his mathematical papers in the Stanhope estate's library, and even subscribed to his series of sermons.

Another incendiary mix of religion and mathematics exploded over England in 1748, when the Scottish philosopher David Hume published an essay attacking some of Christianity's fundamental narratives. Hume believed that we can't be absolutely certain about anything that is based only on traditional beliefs, testimony, habitual relationships, or cause and effect. In short, we can rely only on what we learn from experience.

Because God was regarded as the First Cause of everything, Hume's skepticism about cause-and-effect relationships was especially unsettling. Hume argued that certain objects are constantly associated with each other. But the fact that umbrellas and rain appear together does not mean that umbrellas cause rain. The fact that the sun has risen thousands of times does not guarantee that it will do so the next day. And, most important, the "design of the world" does not prove the existence of a creator, an ultimate cause. Because we can seldom be certain that a particular cause will have a particular effect, we must be content with finding only probable causes and probable effects. In criticizing concepts about cause and effect Hume was undermining Christianity's core beliefs.

Hume's essay was nonmathematical, but it had profound scientific implications. Many mathematicians and scientists believed fervently that natural laws did indeed prove the existence of God, their First Cause. As an eminent mathematician Abraham de Moivre wrote in his influential book *Doctrine of Chances*, calculations about natural events would eventually reveal the underlying order of the universe and its exquisite "Wisdom and Design."

With Hume's doubts about cause and effect swirling about, Bayes began to consider ways to treat the issue mathematically. Today, probability, the mathematics of uncertainty, would be the obvious tool, but during the early 1700s probability barely existed. Its only extensive application was to gambling, where it dealt with such basic issues as the odds of getting four aces in one poker hand. De Moivre, who had spent several years in French prisons because he was a Protestant, had already solved that problem by working from cause to effect. But no one had figured out how to turn his work around backward to ask the so-called inverse question from effect to cause: what if a poker player deals himself four aces in each of three consecutive hands? What is the underlying chance (or cause) that his deck is loaded?

We don't know precisely what piqued Bayes' interest in the inverse probability problem. He had read de Moivre's book, and the Earl of Stanhope was interested in probability as it applied to gambling. Alternatively, it could have been the broad issues raised by Newton's theory of gravitation. Newton, who had died 20 years before, had stressed the importance of relying on observations, developed his theory of gravitation to explain them, and then used his theory to predict new observations. But Newton had not explained the *cause* of gravity or wrestled with the problem of how true his theory might be. Finally, Bayes' interest may have been stimulated by Hume's philosophical essay. In any event, problems involving cause and effect and uncertainty filled the air, and Bayes set out to deal with them quantitatively.

Crystallizing the essence of the inverse probability problem in his mind, Bayes decided that his goal was to learn the approximate probability of a future event he knew nothing about except its past, that is, the number of times it had occurred or failed to occur. To quantify the problem, he needed a number, and sometime between 1746 and 1749 he hit on an ingenious solution. As a starting point he would simply invent a number—he called it a guess—and refine it later as he gathered more information.

Next, he devised a thought experiment, a 1700s version of a computer simulation. Stripping the problem to its basics, Bayes imagined a square table so level that a ball thrown on it would have the same chance of landing on one spot as on any other. Subsequent generations would call his construction a billiard table, but as a Dissenting minister Bayes would have disapproved of such games, and his experiment did not involve balls bouncing off table edges or colliding with one another. As he envisioned it, a ball rolled randomly on the table could stop with equal probability anywhere.

We can imagine him sitting with his back to the table so he cannot see anything on it. On a piece of paper he draws a square to represent the surface of the table. He begins by having an associate toss an imaginary cue ball onto the pretend tabletop. Because his back is turned, Bayes does not know where the cue ball has landed.

Next, we picture him asking his colleague to throw a second ball onto the table and report whether it landed to the right or left of the cue ball. If to the left, Bayes realizes that the cue ball is more likely to sit toward the right side of the table. Again Bayes' friend throws the ball and reports only whether it lands to the right or left of the cue ball. If to the right, Bayes realizes that the cue can't be on the far right-hand edge of the table.

He asks his colleague to make throw after throw after throw; gamblers and mathematicians already knew that the more times they tossed a coin, the more trustworthy their conclusions would be. What Bayes discovered is that, as more and more balls were thrown, each new piece of information made his imaginary cue ball wobble back and forth within a more limited area.

As an extreme case, if all the subsequent tosses fell to the right of the first ball, Bayes would have to conclude that it probably sat on the far left-hand margin of his table. By contrast, if all the tosses landed to the left of the first ball, it probably sat on the far right. Eventually, given enough tosses of the ball, Bayes could narrow the range of places where the cue ball was apt to be.

Bayes' genius was to take the idea of narrowing down the range of positions for the cue ball and—based on this meager information—infer that it had landed somewhere between two bounds. This approach could not produce a right answer. Bayes could never know precisely where the cue ball landed, but he could tell with increasing confidence that it was most probably within a particular range. Bayes' simple, limited system thus moved from observations about the world back to their probable origin or cause. Using his knowledge of the present (the left and right positions of the tossed balls), Bayes had figured out how to say something about the past (the position of the first ball). He could even judge how confident he could be about his conclusion.

Conceptually, Bayes' system was simple. We modify our opinions with objective information: Initial Beliefs (our guess where the cue ball landed) + Recent Objective Data (whether the most recent ball landed to the left or right of our original guess) = A New and Improved Belief. Eventually, names were assigned to each part of his method: Prior for the probability of the initial belief; Likelihood for the probability of other hypotheses with objective new data; and Posterior for the probability of the newly revised belief. Each time the system is recalculated, the posterior becomes the prior of the new iteration. It was an evolving system, which each new bit of information pushed closer and closer to certitude. In short:

Prior times likelihood is proportional to the posterior.

(In the more technical language of the statistician, the likelihood is the probability of the data for a given hypothesis. However, Andrew Dale, a South African historian of statistics, simplified the matter considerably when he observed, "Put somewhat rudely, the likelihood is what remains of Bayes's Theorem once the prior is removed from the discussion.")[2]

As a special case about balls thrown randomly onto a flat table, Bayes' rule is uncontroversial. But Bayes wanted to cover *every* case involving uncertainty, even cases where nothing whatsoever was known about their history—in his words, where we "absolutely know nothing antecedently to any trials."[3] This expansion of his table experiment to cover any uncertain situation would trigger 150 years of misunderstanding and bitter attacks.

Two especially popular targets for attack were Bayes' guesswork and his suggested shortcut.

First, Bayes *guessed* the likely value of his initial belief (the cue ball's position, later known as the prior). In his own words, he decided to make "a guess whereabouts it's [sic] probability is, and . . . [then] see the chance that the guess is right." Future critics would be horrified at the idea of using a mere hunch—a subjective belief—in objective and rigorous mathematics.

Even worse, Bayes added that if he *did not know enough* to distinguish the position of the balls on his table, he would assume they were equally likely to fall anywhere on it. Assuming equal probabilities was a pragmatic approach for dealing with uncertain circumstances. The practice was rooted in traditional Christianity and the Roman Catholic Church's ban on usury. In uncertain situations such as annuities or marine insurance policies, all parties were assigned equal shares and divided profits equally. Even prominent mathematicians assigned equal probabilities to gambling odds by assuming, with a remarkable lack of realism, that all tennis players or fighting cocks were equally skillful.

In time, the practice of assigning equal probabilities acquired a number of names, including equal priors, equal *a priori*'s, equiprobability, uniform distribution probability, and the law of insufficient reason (meaning that without enough data to assign specific probabilities, equal ones would suffice). Despite their venerable history, equal probabilities would become a lightning rod for complaints that Bayes was quantifying ignorance.

Today, some historians try to absolve him by saying he may have applied equal probabilities to his data (the subsequent throws) rather than to the initial, so-called prior toss. But this is also guesswork. And for many working statisticians, the question is irrelevant because in the tightly circumscribed case of balls that can roll anywhere on a carefully leveled surface both produce the same mathematical results.

Whatever Bayes meant, the damage was done. For years to come, the message seemed clear: priors be damned. At this point, Bayes ended his discussion.

He may have mentioned his discovery to others. In 1749 someone told a physician named David Hartley something that sounds suspiciously like Bayes' rule. Hartley was a Royal Society member who believed in cause-and-effect relationships. In 1749 he wrote that "an ingenious friend has communicated to me a Solution of the inverse problem . . . which shews that we may hope to determine the Proportions, and by degrees, the whole Nature, of unknown Causes, by a sufficient Observation of their Effects." Who was this ingenious friend? Modern-day sleuths have suggested Bayes or Stanhope, and in 1999 Stephen M. Stigler of the University of Chicago suggested that Nicholas Saunderson, a blind Cambridge mathematician, made the discovery instead of Bayes. No matter who talked about it, it seems highly unlikely that anyone other than Bayes made the breakthrough. Hartley used terminology that is almost identical to Bayes' published essay, and no one who read the article between its publication in 1764 and 1999 doubted Bayes' authorship. If there had been any question about the author's identity, it is hard to imagine Bayes' editor or his publisher not saying something publicly. Thirty years later Price was still referring to the work as that of Thomas Bayes.

Although Bayes' idea was discussed in Royal Society circles, he himself seems not to have believed in it. Instead of sending it off to the Royal Society for publication, he buried it among his papers, where it sat for roughly a decade. Only because he filed it between memoranda dated 1746 and 1749 can we conclude that he achieved his breakthrough sometime during the late 1740s, perhaps shortly after the publication of Hume's essay in 1748.

Bayes' reason for suppressing his essay can hardly have been fear of controversy; he had plunged twice into Britain's pamphlet wars. Perhaps he thought his discovery was useless; but if a pious clergyman like Bayes thought his work could prove the existence of God, surely he would have published it. Some thought Bayes was too modest. Others wondered whether he was unsure about his mathematics. Whatever the reason, Bayes made an important contribution to a significant problem—and suppressed it. It was the first of several times that "Bayes' rule" would spring to life only to disappear again from view.

Bayes' discovery was still gathering dust when he died in 1761. At that point relatives asked Bayes' young friend Richard Price to examine Bayes' mathematical papers.

Price, another Presbyterian minister and amateur mathematician, achieved fame later as an advocate of civil liberties and of the American and French revolutions. His admirers included the Continental Congress, which asked him to emigrate and manage its finances; Benjamin Franklin, who nominated him for the Royal Society; Thomas Jefferson, who asked him to write to Virginia's youths about the evils of slavery; John Adams and the feminist Mary Wollstonecraft, who attended his church; the prison reformer John Howard, who was his best friend; and Joseph Priestley, the discoverer of oxygen, who said, "I question whether Dr. Price ever had a superior." When Yale University conferred two honorary degrees in 1781, it gave one to George Washington and the other to Price. An English magazine thought Price would go down in American history beside Franklin, Washington, Lafayette, and Paine. Yet today Price is known primarily for the help he gave his friend Bayes.

Sorting through Bayes' papers, Price found "an imperfect solution of one of the most difficult problems in the doctrine of chances." It was Bayes' essay on the probability of causes, on moving from observations about the real world back to their most probable cause.

At first Price saw no reason to devote much time to the essay. Mathematical infelicities and imperfections marred the manuscript, and it looked impractical. Its continual iterations—throwing the ball over and over again and recalculating the formula each time—produced large numbers that would be difficult to calculate.

But once Price decided Bayes' essay was the answer to Hume's attack on causation, he began preparing it for publication. Devoting "a good deal of labour" to it on and off for almost two years, he added missing references and citations and deleted extraneous background details in Bayes' derivations. Lamentably, he also threw out his friend's introduction, so we'll never know precisely how much the edited essay reflects Bayes' own thinking.

In a cover letter to the Royal Society, Price supplied a religious reason for publishing the essay. In moving mathematically from observations of the natural world inversely back to its ultimate cause, the theorem aimed to show that "the world must be the effect of the wisdom and power of an intelligent cause; and thus to confirm . . . from final causes . . . the existence of the Deity." Bayes himself was more reticent; his part of the essay does not mention God.

A year later the Royal Society's *Philosophical Transactions* published "An Essay toward solving a Problem in the Doctrine of Chances." The title avoided religious controversy by highlighting the method's gambling applications. Critiquing Hume a few years later, Price used Bayes for the first and only time. As far as we know, no one else mentioned the essay for the next 17 years, when Price would again bring Bayes' rule to light.

By modern standards, we should refer to the Bayes-Price rule. Price discovered Bayes' work, recognized its importance, corrected it, contributed to the article, and found a use for it. The modern convention of employing Bayes' name alone is unfair but so entrenched that anything else makes little sense.

Although it was ignored for years, Bayes' solution to the inverse probability of causes was a masterpiece. He transformed probability from a gambler's measure of frequency into a measure of informed belief. A card player could start by believing his opponent played with a straight deck and then modify his opinion each time a new hand was dealt. Eventually, the gambler could wind up with a better assessment of his opponent's honesty.

Bayes combined judgments based on prior hunches with probabilities based on repeatable experiments. He introduced the signature features of Bayesian methods: an initial belief modified by objective new information. He could move from observations of the world to abstractions about their probable cause. And he discovered the long-sought grail of probability, what future mathematicians would call the probability of causes, the principle of inverse probability, Bayesian statistics, or simply Bayes' rule.

Given the revered status of his work today, it is also important to recognize what Bayes did not do. He did not produce the modern version of Bayes' rule. He did not even employ an algebraic equation; he used Newton's old-fashioned geometric notation to calculate and add areas. Nor did he develop his theorem into a powerful mathematical method. Above all, unlike Price, he did not mention Hume, religion, or God.

Instead, he cautiously confined himself to the probability of events and did not mention hypothesizing, predicting, deciding, or taking action. He did not suggest possible uses for his work, whether in theology, science, or social science. Future generations would extend Bayes' discovery to do all these things and to solve a myriad of practical problems. Bayes did not even name his breakthrough. It would be called the probability of causes or inverse probability for the next 200 years. It would not be named Bayesian until the 1950s.

In short, Bayes took the first steps. He composed the prelude for what was to come.

For the next two centuries few read the Bayes-Price article. In the end, this is the story of two friends, Dissenting clergymen and amateur mathematicians, whose labor had almost no impact. Almost, that is, except on the one person capable of doing something about it, the great French mathematician Pierre Simon Laplace.

the man who
did everything

Just across the English Channel from Tunbridge Wells, about the time that Thomas Bayes was imagining his perfectly smooth table, the mayor of a tiny village in Normandy was celebrating the birth of a son, Pierre Simon Laplace, the future Einstein of his age.

Pierre Simon, born on March 23, 1749, and baptized two days later, came from several generations of literate and respected dignitaries. His mother's relatives were well-to-do farmers, but she died when he was young, and he never referred to her. His father kept the stagecoach inn in picturesque Beaumont-en-Auge, was a leader of the community's 472 inhabitants, and served 30 years as mayor. By the time Pierre Simon was a teenager his father seems to have been his only close relative. In years to come Pierre Simon's decision to become a mathematician would shatter their relationship almost irretrievably.[1]

Fortunately for the boy there was never any question about his getting an education. Attending school was becoming the norm in France in the 1700s, an enormous revolution fueled by the Catholic Church's fight against Protestant heresy and by parents convinced that education would enrich their children spiritually, intellectually, and financially. The question was, what kind of schooling?

Decades of religious warfare between Protestants and Catholics and several horrendous famines caused by cold weather had made France a determinedly secular country intent on developing its resources. Pierre Simon could have studied modern science and geometry in one of the country's many new secular schools. Instead, the elder Laplace enrolled his son in a local primary and secondary school where Benedictine monks produced clergy for the church and soldiers, lawyers, and bureaucrats for the crown. Thanks to the patronage of the Duke of Orleans, local day students like Pierre Simon attended free. The curriculum was conservative and Latin-based, heavy on copying, memorization, and philosophy. But it left Laplace with a fabulous memory and almost unbelievable perseverance.

Although the monks probably did not know it, they were competing with the French Enlightenment for the child's attention. Contemporaries called it the Century of Lights and the Age of Science and Reason, and the popularization of science was its most important intellectual phenomenon. Given the almost dizzying curiosity of the times, it is not surprising that, shortly after his tenth birthday, Pierre Simon was profoundly affected by a spectacular scientific prediction.[2]

Decades before, the English astronomer Edmond Halley had predicted the reappearance of the long-tailed comet that now bears his name. A trio of French astronomers, Alexis Claude Clairaut, Joseph Lalande, and Nicole-Reine Lepaute, the wife of a celebrated clockmaker, solved a difficult three-body problem and discovered that the gravitational pull of Jupiter and Saturn would delay the arrival of Halley's comet. The French astronomers accurately pinpointed the date—mid-April 1759 plus or minus a month—when Europeans would be able to see the comet returning from its orbit around the sun. The comet's appearance on schedule and on course electrified Europeans. Years later Laplace said it was the event that made his generation realize that extraordinary events like comets, eclipses, and severe droughts were caused not by divine anger but by natural laws that mathematics could reveal.

Laplace's extraordinary mathematical ability may not yet have been apparent when he turned 17 in 1766, because he did not go to the University of Paris, which had a strong science faculty. Instead he went to the University of Caen, which was closer to home and had a solid theological program suitable for a future cleric.

Yet even Caen had mathematical firebrands offering advanced lectures on differential and integral calculus. While English mathematicians were getting mired in Newton's awkward geometric version of calculus, their rivals on the Continent were using Gottfried Leibniz's more supple algebraic calculus. With it, they were forming equations and discovering a fabulous wealth of enticing new information about planets, their masses and details of their orbits. Laplace emerged from Caen a swashbuckling mathematical virtuoso eager to take on the scientific world. He had also become, no doubt to his father's horror, a religious skeptic.

At graduation Laplace faced an anguishing dilemma. His master's degree permitted him to take either the priestly vows of celibacy or the title of abbé, signifying a low-ranking clergyman who could marry and inherit property. Abbés did not have good reputations; Voltaire called them "that indefinable being which is neither ecclesiastic nor secular . . . young men, who are known for their debauchery."[3] An engraving of the period, "What Does the Abbé Think of It?" shows the clergyman peering appreciatively down a lady's bosom as she dresses.[4] Still, the elder Laplace wanted his son to become a clergyman.

If Laplace had been willing to become an abbé, his father might have helped him financially, and Laplace could have combined church and science. A number of abbés supported themselves in science, the most famous being Jean Antoine Nollet, who demonstrated spectacular physics experiments to the paying public. For the edification of the king and queen of France, Nollet sent a charge of static electricity through a line of 180 soldiers to make them leap comically into the air. Two abbés were even elected to the prestigious Royal Academy of Sciences. Still, the lot of most abbé-scientists was neither lucrative nor intellectually challenging. The majority found low-level jobs tutoring the sons of rich nobles or teaching elementary mathematics and science in secondary schools. University-level opportunities were limited because during the 1700s professors transmitted knowledge from the past instead of doing original research.

But Caen had convinced Laplace that he wanted to do something quite new. He wanted to be a full-time, professional, secular, mathematical researcher. And he wanted to explore the new algebra-generated, data-rich world of science. To his father, an ambitious man in bucolic France, a career in mathematics must have seemed preposterous.

Young Laplace made his move in the summer of 1769, shortly after completing his studies at Caen. He left Normandy and traveled to Paris, clutching a letter of recommendation to Jean Le Rond d'Alembert, the most powerful mathematician of the age, one of Europe's most notorious anticlerics, and the object of almost incessant Jesuit attacks. D'Alembert was a star of the Enlightenment and the chief spokesman for the *Encyclopédie*, which was making an enormous body of empirical knowledge universally available, scientific, and free of religious dogma. By throwing in his lot with d'Alembert, Laplace effectively cut his ties to the Catholic Church. We can only imagine his father's reaction, but we know that Laplace did not return home for 20 years and did not attend the old man's funeral.

Once in Paris, Laplace immediately approached the great d'Alembert and showed him a four-page student essay on inertia. Years later Laplace could still recite passages from it. Although besieged by applicants, d'Alembert was so impressed that within days he had arranged a paying job for Laplace as an instructor of mathematics at the new secular, mathematics-based Royal Military School for the younger sons of minor nobles. The school, located behind Les Invalides in Paris, provided Laplace with a salary, housing, meals, and money for wood to heat his room in winter. It was precisely the kind of job he had hoped to avoid.

Laplace could have tried to find work applying mathematics to practical problems in one of the monarchy's numerous research establishments or manufacturing plants. Many mathematically talented young men from modest families were employed in such institutions. But Laplace and his mentor were aiming far higher. Laplace wanted the challenge of doing basic research full time. And to do that, as d'Alembert must have told him, he had to get elected to the Royal Academy of Sciences.

In striking contrast to the amateurism of the Royal Society of London, the French Royal Academy of Sciences was the most professional scientific institution in Europe. Although aristocratic amateurs could become honorary members, the organization's highest ranks were composed of working scientists chosen by merit and paid to observe, collect, and investigate facts free of dogma; to publish their findings after peer review; and to advise the government on technical issues like patents. To augment their low salaries, academicians could use their prestige to cobble together various part-time jobs.

Without financial support from the church or his father, however, Laplace had to work fast. Since most academy members were chosen on the basis of a long record of solid accomplishment, he would have to be elected over the heads of more senior men. And for that to happen, he needed to make a spectacular impact.

D'Alembert, who had made Newton's revolution the focus of French mathematics, urged Laplace to concentrate on astronomy. D'Alembert had a clear problem in mind.

Over the previous two centuries mathematical astronomy had made great strides. Nicolaus Copernicus had moved Earth from the center of the solar system to a modest but accurate position among the planets; Johannes Kepler had connected the celestial bodies by simple laws; and Newton had introduced the concept of gravity. But Newton had described the motions of heavenly bodies roughly and without explanation. His death in 1727 left Laplace's generation an enormous challenge: showing that gravitation was not a hypothesis but a fundamental law of nature.

Astronomy was the era's most quantified and respected science, and only it could test Newton's theories by explaining precisely how gravitation affects the movements of tides, interacting planets and comets, our moon, and the shape of Earth and other planets. Forty years of empirical data had been collected, but, as d'Alembert warned, a single exception could bring the entire edifice tumbling down.

The burning scientific question of the day was whether the universe was stable. If Newton's gravitational force operates throughout the universe, why don't the planets collide with each other and cause the cosmic Armageddon described in the biblical book of Revelation? Was the end of the world at hand?

Astronomers had long been aware of alarming evidence suggesting that the solar system was inherently unstable. Comparing the actual positions of the most remote known planets with centuries-old astronomical observations, they could see that Jupiter was slowly accelerating in its orbit around the sun while Saturn was slowing down. Eventually, they thought, Jupiter would smash into the sun, and Saturn would spin off into space. The problem of predicting the motions of many interacting bodies over long periods of time is complex even today, and Newton concluded that God's miraculous intervention kept the heavens in equilibrium. Responding to the challenge, Laplace decided to make the stability of the universe his lifework. He said his tool would be mathematics and it would be like a telescope in the hands of an astronomer.

For a short time Laplace actually considered modifying Newton's theory by making gravity vary with a body's velocity as well as with its mass and distance. He also wondered fleetingly whether comets might be disturbing the orbits of Jupiter and Saturn. But he changed his mind almost immediately. The problem was not Newton's theory. The problem was the data astronomers used.

Newton's system of gravitation could be accepted as true only if it agreed with precise measurements, but observational astronomy was awash with information, some of it uncertain and inadequate. Working on the problem of Jupiter and Saturn, for example, Laplace would use observations made by Chinese astronomers in 1100 BC, Chaldeans in 600 BC, Greeks in 200 BC, Romans in AD 100, and Arabs in AD 1000. Obviously, not all data were equally valuable. How to resolve errors, known delicately as discrepancies, was anybody's guess.

The French academy was tackling the problem by encouraging the development of more precise telescopes and graduated arcs. And as algebra improved instrumentation, experimentalists were producing more quantitative results. In a veritable information explosion, the sheer collection and systemization of data accelerated through the Western world. Just as the number of known plant and animal species expanded enormously during the 1700s, so did knowledge about the physical universe. Even as Laplace arrived in Paris, the French and British academies were sending trained observers with state-of-the-art instrumentation to 120 carefully selected locations around the globe to time Venus crossing the face of the sun; this was a critical part of Capt. James Cook's original mission to the South Seas. By comparing all the measurements, French mathematicians would determine the approximate distance between the sun and Earth, a fundamental natural constant that would tell them the size of the solar system. But sometimes even up-to-date expeditions provided contradictory data about whether, for instance, Earth was shaped like an American football or a pumpkin.

Dealing with large amounts of complex data was emerging as a major scientific problem. Given a wealth of observations, how could scientists evaluate the facts at their disposal and choose the most valid? Observational astronomers typically averaged their three best observations of a particular phenomenon, but the practice was as straightforward as it was ad hoc; no one had ever tried to prove its validity empirically or theoretically. The mathematical theory of errors was in its infancy.

Problems were ripe for the picking and, with his eye on membership in the Royal Academy, Laplace bombarded the society with 13 papers in five years. He submitted hundreds of pages of powerful and original mathematics needed in astronomy, celestial mechanics, and important related issues. Astutely, he timed his reports to appear when openings occurred in the academy's membership. The secretary of the academy, the Marquis de Condorcet, wrote that never before had the society seen "anyone so young, present to it in so little time, so many important Mémoires, and on such diverse and such difficult matters."[5]

Academy members considered Laplace for membership six times but rejected him repeatedly in favor of more senior scientists. D'Alembert complained furiously that the organization refused to recognize talent. Laplace considered emigrating to Prussia or Russia to work in their academies.

During this frustrating period Laplace spent his free afternoons digging in the mathematical literature in the Royal Military School's 4,000-volume library. Analyzing large amounts of data was a formidable problem, and Laplace was already beginning to think it would require a fundamentally new way of thinking. He was beginning to see probability as a way to deal with the uncertainties pervading many events and their causes. Browsing in the library's stacks, he discovered an old book on gambling probability, *The Doctrine of Chances*, by Abraham de Moivre. The book had appeared in three editions between 1718 and 1756, and Laplace may have read the 1756 version. Thomas Bayes had studied an earlier edition.

Reading de Moivre, Laplace became more and more convinced that probability might help him deal with uncertainties in the solar system. Probability barely existed as a mathematical term, much less as a theory. Outside of gambling, it was applied in rudimentary form to philosophical questions like the existence of God and to commercial risk, including contracts, marine and life insurance, annuities, and money lending.

Laplace's growing interest in probability created a diplomatic problem of some delicacy because d'Alembert believed probability was too subjective for science. Young as he was, Laplace was confident enough in his mathematical judgment to disagree with his powerful patron. To Laplace, the movements of celestial bodies seemed so complex that he could not hope for precise solutions. Probability would not give him absolute answers, but it might show him which data were more likely to be correct. He began thinking about a method for deducing the probable causes of divergent, error-filled observations in astronomy. He was feeling his way toward a broad general theory for moving mathematically from known events back to their most probable causes. Continental mathematicians did not know yet about Bayes' discovery, so Laplace called his idea "the probability of causes" and "the probability of causes and future events, derived from past events."[6]

Wrestling with the mathematics of probability in 1773, he reflected on its philosophical counterpoint. In a paper submitted and read to the academy in March, the former abbé compared ignorant mankind, not with God but with an imaginary intelligence capable of knowing All. Because humans can never know everything with certainty, probability is the mathematical expression of our ignorance: "We owe to the frailty of the human mind one of the most delicate and ingenious of mathematical theories, namely the science of chance or probabilities."[7]

The essay was a grand combination of mathematics, metaphysics, and the heavens that Laplace held to his entire life. His search for a probability of causes and his view of the deity were deeply congenial. Laplace was all of one piece and for that reason all the more formidable. He often said he did not believe in God, and not even his biographer could decide whether he was an atheist or a deist. But his probability of causes was a mathematical expression of the universe, and for the rest of his days he updated his theories about God and the probability of causes as new evidence became available.

Laplace was struggling with probability when one day, ten years after the publication of Bayes' essay, he picked up an astronomy journal and was shocked to read that others might be hot on the same trail. They were not, but the threat of competition galvanized him. Dusting off one of his discarded manuscripts, Laplace transformed it into a broad method for determining the most likely causes of events and phenomena. He called it "Mémoire on the Probability of the Causes Given Events."

It provided the first version of what today we call Bayes' rule, Bayesian probability, or Bayesian statistical inference. Not yet recognizable as the modern Bayes' rule, it was a one-step process for moving backward, or inversely, from an effect to its most likely cause. As a mathematician in a gambling-addicted culture, Laplace knew how to work out the gambler's future odds of an event knowing its cause (the dice). But he wanted to solve scientific problems, and in real life he did not always know the gambler's odds and often had doubts about what numbers to put into his calculations. In a giant and intellectually nimble leap, he realized he could inject these uncertainties into his thinking by considering *all* possible causes and then choosing among them.

Laplace did not state his idea as an equation. He intuited it as a principle and described it only in words: the probability of a cause (given an event) is proportional to the probability of the event (given its cause). Laplace did not translate his theory into algebra at this point, but modern readers might find it helpful to see what his statement would look like today:

$$P(C|E) = \frac{P(E|C)}{\Sigma P\ (E|C')}$$

where $P(C|E)$ is the probability of a particular cause (given the data), and $P(E|C)$ represents the probability of an event or datum (given that cause). The sign in the denominator represented with Newton's sigma sign makes the total probability of all possible causes add up to one.

Armed with his principle, Laplace could do everything Thomas Bayes could have done—as long as he accepted the restrictive assumption that all his possible causes or hypotheses were equally likely. Laplace's goal, however, was far more ambitious. As a scientist, he needed to study the various possible causes of a phenomenon and then determine the best one. He did not yet know how to do that mathematically. He would need to make two more major breakthroughs and spend decades in thought.

Laplace's principle, the proportionality between probable events and their probable causes, seems simple today. But he was the first mathematician to work with large data sets, and the proportionality of cause and effect would make it feasible to make complex numerical calculations using only goose quills and ink pots.

In a mémoire read aloud to the academy, Laplace first applied his new probability of causes to two gambling problems. In each case he understood intuitively what should happen but got bogged down trying to prove it mathematically. First, he imagined an urn filled with an unknown ratio of black and white tickets (his cause). He drew a number of tickets from the urn and, based on that experience, asked for the probability that his next ticket would be white. Then in a frustrating battle to prove the answer he wrote no fewer than 45 equations covering four quarto-sized pages.

His second gambling problem involved piquet, a game requiring both luck and skill. Two people start playing but stop midway through the game and have to figure out how to divide the kitty by estimating their relative skill levels (the cause). Again, Laplace understood instinctively how to solve the problem but could not yet do so mathematically.

After dealing with gambling, which he loathed, Laplace moved happily on to the critical scientific problem faced by working astronomers. How should they deal with different observations of the same phenomenon? Three of the era's biggest scientific problems involved gravitational attraction on the motions of our moon, the motions of the planets Jupiter and Saturn, and the shape of the Earth. Even if observers repeated their measurements at the same time and place with the same instrument, their results could be slightly different each time. Trying to calculate a midvalue for such discrepant observations, Laplace limited himself to three observations but still needed seven pages of equations to formulate the problem. Scientifically, he understood the right answer—average the three data points—but he would have no mathematical justification for doing so until 1810, when, without using the probability of causes, he invented the central limit theorem.

Although Bayes originated the probability of causes, Laplace clearly discovered his version on his own. Laplace was 15 when the Bayes-Price essay was published; it appeared in an English-language journal for the English gentry and was apparently never mentioned again. Even French scientists who kept up with foreign journals thought Laplace was first and congratulated him wholeheartedly on his originality.

Mathematics confirms that Laplace discovered the principle independently. Bayes solved a special problem about a flat table using a two-step process that involved a prior guess and new data. Laplace did not yet know about the initial guess but dealt with the problem generally, making it useful for a variety of problems. Bayes laboriously explained and illustrated why uniform probabilities were permissible; Laplace assumed them instinctively. The Englishman wanted to know the range of probabilities that something will happen in light of previous experience. Laplace wanted more: as a working scientist, he wanted to know the probability that certain measurements and numerical values associated with a phenomenon were realistic. If Bayes and Price searched for the probability that, on the basis of today's puddles, it had rained yesterday and would rain tomorrow, Laplace asked for the probability that a particular amount of rain would fall and then refined his opinion over and over with new information to get a better value. Laplace's method was immensely influential; scientists did not pay Bayes serious heed until the twentieth century.

Most strikingly of all, Laplace at 25 was already steadfastly determined to develop his new method and make it useful. For the next 40 years he would work to clarify, simplify, expand, generalize, prove, and apply his new rule. Yet while Laplace became the indisputable intellectual giant of Bayes' rule, it represented only a small portion of his career. He also made important advances in celestial mechanics, mathematics, physics, biology, Earth science, and statistics. He juggled projects, moving from one to another and then back to the first. Happily blazing trails through every field of science known to his age, he transformed and mathematized everything he touched. He never stopped being thrilled by examples of Newton's theory.

Although he was fast becoming the leading scientist of his era, the academy waited five years before electing him a member on March 31, 1773. A few weeks later he was formally inducted into the world's leading scientific organization. His mémoire on the probability of causes was published a year later, in 1774. At the age of 24, Laplace was a professional researcher. The academy's annual stipend, together with his teaching salary, would help support him while he refined his research on celestial mechanics and the probability of causes.

Laplace was still grappling with probability in 1781, when Richard Price visited Paris and told Condorcet about Bayes' discovery. Laplace immediately latched onto the Englishman's ingenious invention, the starting guess, and incorporated it into his own, earlier version of the probability of causes. Strictly speaking, he did not produce a new formula but rather a statement about the first formula assuming equal probabilities for the causes. The statement gave him confidence that he was on the right track and told him that as long as all his prior hypotheses were equally probable, his earlier principle of 1774 was correct.[8]

Laplace could now confidently marry his intuitive grasp of a scientific situation with the eighteenth century's passion for new and precise scientific discoveries. Every time he got new information he could use the answer from his last solution as the starting point for another calculation. And by assuming that all his initial hypotheses were equally probable he could even derive his theorem.

As Academy secretary, Condorcet wrote an introduction to Laplace's essay and explained Bayes' contribution. Laplace later publicly credited Bayes with being first when he wrote, "The theory whose principles I explained some years after, . . . he accomplished in an acute and very ingenious, though slightly awkward, manner."[9]

Over the next decade, however, Laplace would realize with increasing clarity and frustration that his mathematics had shortcomings. It limited him to assigning equal probabilities to each of his initial hypotheses. As a scientist, he disapproved. If his method was ever going to reflect the actual state of affairs, he needed to be able to differentiate dubious data from more valid observations. Calling all events or observations equally probable could be true only theoretically. Many dice, for example, that appeared perfectly cubed were actually skewed. In one case he started by assigning players equal probabilities of winning, but with each round of play their respective skills emerged and their probabilities changed. "The science of chances must be used with care and must be modified when we pass from the mathematical case to the physical," he counseled.[10]

Moreover, as a pragmatist, he realized he had to confront a serious technical difficulty. Probability problems require multiplying numbers over and over, whether tossing coin after coin or measuring and remeasuring an observation. The process generated huge numbers—nothing as large as those common today but definitely cumbersome for a man working alone without mechanical or electronic aids. (He did not even get an assistant to help with calculations until about 1785.)

Laplace was never one to shrink from difficult computations, but, as he complained, probability problems were often impossible because they presented great difficulties and numbers raised to "very high powers."[11] He could use logarithms and an early generating function that he considered inadequate. But to illustrate how tedious calculations with big numbers could be, he described multiplying $20,000 \times 19,999 \times 19,998 \times 19,997$, etc. and then dividing by $1 \times 2 \times 3 \times 4$ up to 10,000. In another case he bet in a lottery only to realize he could not calculate its formula numerically; the French monarchy's winning number had 90 digits, drawn five at a time.

Such big-number problems were new. Newton had calculated with geometry, not numbers. Many mathematicians, like Bayes, used thought experiments to separate real problems from abstract and methodological issues. But Laplace wanted to use mathematics to illuminate natural phenomena, and he insisted that theories had to be based on actual fact. Probability was propelling him into an unmanageable world.

Armed with the Bayes–Price starting point, Laplace broke partway through the logjam that had stymied him for seven years. So far he had concentrated primarily on probability as a way to resolve error-prone astronomical observations. Now he switched gears to concentrate on finding the most probable causes of known events. To do so, he needed to practice with a big database of real and reliable values. But astronomy seldom provided extensive or controlled data, and the social sciences often involved so many possible causes that algebraic equations were useless.

Only one large amalgamation of truly trustworthy numbers existed in the 1700s: parish records of births, christenings, marriages, and deaths. In 1771 the French government ordered all provincial officials to report birth and death figures regularly to Paris; and three years later, the Royal Academy published 60 years of data for the Paris region. The figures confirmed what the Englishman John Graunt had discovered in 1662: slightly more boys than girls were born, in a ratio that remained constant over many years. Scientists had long assumed that the ratio, like other newly discovered regularities in nature, must be the result of "Divine Providence." Laplace disagreed.

Soon he was assessing not gambling or astronomical statistics but infants. For anyone interested in large numbers, babies were ideal. First, they came in binomials, either boys or girls, and eighteenth-century mathematicians already knew how to treat binomials. Second, infants arrived in abundance and, as Laplace emphasized, "It is necessary in this delicate research to employ sufficiently large numbers in view of the small difference that exists between . . . the births of boys and girls."[12] When the great naturalist Comte de Buffon discovered a small village in Burgundy where, for five years running, more girls had been born than boys, he asked whether this village invalidated Laplace's hypotheses. Absolutely not, Laplace replied firmly. A study based on a few facts cannot overrule a much larger one.

The calculations would be formidable. For example, if he had started with a 52:48 ratio of newborn boys to girls and a sample of 58,000 boys, Laplace would have had to multiply .52 by itself 57,999 times—and then do a similar calculation for girls. This was definitely not something anyone, not even the indomitable Laplace, wanted to do by hand.

He started out, however, as Bayes had suggested, by pragmatically assigning equal probabilities to all his initial hunches, whether 50–50, 33–33–33, or 25–25–25–25. Because their sums equal one, multiplication would be easier. He employed equal probabilities only provisionally, as a starting point, and his final hypotheses would depend on all the observational data he could add.

Next, he tried to confirm that Graunt was correct about the probability of a boy's birth being larger than 50%. He was building the foundation of the modern theory of testing statistical hypotheses. Poring over records of christenings in Paris and births in London, he was soon willing to bet that boys would outnumber girls for the next 179 years in Paris and for the next 8,605 years in London. "It would be extraordinary if it was the effect of chance," he wrote, tut-tutting that people really should make sure of their facts before theorizing about them.[13]

To transform probability's large numbers into smaller, more manageable terms Laplace invented a multitude of mathematical shortcuts and clever approximations. Among them were new generating functions, transforms, and asymptotic expansions. Computers have made many of his shortcuts unnecessary, but generating functions remain deeply embedded in mathematical analyses used for practical applications. Laplace used generating functions as a form of mathematical wizardry to trick a function he could deal with into providing him with the function he really wanted.

To Laplace, these mathematical pyrotechnics seemed as obvious as common sense. To students' frustration, he sprinkled his reports with phrases like, "It is easy to see, it is easy to extend, it is easy to apply, it is obvious that. . . ."[14] When a confused student once asked how he had jumped intuitively from one equation to another, Laplace had to work hard to reconstruct his thought process.

He was soon asking whether boys were more apt to be born in certain geographic regions. Perhaps "climate, food or customs . . . facilitates the birth of boys" in London.[15] Over the next 30-odd years Laplace collected birth ratios from Naples in the south, St. Petersburg in the north, and French provinces in between. He concluded that climate could not explain the disparity in births. But would more boys than girls always be born? As each additional piece of evidence appeared, Laplace found his probabilities approaching certainty "at a dramatically increasing rate."

He was refining hunches with objective data. In building a mathematical model of scientific thinking, where a reasonable person could develop a hypothesis and then evaluate it relentlessly in light of new knowledge, he became the first modern Bayesian. His system was enormously sensitive to new information. Just as each throw of a coin increases the probability of its being fair or rigged, so each additional birth record narrowed the range of uncertainties. Eventually, Laplace decided that the probability of boys exceeding girls was as "certain as any other moral truth" with an extremely tiny margin of being wrong.[16]

Generalizing from babies, he found a way to determine not just the probability of simple events, like the birth of one boy, but also the probability of future composite events like an entire year of births—even when the probability of simple events (whether the next newborn will be male) was uncertain. By 1786 he was determining the influence of past events on the probability of future events and wondering how big his sample of newborns had to be. By then Laplace saw probability as the primary way to overcome uncertainty. Pounding the point home in one short paragraph, he wrote, "Probability is relative in part to this *ignorance*, in part to our knowledge . . . a state of indecision, . . . it's impossible to announce with certainty."[17]

Persevering for years, he used insights gained in one science to shed light on others, researching a puzzle and inventing a mathematical technique to resolve it, integrating, approximating, and generalizing broadly when there was no other way to proceed. Like a modern researcher, he competed and collaborated with others and published reports on his interim progress as he went. Above all, he was tenacious. Twenty-five years later he was still eagerly testing his probability of causes with new information. He combed 65 years' worth of orphanage registries, asked friends in Egypt and Alexander von Humboldt in Central America about birth ratios there, and called on naturalists to check the animal kingdom. Finally, in 1812, after decades of work, he cautiously concluded that the birth of more boys than girls seemed to be "a general law for the human race."[18]

To test his rule on a larger sample Laplace decided in 1781 to determine the size of the French population, the thermometer of its health and prosperity. A conscientious administrator in eastern France had carefully counted heads in several parishes; to estimate the population of the entire nation, he recommended multiplying the annual number of births in France by 26. His proposal produced what was thought to be France's population, approximately 25.3 million. But no one knew how accurate his estimate was. Today's demographers believe that France's population had actually grown rapidly, to almost 28 million, because of fewer famines and because a government-trained midwife was touring the countryside promoting the use of soap and boiling water during childbirth.

Using his probability of causes, Laplace combined his prior information from parish records about births and deaths throughout France with his new information about headcounts in eastern France. He was adjusting estimates of the nation's population with more precise information from particular regions. In 1786 he reached a figure closer to modern estimates and calculated odds of 1,000 to 1 that his estimate was off by less than half a million. In 1802 he was able to advise Napoleon Bonaparte that a new census should be augmented with detailed samples of about a million residents in 30 representative departments scattered equally around France.

As he worked on his birth and census studies during the monarchy's last years, Laplace became involved in an inflammatory debate about France's judicial system. Condorcet believed the social sciences should be as quantifiable as the physical sciences. To help transform absolutist France into an English-style constitutional monarchy, he wanted Laplace to use mathematics to explore a variety of issues. How confident can we be in a sentence handed down by judge or jury? How probable is it that voting by an assembly or judicial tribunal will establish the truth? Laplace agreed to apply his new theory of probability to questions about electoral procedures, the credibility of witnesses, decision making by judicial panels and juries, and procedures of representative bodies and judicial panels.

Laplace took a dim view of most court judgments in France. Forensic science did not exist, so judicial systems everywhere relied on witness testimony. Taking a witness's statement for an event, Laplace asked the probability that the witness or the judge might be truthful, misled, or simply mistaken. He estimated the prior odds of an accused person's guilt at 50–50 and the probability that a juror was being truthful somewhat higher. Even at that, if a jury of eight voted by simple majority, the chance that they judged the accused's guilt wrong would be 65/256, or more than one in four. Thus for both mathematical and religious reasons Laplace sided with the Enlightenment's most radical demand, the abolition of capital punishment: "The possibility of atoning for these errors is the strongest argument of philosophers who have wanted to abolish the death penalty."[19] Laplace also used his rule for more complicated cases where a court must decide among contradictory witnesses or where the reliability of testimony decreases with each telling. For Laplace, these questions demonstrated that ancient biblical accounts by the Apostles lacked credibility.

While still counting babies, Laplace returned to study the seeming instability of Saturn and Jupiter's orbits, the problem that had helped sensitize him early in his career to uncertain data. He did not, however, use his new knowledge of probability to solve this important problem. He used other methods between 1785 and 1788 to determine that Jupiter and Saturn oscillate gently in an 877-year cycle around the sun and that the moon orbits Earth in a cycle millions of years long. The orbits of Jupiter, Saturn, and the moon were not exceptions to Newton's gravitation but thrilling examples of it. The solar system was in equilibrium, and the world would not end. This discovery was the biggest advance in physical astronomy since Newton's law of gravity.

Despite Laplace's astounding productivity, his life as a professional scientist was financially precarious. Fortunately, Paris in the 1700s had more educational institutions and scientific opportunities than anywhere else on Earth, and academy members could patch jobs together to make a respectable living. Laplace tripled his income by examining artillery and naval engineering students three or four months a year and serving as a scientist in the Duke of Orleans' entourage. His increasingly secure position also gave him access to the government statistics he needed to develop and test his probability of causes.

At the age of 39, with a bright future ahead of him, Laplace married 18-year-old Marie Anne Charlotte Courty de Romange. The average age of marriage for French women was 27, but Marie Anne came from a prosperous and recently ennobled family with multiple ties to his financial and social circle. A small street off the Boulevard Saint-Germain is named Courty for her family. The Laplaces would have two children; contraception, whether coitus interruptus or pessaries, was common, and the church itself campaigned against multiple childbirths because they endangered the lives of mothers. Some 16 months after the wedding a Parisian mob stormed the Bastille, and the French Revolution began.

After the revolutionary government was attacked by foreign monarchies, France spent a decade at war. Few scientists or engineers emigrated, even during the Reign of Terror. Mobilized for the national defense, they organized the conscription of soldiers, collected raw materials for gunpowder, supervised munitions factories, drew military maps, and invented a secret weapon, reconnaissance balloons. Laplace worked throughout the upheaval and served as the central figure in one of the Revolution's most important scientific projects, the metric reform to standardize weights and measures. It was Laplace who named the meter, centimeter, and millimeter.

Nevertheless, during the 18 months of the Terror, as almost 17,000 French were executed and half a million imprisoned, his position became increasingly precarious. Radicals attacked the elite Royal Academy of Sciences, and publications denounced him as a modern charlatan and a "Newtonian idolator." A month after the Royal Academy was abolished Laplace was arrested on suspicion of disloyalty to the Revolution but neighbors interceded and he was released the next day at 4 a.m. A few months later he was purged from the metric system commission as not "worthy of confidence as to [his] republican virtues and [his] hatred of kings."[20] His assistant, Jean-Baptiste Delambre, was arrested while measuring the meridian for the meter and then released. At one point Laplace was relieved of his part-time job examining artillery students, only to be given the same job at the École Polytechnique. Seven scientists, including several of Laplace's closest friends and supporters, died during the Terror. Unlike Laplace, who took no part in radical politics, they had identified themselves with particular political factions. The most famous was Antoine Lavoisier, guillotined because he had been a royal tax collector. Condorcet, trying to escape from Paris, died in jail.

The Revolution, however, transformed science from a popular hobby into a full-fledged profession. Laplace emerged from the chaos as a dean of French science, charged with building new secular educational institutions and training the next generation of scientists. For almost 50 years—from the

1780s until his death in 1827—France led world science as no other country has before or since. And for 30 of those years Laplace was among the most influential scientists of all time.

As the best-selling author of books about the celestial system and the law of gravity, Laplace dedicated two volumes to a rising young general, Napoleon Bonaparte. Laplace had launched Napoleon on his military career by giving him a passing exam grade in military school. The two never became personal friends, but Napoleon appointed Laplace minister of the interior for a short time and then appointed him to the largely honorary Senate with a handsome salary and generous expense account that made him quite a rich man. Mme Laplace became a lady-in-waiting to Napoleon's sister and received her own salary. With additional financing from Napoleon, Laplace and his friend the chemist Claude Berthollet turned their country homes in Arceuil, outside Paris, into the world's only center for young postdoctoral scientists.

At a reception in Josephine Bonaparte's rose garden at Malmaison in 1802, the emperor, who was trying to engineer a rapprochement with the papacy, started a celebrated argument with Laplace about God, astronomy, and the heavens.

"And who is the author of all this?" Napoleon demanded.

Laplace replied calmly that a chain of natural causes would account for the construction and preservation of the celestial system.

Napoleon complained that "Newton spoke of God in his book. I have perused yours but failed to find His name even once. Why?"

"Sire," Laplace replied magisterially, "I have no need of that hypothesis."[21]

Laplace's answer, so different from Price's idea that Bayes' rule could prove the existence of God, became a symbol of a centuries-long process that would eventually exclude religion from the scientific study of physical phenomena. Laplace had long since separated his probability of causes from religious considerations: "The true object of the physical sciences is not the search for primary causes [that is, God] but the search for laws according to which phenomena are produced."[22] Scientific explanations of natural phenomena were triumphs of civilization whereas theological debates were fruitless because they could never be resolved.

Laplace continued his research throughout France's political upheavals. In 1810 he announced the central limit theorem, one of the great scientific and statistical discoveries of all time. It asserts that, with some exceptions, any average of a large number of similar terms will have a normal, bell-shaped distribution. Suddenly, the easy-to-use bell curve was a real mathematical construct. Laplace's probability of causes had limited him to binomial problems, but his final proof of the central limit theorem let him deal with almost any kind of data.

In providing the mathematical justification for taking the mean of many data points, the central limit theorem had a profound effect on the future of Bayes' rule. At the age of 62, Laplace, its chief creator and proponent, made a remarkable about-face. He switched allegiances to an alternate, frequency-based approach he had also developed. From 1811 until his death 16 years later Laplace relied primarily on this approach, which twentieth-century theoreticians would use to almost obliterate Bayes' rule.

Laplace made the change because he realized that where large amounts of data were concerned, both approaches generally produce much the same results. The probability of causes was still useful in particularly uncertain cases because it was more powerful than frequentism. But science matured during Laplace's lifetime. By the 1800s mathematicians had much more reliable data than they had had in his youth and dealing with trustworthy data was easier with frequentism. Mathematicians did not learn until the mid-twentieth century that, even with great amounts of data, the two methods can sometimes seriously disagree.

Looking back in 1813 on his 40-year quest to develop the probability of causes, Laplace described it as the primary method for researching unknown or complicated causes of natural phenomena. He referred to it fondly as his source of large numbers and the inspiration behind his development and use of generating functions.

And finally, in the climax of one small part of his career, he proved the elegant, general version of his theorem that we now call Bayes' rule. He had intuited its principle as a young man in 1774. In 1781 he found a way to use Bayes' two-step process to derive the formula by making certain restrictive assumptions. Between 1810 and 1814 he finally realized what the general theorem had to be. It was the formula he had been dreaming about, one broad enough to allow him to distinguish highly probable hypotheses from less valid ones. With it, the entire process of learning from evidence was displayed:

$$P(C|E) = \frac{P(E|C)\, P_{prior}(C)}{\Sigma P(E|C')\, P_{prior}(C')}$$

In modern terms, the equation says that $P(C|E)$, the probability of a hypothesis (given information), equals $P_{prior}(C)$, our initial estimate of its probability, times $P(E|C)$, the probability of each new piece of information (under the hypothesis), divided by the sum of the probabilities of the data in all possible hypotheses.

Undergraduates today study Laplace's first version of the equation, which deals with discrete events such as coin tosses and births. Advanced and graduate students and researchers use calculus with his later equation to work with observations on a continuous range between two values, for example, all the temperatures between 32 and 33 degrees. With it, Laplace could estimate a value as being within such and such a range with a particular degree of probability.

Laplace had owned Bayes' rule in all but name since 1781. The formula, the method, and its masterful utilization all belong to Pierre Simon Laplace. He made probability-based statistics commonplace. By transforming a theory of gambling into practical mathematics, Laplace's work dominated probability and statistics for a century. "In my mind," Glenn Shafer of Rutgers University observed, "Laplace did everything, and we just read stuff back into Thomas Bayes. Laplace put it into modern terms. In a sense, everything is Laplacean."[23]

If advancing the world's knowledge is important, Bayes' rule should be called Laplace's rule or, in modern parlance, BPL for Bayes-Price-Laplace. Sadly, a half century of usage forces us to give Bayes' name to what was really Laplace's achievement.

Since discovering his first version of Bayes' rule in 1774, Laplace had used it primarily to develop new mathematical techniques and had applied it most extensively to the social sciences, that is, demography and judicial reform. Not until 1815, at the age of 66, did he apply it to his first love, astronomy. He had received some astonishingly accurate tables compiled by his assistant Alexis Bouvard, the director of the Paris Observatory. Using Laplace's probability of causes, Bouvard had calculated a large number of observations about the masses of Jupiter and Saturn, estimated the possible error for each entry, and then predicted the probable masses of the planets. Laplace was so delighted with the tables that, despite his aversion to gambling, he used Bayes' rule to place a famous bet with his readers: odds were 11,000 to 1 that Bouvard's results for Saturn were off by less than 1%. For Jupiter, the odds were a million to one. Space-age technology confirms that Laplace and Bouvard should have won both bets.

Late in his career, Laplace also applied his probability of causes to a variety of calculations in Earth science, notably to the tides and to changes in barometric pressure. He used a nonnumerical common-sense version of his probability of causes to advance his famous nebular hypothesis: that the planets and their satellites in our solar system originated in a swirl of dust. And he compared three hypotheses about the orbits of 100 comets to confirm what he already knew: that the comets most probably originate within the sun's sphere of influence.

After the fall of Napoleon, France's new king, Louis XVIII, bestowed the hereditary title of marquis on Laplace, the son of a village innkeeper. And on March 5, 1827, at the age of 78, Laplace died, almost exactly 100 years after his idol, Isaac Newton.

Eulogies hailed Laplace as the Newton of France. He had brought modern science to students, governments, and the reading public and had developed probability into a formidable method for handling unknown and complex causes of natural phenomena. And in one small, relatively insignificant portion of his lifework he became the first to express and use what is now called Bayes' rule. With it, he updated old knowledge with new, explained phenomena that previous centuries had ascribed to chance or to God's will, and opened the way for future scientific exploration.

Yet Laplace had built his probability theory on intuition. As far as he was concerned, "essentially, the theory of probability is nothing but good common sense reduced to mathematics. It provides an exact appreciation of what sound minds feel with a kind of instinct, frequently without being able to account for it."[24] Soon, however, scientists would begin confronting situations that intuition could not easily explain. Nature would prove to be far more complicated than even Laplace had envisioned. No sooner was the old man buried than critics began complaining about Laplace's rule.

3.

many doubts,
few defenders

With Laplace gone, Bayes' rule entered a tumultuous period when it was disdained, reformed, grudgingly tolerated, and finally nearly obliterated by battling theorists. Yet through it all the rule chugged sturdily along, helping to resolve practical problems involving the military, communications, social welfare, and medicine in the United States and Europe.

The backdrop to the drama was a set of unsubstantiated but widely circulated charges against Laplace's reputation. The English mathematician Augustus de Morgan wrote in *The Penny Cyclopaedia* of 1839 that Laplace failed to credit the work of others; the accusation was repeated without substantiation for 150 years until a detailed study by Stigler concluded it was groundless. During the 1880s an anti-Napoleonic and antimonarchical Frenchman named Maximilien Marie painted Laplace as a reactionary ultraroyalist; several English and American authors adopted Marie's version unquestioningly. The *Encyclopaedia Britannica* of 1912 asserted that Laplace "aspired to the role of a politician, and . . . degraded to servility for the sake of a riband and a title."[1] In his long-lived but rather fanciful bestseller *Men of Mathematics* the American E. T. Bell titled the chapter on Laplace "From Peasant to Snob . . . Humble as Lincoln, proud as Lucifer." Bell described Laplace as "grandiose," "snobbish," "pompous," "coarse," "smug," "intimate with Napoleon," and "perhaps the most conspicuous refutation of the pedagogical superstition that noble pursuits necessarily ennoble a man's character."[2] Bell's book, published in 1937, influenced an entire generation of mathematicians and scientists. In the 1960s an Anglo-American statistician Florence Nightingale David wrote without verification that Laplace met with "almost universal condemnation."[3] A scholarly American biography by historian Charles Coulston Gillispie and two collaborators, Robert Fox and Ivor Grattan-Guinness, hemmed and hawed. It began by stating categorically that "not a single testimonial bespeaking congeniality survives" but ended by listing Laplace's "close personal attachments with other French scientists," "warm and tranquil family life," and the help he gave even to critics of his research.[4]

The realization that Laplace was one of the world's first modern professional scientists emerged slowly. The statistician Karl Pearson, no shrinking violet, called the author of the *Britannica* article "one of the most superficial writers that ever obscured the history of science. . . . Such statements published by a writer of one nation about one of the most distinguished men of a second nation, and wholly unsubstantiated by references, are in every way deplorable."[5] Modern historians have shown that many of the disparaging comments about Laplace's life and work were false.

Personal insults aside, Laplace launched a craze for statistics that would ultimately inundate both Bayes' original rule and Laplace's own version of it. He did so by publicizing in 1827 the then-extraordinary fact that the number of dead letters in the Parisian postal system remained roughly constant from year to year. After the French government published a landmark series of statistics about the Paris region, it appeared that many irrational and godless criminal activities, including thefts, murders, and suicides, were also constants. By 1830 stable statistical ratios were firmly dissociated from divine providence, and Europe was swept by a veritable mania for the objective numbers needed by good government.

Unsettled by rapid urbanization, industrialization, and the rise of a market economy, early Victorians formed private statistical societies to study filth, criminality, and numbers. The chest sizes of Scottish soldiers, the number of Prussian officers killed by kicking horses, the incidence of cholera victims—statistics were easy to collect. Even women could do it. No mathematical analysis was necessary or expected. That most of the government bureaucrats collecting statistics were ignorant of and even hostile to mathematics did not matter. Facts, pure facts, were the order of the day.

Gone was the idea that we can use probability to quantify our lack of knowledge. Gone was the search for causes conducted by Bayes, Price, and Laplace. A correspondent admonished the hospital reformer Florence Nightingale in 1861, "Again I must repeat my objections to intermingling Causation with Statistics. . . . The statistician has nothing to do with causation."[6]

"Subjective" also became a naughty word. The French Revolution and its aftermath shattered the idea that all rational people share the same beliefs. The Western world split between Romantics, who rejected science outright, and those who sought certainty in natural science and were enthralled by the objectivity of numbers, whether the number of knifings or of marriages at a particular age.

During the decade after Laplace's death, four European revisionists led the charge against Laplace and probability, the mathematics of uncertainty. John Stuart Mill denounced probability as "an aberration of the intellect" and "ignorance . . . coined into science."[7] Objectivity became a virtue, subjectivity an insult, and the probability of causes a target of skepticism, if not hostility. Awash in newly collected data, the revisionists preferred to judge the probability of an event according to how frequently it occurred among many observations. Eventually, adherents of this frequency-based probability became known as frequentists or sampling theorists.

To frequentists Laplace was such a towering target that Thomas Bayes' existence barely registered. When critics thought of Bayes' rule, they thought of it as Laplace's rule and focused their criticism on him and his followers. Arguing that probabilities should be measured by objective frequencies of events rather than by subjective degrees of belief, they treated the two approaches as opposites, although Laplace had considered them basically equivalent.

The reformers denounced Laplace's pragmatic simplifications as gross abuses. Two of his most popular applications of probability were condemned wholesale. Laplace had asked: given that the sun has risen thousands of times in the past, will it rise tomorrow? and given that the planets revolve in similar ways around the sun, is there a single cause of the solar system? He did not actually use Bayes' rule for either project, only simple gambling odds. Sometimes, though, he and his followers began answering these questions by assuming 50–50 odds. The simplification would have been defensible had Laplace known nothing about the heavens. But he was the world's leading mathematical astronomer, and he understood better than anyone that sunrises and nebulae were the result of celestial mechanics, not gambling odds. He had also started his study of male and female birthrates with 50–50 odds, although scientists already knew that the likelihood of a male birth is approximately 0.52.

Laplace agreed that reducing scientific questions to chance boosted the odds in favor of his deep conviction that physical phenomena have natural causes rather than religious ones. He warned his readers about it. His followers also sometimes weighted their initial odds heavily in favor of natural laws and weakened counterexamples. Critics pounded away at the fact that chance was irrelevant to the questions at hand. They identified Bayes' rule with equal priors and damned the entire rule because of them. Few of the critics tried to even imagine other kinds of priors.

Years later John Maynard Keynes studied the complaints made about Laplace's assessment, based on 5,000 years of history, that "it is a bet of 1,825,214 to 1 that [the sun] will rise tomorrow." Summarizing the arguments, Keynes wrote that Laplace's reasoning "has been rejected by [George] Boole on the ground that the hypotheses on which it is based are arbitrary, by [John] Venn on the ground that it does not accord with experience, by [Joseph] Bertrand because it is ridiculous, and doubtless by others also. But it has been very widely accepted—by [Augustus] de Morgan, by [William] Jevons, by [Rudolf] Lotze, by [Emanuel] Czuber, and by Professor [Karl] Pearson—to name some representative writers of successive schools and periods."[8]

Amid the dissension, Laplace's delicate balance between subjective beliefs and objective frequencies collapsed. He had developed two theories of probability and shown that when large numbers are involved they lead to more or less the same results. But if natural science was the route to certain knowledge, how could it be subjective? Soon scientists were treating the two approaches as diametric opposites. Lacking a definitive experiment to decide the controversy and with Laplace demonstrating that both methods often lead to roughly the same result, the tiny world of probability experts would be hard put to settle the argument.

Research into probability mathematics petered out. Within two generations of his death Laplace was remembered largely for astronomy. By 1850 not a single copy of his massive treatise on probability was available in Parisian bookstores. The physicist James Clerk Maxwell learned about probability from Adolphe Quetelet, a Belgian factoid hunter, not from Laplace, and adopted frequency-based methods for statistical mechanics and the kinetic theory of gases. Laplace and Condorcet had expected social scientists to be the biggest users of Bayes' rule, but they were reluctant to adopt any form of probability. An American scientist and philosopher Charles Sanders Peirce promoted frequency-based probability during the late 1870s and early 1880s. In 1891 a Scottish mathematician, George Chrystal, composed an obituary for Laplace's method: "The laws of . . . Inverse Probability being dead, they should be decently buried out of sight, and not embalmed in text-books and examination papers. . . . The indiscretions of great men should be quietly allowed to be forgotten."[9]

For the third time Bayes' rule was left for dead. The first time, Bayes himself had shelved it. The second time, Price revived it briefly before it again died of neglect. This time theoreticians buried it.

The funeral was a trifle premature. Despite Chrystal's condemnation, Bayes' rule was still taught in textbooks and classrooms and used by astronomers because the anti-Bayesian frequentists had not yet produced a systematic, practical substitute. In scattered niches far from the eyes of disapproving theoreticians, Bayes bubbled along, helping real-life practitioners assess evidence, combine every possible form of information, and cope with the gaps and uncertainties in their knowledge.

Into this breach between theoretical disapproval and practical utility marched the French army, under the baton of a politically powerful mathematician named Joseph Louis François Bertrand. Bertrand reformed Bayes for artillery field officers who dealt with a host of uncertainties: the enemy's precise location; air density; wind direction; variations among their hand-forged cannons; and the range, direction, and initial speed of projectiles. In his widely used textbooks Bertrand preached that Laplace's probability of causes was the only valid method for verifying a hypothesis with new observations. He believed, however, that Laplace's followers had lost their way and must stop their practice of indiscriminately using 50–50 odds for prior causes. To illustrate, he told about the foolish peasants of Britanny who, looking for the possible causes of shipwrecks along their rocky coast, assigned equal odds to the tides and to the far more dangerous northwest winds. Bertrand argued that equal prior odds should be confined to those rare cases when hypotheses really and truly were equally likely or when absolutely nothing was known about their likelihoods.

Following Bertrand's strict standards, artillery officers began assigning equal probabilities only for cannons made in the same factory by roughly the same staff using identical ingredients and processes under identical conditions. For the next 60 years, between the 1880s and the Second World War, French and Russian artillery officers fired their weapons according to Bertrand's textbook.

Bertrand's strict Bayesian reforms figured in the Dreyfus affair, the scandal that rocked France between 1894 and 1906. Alfred Dreyfus, a French Jew and army officer, was falsely convicted of spying for Germany and condemned to life imprisonment. Almost the only evidence against Dreyfus was a letter he was accused of having sold to a German military attaché. Alphonse

Bertillon, a police criminologist who had invented an identification system based on body measurements, testified repeatedly that, according to probability mathematics, Dreyfus had most assuredly written the incriminating letter. Bertillon's notions of probability were mathematical gibberish, and he developed ever more fantastical arguments. As conservative antirepublicans, Roman Catholics, and anti-Semites supported Dreyfus's conviction, a campaign to exonerate him was organized by his family, anticlericals, Jews, and left-wing politicians and intellectuals led by the novelist Émile Zola.

At Dreyfus's military trial in 1899, his lawyer called on France's most illustrious mathematician and physicist, Henri Poincaré, who had taught probability at the Sorbonne for more than ten years. Poincaré believed in frequency-based statistics. But when asked whether Bertillon's document was written by Dreyfus or someone else, he invoked Bayes' rule. Poincaré considered it the only sensible way for a court of law to update a prior hypothesis with new evidence, and he regarded the forgery as a typical problem in Bayesian hypothesis testing.

Poincaré provided Dreyfus's lawyer with a short, sarcastic letter, which the lawyer read aloud to the courtroom: Bertillon's "most comprehensible point [is] false. . . . This colossal error renders suspect all that follows. . . . I do not understand why you are worried. I do not know if the accused will be condemned, but if he is, it will be on the basis of other proofs. Such arguments cannot impress unbiased men who have received a solid scientific education."[10] At that point, according to the court stenographer, the courtroom erupted in a "prolonged uproar." Poincaré's testimony devastated the prosecution; all the judges had attended military schools and studied Bayes in Bertrand's textbook.

The judges issued a compromise verdict, again finding Dreyfus guilty but reducing his sentence to five years. The public was outraged, however, and the president of the Republic issued a pardon two weeks later. Dreyfus was promoted and awarded the Legion of Honor, and government reforms were instituted to strictly separate church and state. Many American lawyers, unaware that probability helped to free Dreyfus, have considered his trial an example of mathematics run amok and a reason to limit the use of probability in criminal cases.

As the First World War approached, a French general and proponent of military aviation and tanks, Jean Baptiste Eugène Estienne, developed elaborate Bayesian tables telling field officers how to aim and fire. Estienne also developed a Bayesian method for testing ammunition. After Germany captured France's industrial base in 1914, ammunition was so scarce that the French could not use wasteful frequency-based methods to test its quality. Mobilized for the national defense, professors of abstract mathematics developed Bayesian testing tables that required destroying only 20 cartridges in each lot of 20,000. Instead of conducting a predetermined number of tests, the army could stop when they were sure about the lot as a whole. During the Second World War, American and British mathematicians discovered similar methods and called them operations research.

Bayes' rule was supposedly still dying on the vine as the First World War approached and the United States faced two emergencies caused by the country's rapid industrialization. In each case self-taught statisticians resorted to Bayes as a tool for making informed decisions, first about telephone communications and second about injured workers.

The first crisis occurred when the financial panic of 1907 threatened the survival of the Bell telephone system owned by American Telephone and Telegraph Company. Alexander Graham Bell's patents had expired a few years earlier, and the company had overexpanded. Only the intervention of a banking consortium led by the House of Morgan prevented Bell's collapse.

At the same time, state regulators were demanding proof of Bell's ability to provide better and cheaper service than local competitors. Unfortunately, Bell telephone circuits were often overloaded in the late morning and early afternoon, when too many customers tried to place calls at the same time. During the rest of the day—80% of the time—Bell's facilities were underutilized. No company could afford to build a system to handle every call that could conceivably be made at peak times.

Edward C. Molina, an engineer in New York City, considered the uncertainties involved. Molina, whose family had emigrated from Portugal via France, was born in New York in 1877. He graduated from a city high school but, with no money for college, got a job first at Western Electric Company and then at AT&T's engineering and research department (later called Bell Laboratories). The Bell System of phone companies was adopting a new mathematical approach to problem solving. Molina's boss, George Ashley Campbell, had studied probability with Poincaré in France but other employees were learning it from the *Encyclopaedia Britannica*. Molina taught himself mathematics and physics and became the nation's leading expert on Bayesian and Laplacean probability.

Unlike many others at the time, he realized that "great confusion exists because many authorities have failed to distinguish clearly between the original Bayes inverse theorem and its subsequent generalization by Laplace. The general theorem embraces, or brings together, both the data obtained from a series of observations and whatever 'collateral' information exists in relation to the observed results."[11] As Molina explained, applied statisticians were often forced to make quick decisions based on meager observational data; in such cases, they had to rely on indirect prior knowledge, called collateral information. This could range from assessments of national or historic trends to an executive's mental health. Methods for utilizing both statistical and nonstatistical types of evidence were needed.

Using Laplace's formula, Molina combined his prior information about the economics of automating Bell telephone systems with data about telephone call traffic, call length, and waiting time. The result was a cost-effective way for Bell to deal with uncertainties in telephone usage.

Molina then worked on automating Bell's labor-intensive system. In many cities the company employed 8 to 20% of the female population as telephone operators, switching wires to route calls through trunking facilities to customers in distant exchanges. Operators were in short supply, annual turnover in some cities was 100% or more annually, and wages doubled between 1915 and 1920. Depending on one's point of view, the work epitomized either opportunities for women or the inhumane pressure of modern technology.

To automate the system Molina conceived of the relay translator, which converted decimally dialed phone numbers into routing instructions. Then he used Bayes to analyze technical information and the economics of various combinations of switches, selectors, and trunking lines at particular exchanges. After women won the right to vote in 1920, Bell feared a backlash if it fired all its operators, so it chose an automating method that merely halved their numbers. Between the world wars, employment of operators dropped from 15 to 7 per 1,000 telephones even as toll call service increased. Probability assumed an important role in the Bell System, and Bayesian methods were used to develop basic sampling theory.

Molina won prestigious awards, but his use of Bayes remained controversial among some Bell mathematicians, and he complained he had trouble publishing his research. Some of his problems may have stemmed from his colorful character. He loved model boats, published articles about Edgar Allan Poe's use of probability, played the piano expertly, and donated to the Metropolitan Opera in New York. He followed the Russo-Japanese war so avidly that his colleagues nicknamed him, not fondly, General Molina. When he independently discovered the Poisson distribution, he called it the Molina distribution until he learned to his embarrassment that Laplace's protégé Siméon Denis Poisson had written about it in the 1830s.

Molina's enthusiasm for Bayes-Laplacean probability did not spread to other American corporations. AT&T often regarded his articles about Bayes as proprietary secrets and published them only in in-house publications years after the fact.

While Bayes' rule was helping to save the Bell System, financiers were rushing to build American railroads and industries. Government safety regulations were nonexistent, however, and 1 out of every 318 industrial workers was killed on the job between 1890 and 1910, and many more were injured. The country's labor force suffered more accidents, sickness, invalidity, premature old age, and unemployment than European workers. Yet, unlike most of Europe, the United States had no system for insuring sick and injured workers, and most blue-collar families lived one paycheck away from needing charity. Federal judges ruled that injured employees could sue only if their bosses were personally at fault. In 1898 a U.S. Department of Labor statistician could think of no other social or legal reform in which the United States lagged so far behind other nations.

The tide turned as growing numbers of workers joined the American Federation of Labor and as local juries started awarding generous settlements to their disabled peers. At that point employers decided it was cheaper to treat occupational health as a predictable business expense than to trust juries and encourage unionization. In an avalanche of no-fault laws passed between 1911 and 1920 all but eight states began requiring employers to insure their workers immediately against occupational injuries and illness. This was the first, and for decades the only, social insurance in the United States.

The legislation triggered an emergency. Normally, the price of an insurance premium reflects years of accumulated data about such factors as accident rates, medical costs, wages, industrywide trends, and particulars about individual companies. No such data existed in the United States. Not even the most industrialized states had enough occupational health statistics to price policies for all their industries. The industrial powerhouse of New York State had only enough experience to price policies for machine printers and garment workers; South Carolina had only enough for cotton spinners and weavers; and St. Louis and Milwaukee for beer brewers. In 1909 Nebraska had only

25 small manufacturers of any kind. As an insurance expert wondered, "When will Nebraska be able to determine its pure premium on 'suspenders without buckles,' or Rhode Island on 'butchers' supplies?' And yet rates must be quoted for either, and moreover they must be adequate and equitable."[12]

Data from other areas were seldom relevant. Germany had collected accident statistics for 30 years, but its industrial conditions were safer, and because its data were collected nationwide, premiums could be based on industrywide information. But in the United States, data were collected by state, and Massachusetts's statistics about shoe- and bootmakers were irrelevant to Nevada's metal mines and their high fatality rates because, as one expert reported, "metal mines are so rare in Massachusetts as snakes in Ireland."[13]

Nevertheless, premiums had to be invented—overnight and out of thin air—for almost every sizeable company in the country. It was a nightmare to keep any mathematically trained statistician awake at night—not that the United States had many. Actuaries were often antagonistic to high-level mathematics, and an official complained that accident and fire premiums were typically priced by untrained clerks who used opinion, what they "euphoniously called underwriting judgment,"[14] rather like "woman's intuition . . . ('I don't know why I think so, but I am sure I am right')."[15] Compounding the crisis, each state legislature mandated its own unique insurance system.

Still, premiums had to be priced accurately: high enough to keep the insurance company solvent for the life of its insurees *and* individualized enough to reward businesses with good safety records. In an extraordinary feat, Isaac M. Rubinow, a physician and statistician for the American Medical Association, organized by hand the analysis, classification, and tabulation of literally millions of insurance claims, primarily from Europe, as a two- or three-year stopgap until each state could accumulate statistics on its occupational casualties. "Every scrap of information," he said, must be used.[16]

Rubinow called together 11 scientifically minded actuaries and formed the Casualty Actuarial Society in 1914. Only seven were college graduates, but their goal was lofty: to set casualty fire and workers' compensation insurance on a sound mathematical basis. Rubinow became the organization's first president but left almost immediately when the insurance industry and the American Medical Association opposed extending social insurance to the sick and aged. Rumors swirled that Rubinow, a Jewish immigrant from Russia, had "socialist tendencies."[17]

Albert Wurts Whitney, a specialist in insurance mathematics from Berkeley, replaced Rubinow on a workers' compensation committee. Whitney was an alumnus of Beloit College and had no graduate degrees but had taught mathematics and physics at the universities of Chicago, Nebraska, and Michigan. At the University of California in Berkeley he had taught probability for future insurance professionals. Though not as steeped in the original mathematical literature as Molina at Bell Labs, Whitney was familiar with the theorems of both Laplace and Bayes and knew he should use one of them. He understood something else too. The equations were too complicated for the fledgling workers' compensation movement.

One afternoon in the spring of 1918, during the First World War, Whitney and his committee worked for hours stripping away every possible mathematical complication and substituting dubious simplifications. They agreed to assume that each business in a particular industrial class (for example, all residential roofers) faced equal risks. They would also consider every actuary equally skilled at supplementing injury data with subjective judgments about "non-statistical" or "exogenous material," such as a business owner's drinking habits. This was Bayes' rule where industrywide experience was used as the basis for the prior and the local business's history for the new data. Whitney cautioned, "We know that the [subjective] rates for some classifications are more reliable than for others. [But] it is doubtful whether it is expedient in practice to recognize this fact."[18]

By the end of the afternoon the committee had decided to base the price of a customer's premium almost exclusively on the experience of the client's broad classification. Thus a machine shop's premium could be based on data from other, similar businesses or, if it was large enough, its own experience. Combining data from related businesses concentrated the numbers, pulling them closer to the mean and making them more accurate, a subtle "shrinking" effect that Charles Stein would explain in the 1950s. What remained was a stunningly simple formula that a clerk could compute, an underwriter could understand, and a salesman could explain to his customers. The committee proudly named its creation Credibility.

For the next 30 years the first social insurance system in the United States relied on this simplified Bayesian system. In a classic understatement an actuary admitted, "Of course [Credibility's] $Z = P/(P+K)$ is not so great a discovery as $E = mc^2$ nor as unalterably true, but it has made life much easier for insurance men for generations."[19] Some 50 years later statisticians and actuaries would be surprised to discover the Bayesian roots of Credibility.

Next, Whitney worked out methods for weighting each datum as to its subjective believability. Soon actuaries were adventuring "beyond anything that has been proven mathematically. The only demonstration they can make," an actuary reported later, "is that, in actual practice, it works."[20]

Skeptical state officials and insurance underwriters sometimes wondered where these strange Credibility figures came from. One insurance commissioner demanded, "You have supported everything else in the filing with actual experience, where is the experience supporting your Credibility factor?"[21] Actuaries hastily changed the subject. When Whitney was asked where he got the mathematical principles underlying Credibility, he pointed flippantly to a colleague's home. "In Michelbacher's dining room," he said.

Credibility theory was a practical American response to a uniquely American problem, and it became a cornerstone of casualty and property insurance. As claims accumulated, actuaries could check the accuracy of their premiums by comparing them with actual claims. In 1922 actuaries gained access to an enormous pool of occupational data compiled by the National Council for Compensation Insurance. As the years passed, practicing actuaries had less and less need to understand the relationship between Credibility and Bayes.

While the United States was using Bayes' theorem for business decisions and France was adapting it for the military, eugenics was shifting Bayes' story back to its birthplace in Great Britain. There Karl Pearson and Ronald Fisher were developing statistics—the mathematics of uncertainties—into the first information science. Early in the twentieth century, as they created new ways to study biology and heredity, theoreticians would change their attitudes toward Bayes' rule from tepid toleration to outright hostility.

Karl Pearson (I repeat his first name because his son Egon figures in Bayes' story too) was a zealous atheist, socialist, feminist, Darwinist, Germanophile, and eugenicist. To save the British Empire, he believed the government should encourage the upper middle class to procreate and the poor to abstain. Karl Pearson ruled Britain's 30-odd statistical theorists for years. In the process he introduced two generations of applied mathematicians to the kind of feuding and professional bullying generally seen only on middle-school playgrounds.

Contentious, unquenchably ambitious, and craggily determined, Karl Pearson was ambivalent about few things, but Bayes' rule was one of them. Uniform priors and subjectivity made him nervous. With few other tools available to statisticians, however, he concluded sadly that "the practical man will . . . accept the results of inverse probability of the Bayes-Laplace brand till better are forthcoming."[22] As Keynes said in *A Treatise on Probability* in 1921, "There is still about it for scientists a smack of astrology, of alchemy." Four years later the American mathematician Julian L. Coolidge agreed: "We use Bayes' formula with a sigh, as the only thing available under the circumstances."[23]

Another geneticist, Ronald Aylmer Fisher, eventually contested Karl Pearson's statistical crown and dealt Bayes' rule a near-lethal blow. If Bayes' story were a TV melodrama, it would need a clear-cut villain, and Fisher would probably be the audience's choice by acclamation.

He didn't look the part. Even with thick glasses he could barely see three feet and had to be rescued from oncoming buses. His clothes were so rumpled that his family thought he looked like a tramp; he smoked a pipe even while swimming; and if a conversation bored him, he sometimes removed his false teeth and cleaned them in public.

Fisher interpreted any question as a personal attack, and even he recognized that his fiery temper was the bane of his existence. A colleague, William Kruskal, described Fisher's life as "a sequence of scientific fights, often several at a time, at scientific meetings and in scientific papers."[24] In a basically sympathetic rendering of Fisher's career, the Bayesian theorist Jimmie Savage said he "sometimes published insults that only a saint could entirely forgive. . . . Fisher burned even more than the rest of us . . . to be original, right, important, famous, and respected. And in enormous measure, he achieved all of that, though never enough to bring him peace."[25] Part of Fisher's frustration may have arisen from the fact that, on many statistical matters, he was correct.

Fisher was 16 when the family business collapsed. At Cambridge on a scholarship, he became the top mathematics student in his class and, in 1911, the founder and chair of the Cambridge University Eugenics Society. A few years later he solved in one page a problem Karl Pearson had struggled over for years; Pearson thought Fisher's solution was rubbish and refused to publish it in his prestigious journal *Biometrika*. The two continued to feud as long as they lived. But in straightening out inconsistencies in Karl Pearson's work, Fisher pioneered the first comprehensive and rigorous theory of statistics and set it on its mathematical and anti-Bayesian course.

The enmity between these two volatile men is striking because both were fervent eugenicists who believed that the careful breeding of British supermen and superwomen would improve the human population and the British Empire. To help support his wife and eight children on a subsistence farm, Fisher accepted funds from a controversial source, Charles Darwin's son Leonard, who, as honorary president of the Eugenics Education Society, advocated the detention of "inferior types, . . . the sexes being kept apart" to prevent them from bearing children.[26] In return for his financial help, Fisher published more than 200 reviews in Darwin's magazine between 1914 and 1934.

In 1919, few jobs were available in statistics or eugenics but Fisher landed a position analyzing fertilizers at Rothamsted Agricultural Experiment Station. Other statistical pioneers worked in breweries, cotton thread and light bulb factories, and the wool industry. Fisher's job was analyzing volumes of data compiled over decades about horse manure, chemical fertilizers, crop rotation, rainfall, temperature, and yields. "Raking over the muck-heap," he called it.[27] At first, like Karl Pearson, Fisher used Bayes. But during afternoon teas at Rothamsted soil scientists confronted Fisher with new kinds of literally down-to-earth problems. Fascinated, Fisher worked out better ways to design experiments.

Over the years he pioneered randomization methods, sampling theory, tests of significance, maximum likelihood estimation, analysis of variance, and experimental design methods. Thanks to Fisher, experimental scientists, who had traditionally ignored statistical methods, learned to incorporate them when designing their projects. As the twentieth century's statistical magistrate, Fisher often ended lengthy discussions with a one-word verdict: "Randomize." In 1925 he published a revolutionary manual of new techniques, *Statistical Methods for Research Workers*. A cookbook of ingenious statistical procedures for nonstatisticians, it turned frequency into the de facto statistical method. His first manual sold 20,000 copies, and a second went through seven editions before Fisher's death in 1962. His analysis of variance, which tells how to separate the effects of various treatments, has become one of the natural sciences' most important tools. His significance test and its p-values would be used millions of times even as, over the years, it became increasingly controversial. No one today can discuss statistics— what he called "mathematics applied to observational data"—without using some of Fisher's vocabulary.[28] Many of his ideas were solutions to computational problems caused by the limitations of the era's desk calculators. Soon statistical departments rang with the music of bells activated by mechanical calculating machines at every step of their Fisherian calculations.

Fisher himself became a superb geneticist who did mathematical statistics on the side. He filled his house with cats, dogs, and thousands of mice for cross-breeding experiments; he could document each animal's pedigree for generations. Unlike Bayes, Price, and Laplace, he did not need to supplement inadequate or conflicting observations with hunches or subjective judgments. His experiments produced small data sets and subsets tightly focused to answer a single question with rigorous mathematics. He dealt with few uncertainties or gaps in data and could compare, manipulate, or repeat his experiments as needed. Thanks to the problems he analyzed, Fisher redefined most uncertainties not by their relative probabilities but by their relative frequencies. He brought to fruition Laplace's frequency-based theories, the methods Laplace himself preferred toward the end of his life.

After 15 years at Rothamsted, Fisher moved, first, to University College London and then to Cambridge as professor of genetics. Today, statisticians regard him as one of the great minds of the twentieth century and a "mythical aura" surrounds both him and Karl Pearson.[29] The aura around Fisher is a trifle tarnished, though. He left his family on the farm with a precarious, barebones allowance. As a colleague wrote, "If only . . . if only . . . RAF had been a nicer man, if only he had taken pains to be clearer and less enigmatic, if only he had not been obsessed with ambition and personal bitternesses. If only. But then we might not have had Fisher's magnificent achievements."[30]

Lashing out at Bayes' rule, Fisher called it an "impenetrable jungle" and "a mistake, perhaps the only mistake to which the mathematical world has so deeply committed itself."[31] Equal priors constituted "staggering falsity."[32] "My personal conviction," he declared, is "that the theory of inverse probability is founded upon an error, and must be wholly rejected."[33] A scholarly statistician, Anders Hald, politely lamented "Fisher's arrogant style of writing."[34] Although Fisher's work had many Bayesian elements, he battled Bayes for decades and rendered it virtually taboo among respectable statisticians. His constant readiness to quarrel made it hard for opponents to engage him. Bayesians were not alone in concluding that Fisher adopted some of his positions "simply to avoid agreeing with his opponents."[35]

Driven by the need to deal with uncertainties and save time and money, frequency-based sampling theorists enjoyed a golden age during the 1920s and 1930s. Fisher liberated scientists to summarize and draw conclusions without having to deal with Bayes' messy prior prejudices and hunches. And thanks to his insistence on mathematical rigor, statistics was becoming, if not quite "real mathematics," at least a distinct mathematical discipline, mathematics as applied to data.

The feud between Karl Pearson and Fisher entered its second generation when Karl's son Egon became another victim of Fisher's wrath. Unlike his father, Egon Pearson was a modest, even self-effacing gentleman. At first, like his father and Fisher earlier in their careers, Egon Pearson frequently used Bayes' rule. In 1925 he published the most extensive exploration of Bayesian methods conducted between Laplace in the 1780s and the 1960s. Using priors for a series of seemingly whimsical experiments, he calculated such probabilities as the fraction of London taxicabs with LX license plates; men smoking pipes on Euston Road; horse-drawn vehicles on Gower Street; chestnut colts born to bay mares; and hounds with fawn-spotted coats. His experiments had a serious purpose, though. He was looking at all sorts of binomial problems, "working backwards" to find "nature's prior," one that anyone with a binomial problem could use. He concluded that more data had to be collected, but no one took up the challenge. Instead, Egon Pearson busied himself trying to make Fisher's work more mathematically rigorous, thereby enraging both Fisher and his father.

Egon Pearson and a Polish mathematician, Jerzy Neyman, teamed up in 1933 to develop the Neyman-Pearson theory of hypothesis testing. Until then, statisticians had tested one hypothesis at a time and either accepted or rejected it without considering alternatives. Egon Pearson's idea was that the only correct reason for rejecting a statistical hypothesis was to accept a more probable one. As he, Neyman, and Fisher developed it, the theory became one of the twentieth century's most influential pieces of applied mathematics. Egon Pearson was afraid of contradicting his father, however. His "fears of K.P. and R.A.F." precipitated a psychological crisis in 1925 and 1926: "I had to go through the painful stage of realizing that K.P. could be wrong . . . and I was torn between conflicting emotions: a. finding it difficult to understand R. A. F., b. hating him for his attacks on my paternal 'god,' c. realizing that in some things at least he was right."[36] To placate his father, Egon gave up the woman he loved; they married many years later. And he was so afraid to submit articles to his father's *Biometrika* that he and Neyman published their own journal, *Statistical Research Memoirs*, for two years between 1936 and 1938 and ceased publication only after Karl Pearson died.

Over the years Fisher, Egon Pearson, and Neyman would develop a host of powerful statistical techniques. Fisher and Neyman became fervent anti-Bayesians who limited themselves to events that could theoretically be repeated many times; regarded samples as their only source of information; and viewed each new set of data as a separate problem, to be used if the data were powerful enough to provide statistically significant conclusions and discarded if not. As anti-Bayesians, they banned subjective priors, although they did not argue with Bayes' theorem when the priors were known; the difficulties and controversies arose when the prior probabilities were unknown. Neyman, for example, denounced Bayes' equal-prior shortcut as "illegitimate."[37]

In addition, in a deep philosophical divide between the two methods, frequentists asked for the probability of a data set given full knowledge of its probable causes, while Bayesians could ask for better knowledge of the causes in light of the data. Bayesians could also consider the probability of a single event, like rain tomorrow; encapsulate subjective information in priors; update their hunches with new information; and include every datum possible because each one might change the answer by a small amount.

In time, however, Fisher and Neyman also split asunder, starting yet another juicy 30-year feud. Their views on testing, which could be an order of magnitude apart, formed the crux of their bitter fight. According to Neyman, though, the argument began because Fisher demanded that Neyman lecture only from Fisher's book. When Neyman refused, Fisher promised to oppose him "in all my capacities."

In an argument during a meeting of the Royal Society on March 28, 1934, a secretary took the customary word-for-word notes for verbatim publication. Neyman presented a paper arguing that the Latin square (a technique invented by Fisher for experimental design) was biased. Fisher immediately marched to a blackboard, drew a Latin square and, using a simple argument, showed that Neyman was wrong. But Fisher was far from polite. He complained sarcastically that "he had hoped that Dr. Neyman's paper would be on a subject with which the author was fully acquainted, and on which he could speak with authority. . . . Dr. Neyman had been somewhat unwise in his choice of topics." Fisher couldn't seem to stop. He kept on: "Dr. Neyman had arrived or thought he had arrived at . . . Apart from its theoretical defects . . . the apparent inability to grasp the very simple argument . . . How had Dr. Neyman been led by his symbolism to deceive himself on such a simple question?" and on and on.[38]

By 1936 the feud between the Neyman camp and the Fisherites was an academic cause célèbre. The two groups occupied different floors of the same building at University College London but they never mixed. Neyman's group met in the common room for India tea between 3:30 and 4:15 p.m. Fisher's group sipped China tea from then on. They were fighting over scraps. The statistics building had no potable water, so little electricity that blackboards were unreadable after dark, and so little heat in winter that overcoats were worn inside.

George Box, who straddled both groups (he studied under Egon Pearson, became a Bayesian, and married one of Fisher's daughters), said that Fisher and Neyman "could both be very nasty and very generous at times." Because Neyman was decision-oriented and Fisher was more interested in scientific inference, their methodologies and types of applications were different. Each was doing what was best for the problems he was working on, but neither side made any attempt to understand what the other was doing. A popular in-house riddle described the situation: "What's the collective noun for a group of statisticians?" "A quarrel."[39]

Shortly before the Second World War, Neyman moved to the University of California at Berkeley and transformed it into an anti-Bayesian powerhouse. The Neyman-Pearson theory of tests became the Berkeley school's glory and emblem. The joke, of course, was that the University of Berkeley's namesake, the Bishop Berkeley, had disapproved of calculus and mathematicians.

The golden age of probability theory had turned into a double-fronted attack by two camps of warring frequentists united in their abhorrence of Bayes. In the maelstrom, the lack of reasoned discourse among the leaders of statistical mathematics delayed the development of Bayes' rule for decades. Caught in the infighting, the rule was left to find its way alone, stymied and disparaged.

Yet even as the frequentists' assault laid it low, the first glimmerings of a revival flickered here and there, quietly and almost unnoticed. In a remarkable confluence of thinking, three men in three different countries independently came up with the same idea about Bayes: knowledge is indeed highly subjective, but we can quantify it with a bet. The amount we wager shows how much we believe in something.

In 1924 a French mathematician, Émile Borel, concluded that a person's subjective degree of belief could be measured by the amount he was willing to bet. Borel argued that applying probability to real problems, such as insurance, biology, agriculture, and physics, was far more important than mathematical theorizing. He believed in rational behavior and lived as he taught. At the height of a scandal over Marie Curie's possible affair with another scientist, Borel sheltered her and her daughters; in reaction, the minister of public instruction threatened to fire him from his teaching post at the École Normale Supérieur, a leading school of mathematics and science.[40]

Between the two world wars, Borel was a member of the French Chamber of Deputies and a minister of the navy and helped direct national policy toward research and education. Imprisoned briefly during the Second World War by the pro-Nazi Vichy government, he later received the Resistance Medal.

Two years after Borel's suggestion a young English mathematician and philosopher named Frank P. Ramsey made the same suggestion. Before he died at the age of 26 following surgeries for jaundice, he wondered how we should make decisions in the face of uncertainty. In an informal talk to students at the Moral Sciences Club at Cambridge University in 1926 Ramsey suggested that probability was based on personal beliefs that could be quantified by a wager. Such extreme subjectivity broke radically with previous thinkers like Mill, who had denounced subjective probabilities as the abhorrent quantification of ignorance.

Ramsey, who in his brief career also contributed to economics, logic, and philosophy, believed uncertainty had to be described in terms of probability rather than by tests and procedures. By talking about a measure of belief as a basis for action and by introducing a utility function and the maximization of expected utility, he showed how to act in the face of uncertainty. Neither Bayes nor Laplace had ventured into the world of decisions and behavior. Because Ramsey worked in Cambridge, England, the history of Bayes' rule might have been quite different had he lived longer.

At almost the same time as Borel and Ramsey, an Italian actuary and mathematics professor, Bruno de Finetti, also suggested that subjective beliefs could be quantified at the racetrack. He called it "the art of guessing."[41] De Finetti had to deliver his first important paper in Paris because the most powerful Italian statistician, Corrado Gini, regarded his ideas as unsound. (In Gini's defense, de Finetti told colleagues he was convinced Gini had "the evil eye.")[42] De Finetti, considered the finest Italian mathematician of the twentieth century, wrote about financial economics and is credited with putting Bayes' subjectivity on a firm mathematical foundation.

Not even probability experts, however, noticed these outbursts of subjective betting. During the 1920s and 1930s the anti-Bayesian trio of Fisher, Egon Pearson, and Neyman attracted all the attention. Ramsey, Borel, and de Finetti worked outside English-speaking statistical circles.

Another outsider carved out a safe haven for Bayes in paternity law, a small and inconspicuous corner of the American judicial system. Paternity law asks, Did this man father this child? And if so, how much should he pay in child support? In 1938 a Swedish professor of genetics and psychiatry named Erik Essen-Möller developed an index of probability that was mathematically equivalent to Bayes' theorem. For 50 years, until DNA profiling became available, American lawyers used the Essen-Möller index without knowing its Bayesian paternity. In the U.S. Uniform Parentage Act, Bayes even became a model for state legislation. Because paternity lawyers began by assigning 50–50 odds to the man's innocence, the index favored fathers even though Essen-Möller believed that "mothers more frequently accuse true than false fathers."[43] Bayesian paternity law was also used in immigration and inheritance cases and in cases where a child was born as a result of rape. Today, DNA evidence typically gives paternity probabilities of 0.999 or more.

Yet another outsider, Lowell J. Reed, a medical researcher at Johns Hopkins University in Baltimore, dramatized the shortcomings of frequentism and the value of Bayes in 1936. Reed, a member of the department of biostatistics, wanted to determine the X-ray dosages that would kill cancerous tumors but leave patients unharmed. He had no precise exposure records, however, and the effects of low doses were not understood. Reed normally used frequency methods and repeated tests on fruit flies, protozoa, and bacteria; but to ascertain doses for humans he would have to use expensive mammals. With Bayes, he determined the most therapeutic doses for human cancer patients by sacrificing a relatively small number of cats, 27. But Reed worked outside the statistical mainstream, used Bayes only occasionally, and had little influence on statistics. Even Ramsey, Borel, de Finetti, and Essen-Möller had to wait decades before the importance of their work was recognized.

It was a geophysicist, Harold Jeffreys, who almost singlehandedly kept Bayes alive during the anti-Bayesian onslaught of the 1930s and 1940s. Cambridge University students liked to joke that they had the world's two greatest statisticians, although one was a professor of astronomy and the other a professor of genetics. Fisher was the geneticist. Jeffreys was an Earth scientist who studied earthquakes, tsunamis, and tides. He said he qualified for the astronomy department "because Earth is a planet."[44]

Thanks in large part to Jeffreys's quiet, gentlemanly personality he and Fisher became friends even though they disagreed, irrevocably and vociferously, over Bayes. Jeffreys said he told Fisher that "on most things we should agree and when we disagreed, we would both be doubtful. After that, Fisher and I were great friends."[45] For example, Jeffreys believed that Fisher's maximum likelihood method was basically Bayesian, and he often used it because with large samples the prior did not matter and the two techniques produced approximately the same results. They differed, however, when small amounts of data were involved. Years later others would dramatize situations where Jeffreys's and Fisher's significance test results can differ by an order of magnitude.

Aside from their views on Bayes, Jeffreys and Fisher had much in common. Both were practicing scientists who manipulated statistical data; neither was a mathematician or statistician. Both were educated at Cambridge; Jeffreys, in fact, never left and was a fellow there for 75 years, longer than any other professor. Neither man was outgoing; both were appalling lecturers, their feeble voices inaudible beyond a few rows, and a student once counted Jeffreys mumbling "er" 71 times in five minutes. Both were knighted for their work.

Of the two, Jeffreys led the richer personal life. At the age of 49 he married his longtime collaborator, the mathematician Bertha Swirles; they proofread their monumental *Methods of Mathematical Physics* during all-night stints as air raid wardens during the Second World War. He enjoyed annotating discrepancies in detective novels, singing tenor in choral societies, botanizing, walking, traveling, and, until he was 91, bicycling to work.

Like Laplace, Jeffreys studied the formation of the Earth and planets in order to understand the origin of the solar system. He became involved in statistics because he was interested in how earthquake waves travel through the Earth. A major earthquake generates seismic waves that can be recorded thousands of miles away. By measuring their arrival times at different stations, Jeffreys could work back to determine the earthquake's likely epicenter and the likely composition of the Earth. It was a classic problem in the inverse probability of causes. In 1926 Jeffreys inferred that Earth's central core is liquid—probably molten iron, probably mixed with traces of nickel.

As one historian said, "Perhaps in no other field were as many remarkable inferences drawn from so ambiguous and indirect data."[46] Signals were often difficult to interpret, and seismograph machines differed greatly. Earthquakes, which often occurred far apart under very different conditions, were hardly repeatable. Jeffreys's conclusions involved far more uncertainty than Fisher's breeding experiments, which were designed to answer precise, repeatable questions. Like Laplace, Jeffreys spent a lifetime updating his observations with new results. He wrote, "The propositions that are in doubt . . . constitute the most interesting part of science; every scientific advance involves a transition from complete ignorance, through a stage of partial knowledge based on evidence becoming gradually more conclusive, to the stage of practical certainty."[47]

Working on his office floor ankle-deep with papers, Jeffreys composed *The Earth: Its Origin, History, and Physical Constitution*, the standard work on the planet's structure until plate tectonics was discovered in the 1960s. (Sadly, while Jeffreys played the hero defending Bayes, he opposed the idea of continental drift as late as 1970, when he was 78, because he thought it meant the continents would have to push their way through viscous liquid.)

While analyzing earthquakes and tsunamis, Jeffreys worked out a new, objective form of Bayes for scientific applications and devised formal rules for selecting priors. As he put it, "Instead of trying to see whether there was any more satisfactory form of the prior probability, a succession of authors have said that the prior probability is nonsense and therefore that the principle of inverse probability, which cannot work without [priors], is nonsense too."[48]

Jeffreys considered probability appropriate for all uncertainty, even something as apparently certain as a scientific law, whereas frequentists usually restricted probability to the uncertainties associated with theoretically repeatable data. As the statistician Dennis Lindley wrote, Jeffreys "would admit a probability for the existence of the greenhouse effect, whereas most [frequentist] statisticians would not and would confine their probabilities to the data on CO_2, ozone, heights of the oceans, etc."[49]

Jeffreys was particularly annoyed by Fisher's measures of uncertainty, his "p-values" and significance levels. The p-value was a probability statement about data, given the hypothesis under consideration. Fisher had developed them for dealing with masses of agricultural data; he needed some way to determine which should be trashed, which filed away, and which followed up on immediately. Comparing two hypotheses, he could reject the chaff and save the wheat.

Technically, p-values let laboratory workers state that their experimental outcome offered statistically significant evidence against a hypothesis if the outcome (or a more extreme outcome) had only a small probability (under the hypothesis) of having occurred by chance alone.

Jeffreys thought it very strange that a frequentist considered possible outcomes that had not occurred. He wanted to know the probability of his hypothesis about the epicenter of a particular earthquake, given his information about the arrival times of tsunamis caused by the earthquake. Why should possible outcomes that had not occurred make anyone reject a hypothesis? Few researchers repeated—or could repeat—an experiment at random many, many more times. "Imaginary repetitions," a critic called them. Bayesians considered data as fixed evidence, not as something that can vary. Jeffreys certainly could not repeat a particular earthquake. Moreover, the p-value is a statement about data, whereas Jeffreys wanted to know about his hypothesis given his data. As a result, Jeffreys proposed using only observed data with Bayes' rule to compute the probability that the hypothesis was true.

Newton, as Jeffreys pointed out, derived his law of gravity 100 years before Laplace proved it by discovering Jupiter's and Saturn's 877-year cycle: "There has not been a single date in the history of the law of gravitation when a modern significance test would not have rejected all laws [about gravitation] and left us with no law."[50]

Bayes, on the other hand, "makes it possible to modify a law that has stood criticism for centuries without the need to suppose that its originator and his followers were useless blunderers."[51]

Jeffreys concluded that p-values fundamentally distorted science. Frequentists, he complained, "appear to regard observations as a basis for possibly rejecting hypotheses, but in no case for supporting them."[52] But odds are that at least some of the hypotheses Fisher rejected were worth investigating or were actually true.

A frequentist who tests a precise hypothesis and obtains a p-value of .04, for example, can consider that significant evidence against the hypothesis. But Bayesians say that even with a .01 p-value (which many frequentists would see as extremely strong evidence against a hypothesis) the odds in its favor are still 1 to 9 or 10—"not earth-shaking," says Jim Berger, a Bayesian theorist at Duke University. P-values still irritate Bayesians. Steven N. Goodman, a distinguished Bayesian biostatistician at Johns Hopkins Medical School, complained in 1999, "The p-value is almost nothing sensible you can think of. I tell students to give up trying."[53]

Jeffreys was making Laplace's probability of causes useful for practicing scientists, even as Fisher was doing the same for Laplace's frequency-based methods. The difference was that Fisher used the word "Bayes" as an insult, while Jeffreys called it the Pythagorean theorem of probability theory. As the first since Laplace to apply formal Bayesian theory to a variety of important scientific problems, Jeffreys became the founder of modern Bayesian statistics.

Statistically, the lines were drawn. Jeffreys and Fisher, two otherwise cordial Cambridge professors, embarked on a two-year debate in the Royal Society's *Proceedings*. Jeffreys may have been shy and uncommunicative, but when he was sure of himself he dug in, placidly but implacably. Fisher remained his usual "volcanic and paranoid" self.[54] Both men were magnificent scientists, the world's leading statisticians, and each used the methods best suited to his respective field. Yet neither could see the other's point of view. Like gladiators of old, they hurled impassioned articles back and forth, criticizing one another and issuing formal rejoinders, probing rebuttals, and brilliant clarifications—until Royal Society editors threw up their hands in exasperation and ordered the warriors to cease and desist.

After the grand debate, Jeffreys wrote a monumental book, *Theory of Probability*, which remained for years the only systematic explanation of how to apply Bayes to scientific problems. Fisher complained publicly that Jeffreys makes "a logical mistake on the first page which invalidates all the 395 formulae in his book."[55] The mistake, of course, was to use Bayes' theorem. Summarizing Jeffreys's books, Lindley said, "De Finetti is a master of theory, Fisher a master of practice, but Jeffreys is brilliant at both."[56]

The Fisher–Jeffreys debate ended inconclusively. Practically speaking, however, Jeffreys lost. For the next decade and for a variety of reasons frequentism almost totally eclipsed Bayes and the inverse probability of causes.

First, Fisher was persuasive in public, while the mild-mannered Jeffreys was not: people joked that Fisher could win an argument even when Jeffreys was right. Another factor was that social scientists and statisticians needed objective methods in order to establish themselves as academically credible in the 1930s. More particularly, physicists developing quantum mechanics were using frequencies in their experimental data to determine the most probable locations of electron clouds in nuclei. Quantum mechanics was new and chic, and Bayes was not.

In addition, Fisher's techniques, written in a popular style with minimal mathematics, were easier to apply than those of Jeffreys. A biologist or psychologist could easily use Fisher's manual to determine whether results were statistically significant. To use Jeffreys's rather opaque and mathematical approach, a scientist had to choose among five nuanced categories: the evidence against the hypothesis is "not worth more than a bare mention" or it is substantial, strong, very strong, or decisive.[57] Characteristically, Jeffreys tucked the five categories into appendix B of his book.

Finally and most important, Jeffreys was interested in making inferences from scientific evidence, not in using statistics to guide future action. To him, decision making—so important for the rise of mathematical statistics during the Second World War and the Cold War—was irrelevant. Others parted at the same divide: a big reason for the feud between Fisher and Neyman, for example, was decision theory.

All these factors left Jeffreys almost totally isolated from statistical theorists. His link with Fisher was their interest in applying statistics to science. Jeffreys knew Ramsey and visited him as he was dying in the hospital, but neither realized the other was working on probability theory; in any case, Jeffreys was interested in scientific inference and Ramsey in decision making. Jeffreys and de Finetti worked on similar probability issues during the 1930s, but Jeffreys did not even know the Italian's name for half a century and would have rejected outright de Finetti's subjectivity. Most statisticians ignored Jeffreys's book on probability theory for years; he said that "they were completely satisfied with frequency theories."[58] Jeffreys accepted a medal from the Royal Statistical Society, but attended none of its meetings. Geophysicists did not know about his probability work; a surprised geologist once asked Lindley, "You mean that your Jeffreys is the same Jeffreys as mine?"[59]

By 1930 Jeffreys was truly a voice in the wilderness. Most statisticians were using the powerful body of ideas developed by the anti-Bayesian trio. Jeffreys's great book Theory of Probability was published as part of a series of physics books, not statistics books. It also appeared in the last year of peace, just before the start of the Second World War and a new opportunity for Bayes' rule.

part II
second world war era

4.

bayes goes to war

By 1939 Bayes' rule was virtually taboo, dead and buried as far as statisticians in the know were concerned. A disturbing question remained, though. How could wartime leaders make the best possible life-and-death decisions swiftly, without waiting for complete information? In deepest secrecy some of the greatest mathematical minds of the century would contribute to rethinking Bayes' role during the uncertain years ahead.

The U-boat peril was the only thing that ever really frightened Winston Churchill during the Second World War, he recalled in his history of the conflict. Britain was self-sufficient in little other than coal; it grew enough food to feed only one in three residents. But after the fall of France in 1940, Germany controlled Europe's factories and farms, and unarmed merchant ships had to deliver to Britain 30 million tons of food and strategic supplies a year from the United States, Canada, Africa, and eventually Russia. During the Battle of the Atlantic, as the fight to supply Britain was called, German U-boats would sink an estimated 2,780 Allied ships, and more than 50,000 Allied merchant seamen would die. For Prime Minister Churchill, feeding and supplying his country was the dominating factor throughout the war.

Hitler said simply, "U-boats will win the war."[1]

U-boat operations were tightly controlled by German headquarters in occupied France. Each submarine went to sea without orders and received them by radio after it was well out in the Atlantic. As a result, an almost endless cascade of coded radio messages—more than 49,000 are still archived —raced back and forth between the U-boats and France. Although the British desperately needed to know where the U-boats were, the messages were unreadable. They had been encrypted by word-scrambling machines, and no one in Germany or Britain thought their codes could be broken.

Strangely enough, the Poles were the first to think otherwise. A few intelligence officers in Poland, sandwiched as they were between Germany and Russia, realized a full decade before the start of the Second World War that mathematics could make eavesdropping on their rapacious neighbors quite informative. The First World War had made the need for machines to encode radio messages painfully obvious. When an alphabet-scrambling machine was exhibited at an international trade show in 1923, Germany bought some and began introducing complexities to make their codes more secure. The machines were named Enigma.

And enigmas they were. The Poles spent three years trying unsuccessfully to crack German messages before realizing that automated cipher machines had transformed cryptography. The science of coding and decoding secret messages had become a game for mathematicians. When the Polish secret service organized a top-secret cryptography class for German-speaking mathematics students, its star pupil was an actuarial mathematician named Marián Rejewski. He used inspired guesswork and group theory—the new mathematics of transformation—to make a crucial discovery: how the wheels on an Enigma were wired. By early 1938 the Poles were reading 75% of Germany's army and air force messages. Shortly before their country was invaded in 1939 they invited French and British agents to a safehouse in the Pyry Forest outside Warsaw, revealed their system, and sent an updated Enigma machine to London.

To an observer, an Enigma looked rather like a complicated typewriter, with a traditional keyboard of 26 letter keys and a second array of 26 lettered lights. Each time a typist pressed a letter key, an electric current passed through a set of three wheels and advanced one of them a notch. The enciphered letter lit up on the lampboard, and the typist's assistant read the letter off to a third aide, who radioed the scramble in Morse code. At its destination, the process was reversed. The recipient typed the coded letters into his Enigma keyboard, and the original message lit up on his lampboard. By changing the wiring, wheels, starting places, and other features, an Enigma operator could churn out millions upon millions of permutations.

Germany standardized its military communications with increasingly complex versions of the machines. Approximately 40,000 military Enigmas were distributed to the German army, air force, navy, paramilitary, and high command as well as to the Spanish and Italian nationalist forces and the Italian navy. When German troops invaded Poland on September 1, 1939, battery-powered Enigmas were the key to their high-speed blitzkrieg as field officers in Enigma-equipped command vehicles coordinated, as never before, a barrage of artillery fire, dive-bombing airplanes, and panzer tanks. Most German naval vessels, particularly battleships, minesweepers, supply ships, weather report boats, and U-boats, had an Enigma.

Unlike the Poles, the British agency charged with cracking German military codes and ciphers clung to the tradition that decryption was a job for gentlemen with linguistic skills. Instead of hiring mathematicians, the Government Code and Cypher School (GC&CS) employed art historians, scholars of ancient Greek and medieval German, crossword puzzlers, and chess players. Mathematicians were regarded as "strange fellows."[2]

The British government and educational systems treated applied mathematics and statistics as largely irrelevant to practical problems. Well-to-do boys in English boarding schools learned Greek and Latin but not science and engineering, which were associated with low-class trade. Britain had no elite engineering schools like MIT or the École Polytechnique. Two years into the war, when government officials went to Oxford to recruit men proficient in both mathematics and modern languages, they found only an undergraduate mathematics major teaching himself beginning German. The government did not even plan to exempt mathematicians from combat. Knowing that their skills would be needed eventually, mathematicians quietly spread word to their colleagues to register with the government as physicists because they at least were considered vital to the nation's defense.

Exacerbating the emergency was the fact that the government regarded statistical data as bothersome details. A few months before war was declared in 1939, the giant retailer Lord Woolton was asked to organize the clothing for Britain's soldiers. He discovered to his horror that "the War Office had no statistical evidence to assist me. . . . I had the greatest difficulty in arriving at any figures that would show how many suits of uniform and how many boots were involved."[3] The Department of Agriculture ignored a study of the fertilizers needed to increase Britain's food and timber supplies because it thought the Second World War was going to be a nonscientific war and no more data would be needed. Government functionaries also seemed to think that applying mathematics to real life would be easy. When the Ministry of Supply needed to assess new rockets, it gave an employee one week to "learn statistics."[4]

Probability experts were scarce. For a small elite the 1930s had been the golden age of probability theory, the language of statistics. But the majority of mathematicians thought of probability as arithmetic for social scientists. Cambridge, the center of British mathematics, was a backwater in probability. Germany, a leader in modern mathematics and quantum physics, produced few statisticians. And one of the greatest probability thinkers of the twentieth century, Wolfgang Doeblin, was a 25-year-old French soldier fighting for his life as France fell to the Germans in June 1940. The Gestapo was hunting his father, and Doeblin, surrounded and without hope of escape, killed himself to avoid any chance of betraying his parent. Doeblin's work would one day be crucially relevant to chaos theory and random mapping transformations.

Oddly, the Allies' top three statisticians were sidelined during the war. Harold Jeffreys was ignored, perhaps because he was an earthquake specialist and astronomy professor. British security apparently considered Ronald Fisher, the anti-Bayesian geneticist, to be politically untrustworthy because he had corresponded with a German colleague. Fisher's offers to help the war effort were ignored, and his application for a visa to the United States was rejected without explanation. A chemist calculating the dangers of poison gas succeeded in arranging a visit to Fisher only by claiming he was collecting a horse nearby. As for Jerzy Neyman, he persisted in carrying out full theoretical studies that could lead to a new theorem even though the military desperately needed quick and dirty advice; one of Neyman's grants was formally terminated.

With applied mathematicians and statisticians in short supply, wartime data were often analyzed not by statisticians but by actuaries, biologists, physicists, and pure mathematicians—few of whom knew that, as far as sophisticated statistics was concerned, Bayes' rule was unscientific. Their ignorance proved fortunate.

Despite the strange reputation of British mathematicians, the operational head of GC&CS prepared for war by quietly recruiting a few nonlinguists—"men of the Professor type"[5]—from Oxford and Cambridge universities. Among that handful of men was Alan Mathison Turing, who would father the modern computer, computer science, software, artificial intelligence, the Turing machine, the Turing test—and the modern Bayesian revival.

Turing had studied pure mathematics at Cambridge and Princeton, but his passion was bridging the gap between abstract logic and the concrete world. More than a genius, Turing had imagination and vision. He had also developed an almost unique set of interests: the abstract mathematics of topology and logic; the applied mathematics of probability; the experimental derivation of fundamental principles; the construction of machines that could think; and codes and ciphers. Turing had already spent hours in the United States discussing cryptography in his high-pitched stammer with a Canadian physicist named Malcolm MacPhail.

After Turing returned to England in the spring of 1939, his name was quietly added to a short "emergency list" of people with orders to report immediately to the GC&CS in the event war was declared. He worked alone that summer, studying both probability theory and Enigma codes. Occasionally he visited GC&CS to talk with a cryptanalyst, Dillwyn Knox, who had already solved a relatively simple Enigma code used by the Italian navy. By the time Germany invaded Poland, Knox and Turing probably understood more about military Enigmas than anyone else in Britain.

On September 4, the day after England declared war on Germany, Turing took a train to the GC&CS research center in Bletchley Park, a small town north of London. He was 27 but looked 16. He was handsome, athletic, shy, and nervous and had been openly homosexual at Cambridge. He cared little about appearances; he wore shabby sports coats and had dirty fingernails and a permanent five-o'clock shadow. He would devote the next six years to Enigma and to other coding and decoding projects.

On his arrival in Bletchley Park, GC&CS analysts divided up the Enigma systems, and Turing worked awhile on army codes. By January the English were reading German air force messages. During the first weeks of the war Turing also designed the "bombe." This was not a weapon in the traditional sense but a high-speed electromechanical machine for testing every possible wheel arrangement in an Enigma. Turing's bombe, a radical redesign and upgrade of the device invented by the Poles, would turn Bletchley Park into a code-breaking factory. Turing's machine tested hunches, 15-letter tidbits suspected of being in the original message. Because it was faster to toss out possibilities than to find one that fit, Turing's bombe simultaneously tested for wheel combinations that could not produce the hunch.

Turing refined the bombe's design with the help of mathematician Gordon Welchman and engineer Harold "Doc" Keen. Their prototype, a metal cabinet roughly 7 by 6 by 2.5 feet, appeared at Bletchley Park in March 1940. Some believe the bombe's design was Turing's biggest contribution to breaking Enigma.

Despite the progress made on breaking German air force and army codes, no one at Bletchley Park wanted to tackle the German naval codes, the key to winning the U-boat war in the Atlantic. Of all the branches of the

Axis military, Hitler's navy operated the most complex Enigma machines and security systems. By war's end, a naval Enigma machine could be set up an astronomical number of ways. According to a Bletchley Park decoder, "All the coolies in China could experiment for months without reading a single message."[6] At any one time the machine could use 1 of 4 reflector combinations (each of which could be set in 26 different ways); 3 of 8 rotors (giving up to 336 permutations); more than 150 billion plugboard combinations; 17,000 possible clip positions around the rotors; and 17,000 possible starting positions (half a million in four-rotor machines). Many of these settings were changed every two days, sometimes every 8 or 24 hours.

According to Frank Birch, head of the GC&CS naval intelligence branch, superior officers informed him that the "German codes were unbreakable. I was told it wasn't worthwhile putting pundits onto them. . . . Defeatism at the beginning of the war, to my mind, played a large part in delaying the breaking of the codes."[7] The naval codes were assigned to one officer and one clerk; not a single cryptanalyst was involved. Birch, however, thought the naval Enigma could be broken because it had to be. The U-boats put Britain's very existence at stake.

Turing had still another attitude. The fact that no one else wanted to work on the naval codes made them doubly attractive. A close friend called Turing "a confirmed solitary."[8] Isolation appealed to him. Announcing that "no one else was doing anything about it and I could have it to myself," Turing decided to attack the German naval code.[9] He began working on naval Enigma with a staff of two "girls" and an Oxford mathematician-physicist, Peter Twinn.[10] Turing thought the code "could be broken because it would be so interesting to break it."[11]

One of Turing's first jobs was to reduce the number of tests a bombe had to conduct. Although it was fast, a bombe took 18 minutes to test a possible wheel setting. Assuming the worst, a bombe would need four days to test all 336 possible wheel permutations on an Enigma. Until more bombes could be built, their workload had to be drastically reduced.

Late one night soon after joining Bletchley Park, Turing invented a manual method for reducing the burden on the bombes. It was a highly labor-intensive, Bayesian system he nicknamed Banburismus for the nearby town of Banbury, where a printing shop would produce needed materials.

"I was not sure that it would work in practice," Turing said.[12] But if it did, it would let him guess a stretch of letters in an Enigma message, hedge his bets, measure his belief in their validity by using Bayesian methods to assess their probabilities, and add more clues as they arrived. If it worked, it would identify the settings for 2 of Enigma's 3 wheels and reduce the number of wheel settings to be tested on the bombes from 336 to as few as 18. At a time when every hour counted, the difference could save lives.

Turing and his slowly growing staff began to comb intelligence reports to collect "cribs," Bletchley-ese for German words predicted to occur in the plain-text, that is, the original, uncoded message. The first cribs came primarily from German weather reports because they were standardized and repeated often: "Weather for the night," "Situation Eastern Channel," and, as one blessed fool radioed nightly, "Beacons lit as ordered." Reports from British meteorologists about weather in the Channel provided more hunches. Knowing the most frequent letter combinations in German words helped too. When a prisoner of war told them the German navy spelled out numbers, Turing realized that the word "ein" ("one," "a," or "an") appeared in 90% of Enigma messages; Bletchley Park clerks catalogued by hand 17,000 ways "ein" could be encrypted, and a special machine was constructed to screen for them.

In a fundamental breakthrough, Turing realized he could not systematize his hunches or compare their probabilities without a unit of measurement. He named his unit a ban for Banburismus and defined it as "about the smallest change in weight of evidence that is directly perceptible to human intuition."[13] One ban represented odds of 10 to 1 in favor of a guess, but Turing normally dealt with much smaller quantities, decibans and even centibans. The ban was basically the same as the bit, the measure of information Claude Shannon discovered by using Bayes' rule at roughly the same time at Bell Telephone Laboratories. Turing's measure of belief, the ban, and its supporting mathematical framework have been called his greatest intellectual contribution to Britain's defense.

To estimate the probability of a guess when information was arriving piecemeal, Turing used bans to discriminate between sequential hypotheses. He was thus one of the first to develop what came to be called sequential analysis. He used bans to quantify how much information was needed to solve a particular problem so that, instead of deciding how many observations to make, he could target the amount of evidence needed and stop when he had it.

Bans involved a manual, paper-and-pencil system far removed from a modern computerized Bayesian calculation. Bans automated the kind of subjective guessing that Émile Borel, Frank Ramsey, and Bruno de Finetti had tried to validate during the anti-Bayesian onslaught of the 1920s and 1930s. Using Bayes' rule and bans, Turing began calculating credibility values for various kinds of hunches and compiling reference tables of bans for technicians to use. It was a statistics-based technique and produced no absolute certainties, but when the odds of a hypothesis added up to 50 to 1, cryptanalysts could be close to certain they were right. Each ban made a hypothesis 10 times more likely.

A top modern-day cryptographer explained Turing's thinking: "When you work day after day, year after year, you need to make a best guess of what's most likely to be breakable with the resources at hand. You may have too many choices, so you pick the more checkable guesses. At every step you hedge bets. . . . Sometimes you make approximations, and other times you have precisely correct numbers with the right formulas, the right numbers, for the decibans."[14]

In operation, Banburismus used 5- or 6-foot-long strips of thin cardboard printed in Banbury. Decoders look for repetitions and coincidences, so Wrens, technicians from the Women's Royal Naval Service, punched each intercepted message by hand, letter by letter, into a Banbury sheet. Then they slipped one strip on top of others so that any two messages could be compared. When enough letter holes showed through both Banburies, the number of repeats was recorded.

As Patrick Mahon, who worked on Banburismus during the war, wrote in his secret history of Bletchley Park, "If by any chance, the two messages have identical content for 4 or 6 or 8 more letters . . . such a coincidence between cipher texts is known as a 'fit.'"

"The game of Banburismus involved putting together large numbers of pieces of probabilistic information somewhat like the reconstruction of DNA sequences," Turing's statistical assistant, I. J. "Jack" Good, explained later.[15] Good, the son of a Jewish watchmaker from tsarist Russia, had studied pure mathematics at Cambridge and waited a year for a defense job before being hired on the strength of his chess playing. Good thought "the game of Banburismus was enjoyable, not easy enough to be trivial, but not difficult enough to cause a nervous breakdown."[16] Bayes' rule was proving to be a natural for cryptography, good for hedging bets when there were prior guesses and decisions to be made with a minimum of time or cost.

Turing was developing a homegrown Bayesian system. Finding the Enigma settings that had encoded a particular message was a classic problem in the inverse probability of causes. No one is sure where Turing picked Bayes up, whether he rediscovered it independently or adapted it from something overheard about Jeffreys, Cambridge's lone defender of Bayes' rule before the war. All we know for sure is that, because Turing and Good had studied pure mathematics and not statistics, neither had been sufficiently poisoned by anti-Bayesian attitudes.

In any event, Turing talked at Bletchley Park about bans, not Bayes.

Once Good asked, "Aren't you essentially using Bayes' theorem?"[17] Turing answered, "I suppose." Good concluded that Turing knew of the theorem's existence. But Turing and Good may have been the only ones at Bletchley Park who realized that Banburismus was Bayesian, and heavily so.

Good met a friend, George A. Barnard, one day in London and—strictly against the rules—"told him that we were using Bayes factors, and their logarithms, sequentially, to discriminate between two hypotheses but of course I did not mention the application. Barnard said that curiously enough a similar method was being used for quality control in the Ministry of Supply for discriminating between lots rather than hypotheses. It was really the same method because the selection of a lot can be regarded as the acceptance of a hypothesis."[18] Sequential analysis differed from frequency-based testing, where the number of items to be tested was fixed from the beginning. In sequential analysis, once several tests or observations strongly cleared or condemned a case of, say, field rations or machine-gun ammunition, the tester could move on to the next box. This almost halved the number of tests required, and the use of logarithms massively simplified calculations by substituting addition for multiplication. Abraham Wald of Columbia University is generally credited with discovering sequential analysis for testing ammunition in the United States later during the war. But Good concluded that Turing had used it first and that Turing, Wald, and Barnard all deserved credit for discovering and applying it. Oddly enough, after the war Barnard would become a prominent anti-Bayesian.

Turing was making progress when, in May 1940, the doldrums hit. He had both the theory and the method for breaking Enigma codes but still could not read U-boat messages. The Germans were building more U-boats, and Adm. Karl Doenitz had formed wolf packs of subs strung across the North Atlantic; when one U-boat spotted a convoy, it radioed the rest. During the first 40 months of the war, U-boats sank 2,177 merchant ships totaling more than 1 million tons, far more than were lost to German aircraft, mines, warships, and other causes.

If the British were going to be able to route supply convoys around the U-boats, Turing needed more information. He needed to see one of the code-books that U-boat Enigma operators used before broadcasting a ciphered message. One of the factors that made breaking the Enigma code so difficult was that the operator *doubly*-enciphered a trio of letters that began each message and that indicated the starting positions of the Enigma's three wheels. The operator enciphered the three letters twice over: once mechanically, with his Enigma machine, and once manually, by selecting one of nine sets of tables in a codebook issued to each sub. The operator learned which table to use each day by consulting a calendar issued with the tables. If a U-boat came under attack, crews had strict orders to destroy the tables either before abandoning ship or as the enemy was about to board.

In a brilliant piece of deduction shortly after war was declared, Turing figured out this double-encipherment system, but he needed a copy of the codebook to make Banburismus work. Enigmas had so many variations that trial-and-error methods were ineffective. A codebook had to be "pinched," as Turing put it. The wait for a pinch would stretch through ten nerve-racking months.

As Turing waited desperately for the navy to get him a codebook, morale at GC&CS sank. Alastair G. Denniston, the head of GC&CS, told Birch, "You know, the Germans don't mean you to read their stuff, and I don't expect you ever will."[19]

Long and bitter arguments broke out about whether more bombes should be built, and if so, how many. In August 1940 Birch wrote, "Turing and Twinn are like people waiting for a miracle, without believing in miracles. . . . Turing has stated categorically that with 10 machines [bombes] he could be sure of breaking Enigma and keeping it broken. Well can't we have 10 machines?"[20]

A second bombe incorporating Welchman's improvements arrived later that month, but the fight for more bombes continued throughout 1940. Birch complained that the British navy was not getting its fair share of the bombes: "Nor is it likely to. It has been argued that a large number of bombes would cost a lot of money, a lot of skilled labour to make and a lot of labour to run, as well as more electric power than is at present available here. Well, the issue is a simple one. Tot up the difficulties and balance them against the value to the Nation of being able to read current Enigma."[21]

To capture a codebook, Lt. Cmdr. Ian Fleming, the future creator of James Bond but at the time an aide to the head of Britain's Directorate of

Naval Intelligence, concocted Operation Ruthless. It was a scheme worthy of his postwar spy. The British would outfit a captured German plane with a crew that was to include a "word-perfect German speaker" (Fleming himself, who had studied German in Austria as a youth).[22] After the plane faked a crash into the Channel and its crew was rescued by a German boat, the British would capture the vessel and bring it and its Enigma equipment home to Turing. The escapade was elaborately planned but canceled, and Turing and Twinn went to Birch looking "like undertakers cheated of a nice corpse . . . , all in a stew."[23] Instead, documents and papers—bits and pieces of clues to the contents of the all-important codebooks—were taken from two weather ships captured off Iceland and, in a commando raid organized specifically to help Turing, from an armed German trawler off the Norwegian coast. With these clues, Turing began trying to deduce the contents of the all-important codebooks.

Turing's group was beginning to break the German naval cipher on the glorious day of May 27, 1941, when the British sank the Bismarck, then the world's largest battleship. By June Turing had succeeded in reconstructing the codebooks from various clues, and for the first time Bletchley Park could read the messages to and from the U-boat wolf packs within an hour of their arrival. Finally, the British could reroute convoys safely around the subs. For 23 blessed days in June 1941, a time when Britain still fought alone, no convoy in the North Atlantic was attacked.

By then, Bletchley Park regarded Turing fondly as its eccentric genius, although some of his unconventional behavior made practical sense. He wore a gas mask while bicycling to work during the June hay fever season. And he managed his bicycle's broken chain by counting pedal strokes and executing a certain maneuver every 17 revolutions. Bicycle parts were scarce, and he liked identifying repeated patterns in his work.

By autumn of 1941, Banburismus was again in trouble, critically short of typists and junior clerks, otherwise known as "girl power." Turing and three other decoders took a direct but unorthodox approach to the problem. Appealing directly to Churchill on October 21, they wrote, "We despair of any early improvement without your intervention." Welchman probably drafted the letter, but Turing signed it first, followed by Welchman, their colleague Hugh Alexander, and P. Stuart Milner-Barry, a Cambridge mathematics graduate who was the chess correspondent for The Times newspaper. Milner-Barry took a train to London, hailed a taxi, and "with a sense of total incredulity (can this really be happening?) invited the taxi driver to take him to 10 Downing Street." There he persuaded a brigadier general to deliver the letter personally to the prime minister and to stress its urgency.

Churchill, who had visited Bletchley Park, had recently been informed that Britain was running out of food and war supplies. He immediately sent a memorandum to his chief of staff: "Action this day: Make sure they have all they want on extreme priority and report to me that this had been done."[24] Turing and company heard nothing directly in response but noticed that work went more smoothly, bombes were built faster, and staff arrived sooner.

As Bletchley Park was beginning to break naval Enigma, Hitler invaded Russia with two-thirds of his forces in June 1941 and launched a merciless bombardment of Moscow. Early in the campaign, Russia's greatest mathematician, Andrei Kolmogorov, was evacuated east to safety in Kazan along with the rest of the Russian Academy of Sciences. Shortly after, Russia's Artillery Command, reeling from Germany's massive bombing raids, asked Kolmogorov to return to the capital for consultations. Amidst the chaos, he was lodged for awhile on a sofa.

In a country that idolized its intelligentsia, Kolmogorov was a famous man. When a professor's wife heard he was going to visit her home, she began frantically cleaning and cooking. When a maid asked why, the hostess replied, "How can I explain it to you? Just imagine that you will be getting a visit from the tsar himself."[25] Kolmogorov's legend began with his mother, an independent woman of "lofty social ideals" who never married and died in childbirth. Her two sisters raised Andrei, ran a small school for him and his friends, and published a newsletter with little problems he had composed, such as "How many different ways can a button with four holes be sewn?"[26] At the age of 19 at Moscow State University he escaped final examinations in his 14 courses by writing 14 original papers. He was more proud of having taught school to pay his way through the university than of winning any of his awards; late in life he volunteered at a school for gifted children, where he introduced them to literature, music, and nature.

Kolmogorov became the world's authority on probability theory. In 1933 he demonstrated that probability is indeed a branch of mathematics, founded on basic axioms and far removed from its indecorous gambling origins. So fundamental was Kolmogorov's approach that any mathematician, frequentist or Bayesian, could legitimately use probability. Kolmogorov himself espoused the frequentist approach.

Now the generals were asking him about using Bayes against the German barrage. Russia's artillery, like that of the French, had used Bayesian firing tables for years, but the generals were split over an esoteric point about aiming. They asked Kolmogorov his opinion.

"Strictly speaking," he told the generals, starting with Bayes' 50–50 prior odds was "not only arbitrary but surely wrong because it contradicts the main requirements of the probability theory."[27] But with Germany on Moscow's doorstep, Kolmogorov felt he had no choice but to start with equal priors. Agreeing with Joseph Bertrand's strictly reformed version of Bayes, Kolmogorov told the generals they should start with 50–50 odds whenever shooting repeatedly at a small area. Because it was sometimes better to shoot randomly than aim precisely, the guns in a battery of weapons should aim slightly wide of the mark, the way a hunter shooting at moving birds uses pellets for wider dispersion.

That same autumn of 1941, Kolmogorov taught a wartime course at Moscow State University on firing dispersion theory and made the class compulsory for probability majors. Surprisingly, on September 15, 1941, three months into the German invasion of Russia, Kolmogorov submitted his theory of firing to a journal for publication. The article was so mathematical and theoretical that Russia's censors, not realizing it could help the Germans as well as the Russians, allowed it to be printed in 1942. Fortunately, the enemy did not understand the theory any better than the censors did. After the war Kolmogorov published two more practical problems of Bayesean artillery that are still in print—in English—for military authorities to study. Years later a general in the Russian artillery recalled that during the invasion Kolmogorov "did a lot of useful things for us as well, we remember it, and appreciate him too."[28]

Shortly after Germany attacked Russia, British radio listening posts intercepted a new kind of German army message. Analysts at Bletchley Park thought it came from a teletype machine. They were right. The Germans were encrypting and decrypting at the speed of typing. The new Lorenz machines and their family of ultrasecret codes were technically far more sophisticated than the Enigmas, which had been built for commercial use in the 1920s. The supreme command in Berlin relied on its new codes to communicate the highest level of strategy to army group commanders across Europe. The messages were so important that Hitler himself signed some of them.

Code-naming the new Lorenz machines Tunny for "tuna fish," a group of Britain's leading mathematicians began a year of desperate struggle. They used Bayes' rule, logic, statistics, Boolean algebra, and electronics. They also began work on designing and building the first of ten Colossi, the world's first large-scale digital electronic computers.

When Good and others started work on the Tunny-Lorenz codes, they incorporated Turing's Bayesian scoring system and his fundamental units of bans, decibans, and centibans. They employed Bayes' theorem and a spectrum of priors: honest priors and improper ones; priors that represented what was known and sometimes not; and in different places both Thomas Bayes' uniform priors and Laplace's unequal ones. To deduce the pattern of cams surrounding the wheels of the Tunny-Lorenz machines Turing invented a highly Bayesian method known as Turingery or Turingismus in July 1942. Turingery was a paper-and-pencil method, "more artistic than mathematical. . . . [You had to rely on what] you felt in your bones," according to Turingery player William T. Tutte.[29] The first step was to make a guess and assume, as Bayes had suggested, that it had a 50% chance of being correct. Add more and more clues, some good, some bad, and "with patience, luck, a lot of rubbing out, and a lot of cycling back and forth," the plain text appeared. When the odds of being correct reached 50 to 1, a pair of wheel settings was declared certain.[30]

As Bletchley Park analysts worked on Tunny's wheel patterns and Russia resisted the German onslaught, Japan attacked the United States at Pearl Harbor on December 7, 1941. Supplying Great Britain immediately became more difficult. When American ships that had protected the convoys supplying Britain were quickly transferred to the Pacific, 15 German U-boats took their places in the shipping lanes off the American East Coast. As convoys of Argentine beef and Caribbean oil hugged the coast, they were silhouetted at night against shore lights that local communities dependent on tourism refused to dim. Miami's neon signs, for example, stretched for six deadly miles. The U-boats, lying in wait at periscope depth, caused three months of devastation until the U.S. military ordered coastal lights turned off at dusk.

Making matters worse, the Atlantic U-boats added a fourth wheel to their Enigmas, and the Turing-Welchman bombes were stymied. For most of 1942 Turing and his coworkers could not read any message to or from German submarines. Bletchley Park called it the Great Blackout. For four months the U-boats ran riot in the Atlantic at large, sinking 43 ships in August and September alone. The average U.S. vessel crossed the Atlantic and back three times before it was sunk on its fourth trip.

Finally, in December 1942, three young British crewmen, Lt. Anthony Fasson, Able-Seaman Colin Grazier, and Tommy Brown, swam from their ship to a sinking German submarine off Egypt to pinch its vital codebook of encrypting tables. Fasson and Grazier drowned in the attempt, but Brown, a 16-year-old canteen assistant, survived to rescue the tables. The tables were raced to Bletchley Park, and within hours of cryptanalysts' receiving them, U-boat messages from the Atlantic were being decrypted and convoys rerouted.

The month before that happened, however, would be the war's most dangerous month for Allied shipping, and during it Turing sailed for the United States on the *Queen Elizabeth*, a fast ship that traveled without convoy. Clearance from the White House made Turing a liaison between Bletchley Park and the U.S. Navy. The British had been teaching the Americans about Enigma in general before Pearl Harbor. Now Turing was to tell U.S. officials everything that had been learned, and the United States would accelerate the production of bombes. Surprisingly, the British planned his trip rather haphazardly. He arrived with inadequate identification, and U.S. immigration authorities almost confined him to Ellis Island. In addition, he had not been told whether he could discuss Tunny code breaking with Americans, and the Americans did not realize he expected to have full access to their voice-scrambling research. Nevertheless, during his stay he held high-level meetings in Dayton, Ohio, Washington, and New York City.

Turing spent at least one afternoon in Dayton, where the National Cash Register Company planned to manufacture 336 bombes. He was dismayed to discover that the U.S. Navy was ignoring Banburismus and its ability to economize on bombe usage. The Americans seemed uninterested in the Enigma outside of their obligation to supply bombes for it.

In Washington, Turing discussed Bletchley Park's methods and bombes with U.S. Navy cryptographers. According to a previous agreement, the United States was concentrating on Japanese navy codes and ciphers while the British worked on Enigma. Bletchley Park had already sent a detailed technical report of its work to the Americans, but a civilian navy cryptographer, Agnes Meyer Driscoll, had sat on it; she had broken many Japanese codes and ciphers before the war and had her own, mistaken notions about how to solve Germany's naval Enigma. Turing's mathematics may also have been too technical for the Americans. At first he was alarmed that no one seemed to be working mathematically "with pencil and paper," and he tried in vain to explain the general principle that confirming inferences suggested by a hypothesis would make the hypothesis itself more probable.[31] Later, he was relieved to meet American mathematicians involved in cryptography.

From Washington, Turing went to the Bell Laboratory in New York City, where he and Claude Shannon met regularly at afternoon tea. Shannon, like Turing and Kolmogorov, was a great mathematician and an original thinker, and he was using Bayes' rule for wartime projects. But Turing and Shannon had more than Bayes in common. Both were shy, unconventional men with deep interests in cryptography and machines that could think. As young men, both had written seminal works combining machines and mathematics. In his master's thesis in mathematics, written at MIT, Shannon showed that Molina's relay circuits could be analyzed using Boolean algebra. Both Turing and Shannon liked cycling. Turing rode a bicycle for transportation and exercise; Shannon avoided social chitchat by riding a unicycle through Bell Labs' hallways, sometimes juggling balls along the way. Both men liked to design equipment, in Shannon's case whimsical machines like a robotic mouse to solve mazes or a computer for Roman numerals. His garage was filled with chess-playing machines. Unlike Turing, though, Shannon had a warm family life. His father was a businessman, his mother a high school principal, and his sister a mathematics professor, and he and his wife had three children.

When Turing visited Bell Labs, the next cryptographic frontier was speech. Britain and the United States wanted their best people, Shannon and Turing, working on it. Shannon was already developing the SigSaly voice scrambler; it had a nonsense nursery-rhyme name but by war's end Franklin D. Roosevelt, Churchill, and their top generals in eight locations around the world could talk together in total secrecy. With naval Enigma reduced to a largely administrative problem, Turing would tackle voice communications when he returned to Britain. When Turing and Shannon met for tea, they probably discussed the SigSaly.

Shannon was also working on a theory of communication and information and its application to cryptography. In a brilliant insight Shannon realized that noisy telephone lines and coded messages could be analyzed by the same mathematics. One problem complemented the other; the purpose of information is to reduce uncertainty while the purpose of encryption is to increase it. Shannon was using Bayesian approaches for both. He said, "Bell Labs were working on secrecy systems. I'd work on communications systems and I was appointed to some of the committees studying cryptanalytic techniques. The work on both the mathematical theory of communications and the cryptography went forward concurrently from about 1941. I worked on both of them together and I had some of the ideas while working on the other. I wouldn't say that one came before the other—they were so close together you couldn't separate them."[32]

Shannon's efforts united telegraph, telephone, radio, and television communication into one mathematical theory of information. Roughly speaking, if the posterior in a Bayesian equation is quite different from the prior, something has been learned; but when a posterior is basically the same as the prior guess, the information content is low.

Communication and cryptography were in this sense the reverse of one another. Shannon called his logarithmic units for measuring information binary dibits, or bits, a word suggested by John W. Tukey of Bell Labs and Princeton University. In a confidential report published in 1949 Shannon used Bayes' theorem and Kolmogorov's theory of probability from 1933 to show that, in a perfectly secret system, nothing is learned because the prior and posterior of Bayes' theorem are equal. Bell Labs communications theorists were still developing extensions of Shannon's theory and using Bayesian techniques extensively in 2007.

Returning home, Turing boarded the *Empress of Scotland* in New York City on March 23, 1943. New York was the world's greatest port during the war: more than 50 vessels streamed in and out of the city's harbors each day. Turing was traveling during what would be the second most dangerous month of the war for Allied shipping. The U-boat offensive reached its peak that month and would sink 108 Allied ships while losing only 14 subs. Germany had broken the convoys' routing cipher. Approximately 1,350 mostly unarmed merchant ships were at sea every day that spring. They joined a long coastal shipping line that stretched from Brazil to the mouth of the St. Lawrence River, where they formed convoys to cross the Atlantic. Allied escort vessels concentrated on protecting convoys carrying troops to Britain for an invasion of Europe, however, so Turing's ship was one of 120 fast-moving ships that traveled unescorted. Speed was no guarantee of safety, though; the week before, U-boats had sunk the *Empress of Scotland's* sister ship. Despite the danger, Turing made it back to England without incident.

Clearly, the Allies had to locate and destroy the U-boats, not just evade them. U-boats were tying up thousands of Allied ships, planes, and troops needed to supply Britain and invade Continental Europe. The hunt for U-boats involved Bayes' rule in still another part of the Battle of the Atlantic.

Applying scientific techniques to the antisubmarine campaign, the British Air Ministry organized a small group of scientists to improve its operational efficiency. This was a new idea, and the British called it O.R., for operational or operations research. Its statistics were fairly elementary but imbued with Bayesian ideas.

O.R. concentrated on boosting the efficiency of torpedo attacks, airplane navigation, and formation flying by squadrons of planes searching for U-boats. Bayes' "a priori Method" played "quite a large role in operational research," especially when comparatively few variables were involved, reported O.R.'s chief, the future developmental biologist Conrad H. Waddington.[33]

Typically, O.R. employed Bayes for small, detailed parts of big problems, such as the number of aircraft needed to protect a convoy, the length of crews' operational tours, and whether an aircraft patrol should deviate from its regular flight pattern. Observing the success of British O.R., Adm. Ernest King, commander in chief of the U.S. Fleet, assigned 40 civilian physicists, chemists, mathematicians, and actuaries to his staff. This Anti-Submarine Warfare Operations Research Group was headed by physicist Philip M. Morse of MIT and chemist George E. Kimball of Columbia University.

The Allies had built a string of high-frequency direction-finding stations along the perimeter of the Atlantic. Much of the system was devoted to capturing encoded radio messages and relaying them to code breakers in the United States and at Bletchley Park. With six or seven listening posts intercepting the same message from a particular U-boat, the position of a submarine in the Atlantic could be determined within about 10,000 square miles. This gave patrol planes a good idea of where to look, but 10,000 square miles still meant a rectangle measuring 50 × 200 miles. The Allies needed an efficient method for narrowing the search.

Since almost every aspect of searching for targets in the open seas involves uncertainties and probabilities, mathematician Bernard Osgood Koopman of Columbia University was assigned the job of finding a workable method. After graduating from Harvard in 1922, Koopman had studied probability in Paris and earned a Ph.D. from Columbia. His dream was to bridge the gap between Bayes' "intuitive probability . . . of a subjective nature" and the "purely objective" frequency-based probability used in quantum physics and statistical mechanics.[34]

A crusty man with a rough frankness and a pungent wit, Koopman saw no reason to be bashful about Bayes or Bayesian priors. He assumed from the very beginning that he was dealing with probabilities: "Every operation involved in search is beset with uncertainties; it can be understood quantitatively only in terms of . . . probability. This may now be regarded as a truism; but it seems to have taken the developments in operational research of the Second World War to drive home its practical implications."[35]

Searching for a U-boat at sea, Koopman first asked what its heading was likely to be. To him, this was a classic Bayesian "probability of causes" problem. Priors would obviously be needed. "No rational prospector would search a region for mineral deposits unless a geological study, or the experience of previous prospectors, showed a sufficiently high probability of their presence," he commented. "Police will patrol localities of high incidence of crime. Public health officials will have ideas in advance of the likely sources of infection and will examine them first."[36]

Koopman started right off by assigning Thomas Bayes' 50–50 odds to the presence of a target U-boat inside the 10,000 square miles. Then he added data that were as objective as possible, as Jeffreys advised. Unlike Turing, Koopman had access to enormous amounts of detailed information that the military had accumulated about U-boat warfare.

Unfortunately, a U-boat could spot a destroyer long before the destroyer's sonar picked up the U-boat. Many U.S. planes were not equipped with windshield wipers, and crews peered through scratched and soiled windows. "The need for keeping the windows clean and clear cannot be overemphasized," Koopman admonished. If a crew was lucky enough to get binoculars, they were standard navy 7 x 50 issue, hazy at best. Unless crew members changed stations frequently to minimize the monotony, they lost focus. And the best angle for watching was generally 3 or 4 degrees below the horizon—"a rough and ready rule for finding this locus," Koopman wrote, "is to extend the fist at arm's length and look about two or three fingers below the horizon."[37] He figured that most aircraft crews were only a quarter as efficient as lookouts working under laboratory conditions.

As a practical problem, Koopman asked how a naval officer could find a U-boat within 10,000 square miles if he had 4 planes, each of which could fly 5 hours at 130 knots up and down 5 search lanes, each 5 miles wide. Although few O.R. investigations required such intricate mathematics, Koopman found a way to answer the question mathematically using logarithmic functions. Knowing only that 3 of the 5 lanes had a 10% probability of success, another had 30%, and a fourth had 40%, Koopman could do the Bayesian math. The officer should assign two planes each to the 40% lane and the 30% lane and none to the least probable areas. He calculated this by hand; his problem was not calculating but getting appropriate observational data. He later said that computers would have been irrelevant.

Applying his theories, Koopman wrote a fat manual of precomputed recipes for conducting a U-boat search. The effort needed for each subsection of the search area equaled the logarithm of the probability at that point. The regions to be searched did not have to be boxes or circles; they could have squiggly, irregular shapes. But using his formulas, he could tell a commander how many hours of search to devote to each squiggly region.

Using Koopman's cookbook, a shipboard officer could lay out the optimal way to search given his limited resources: the expected time needed to find the target; the boundaries beyond which he should not venture; and what he should do every two hours until either the U-boat was found or the search was called off. He could plan an eight-hour day, starting off with an optimum search for the first four hours; then, if a U-boat had not been found, the commander could use Bayes' rule to update the target's probable location and launch a new plan every two hours to maximize his chance of locating it.

All of the commander's planning for two-hour sequential searches could be done ahead of time in his stateroom. Koopman called it a "continuous distribution of effort." His U-boat sea searches were theoretically similar to Kolmogorov's artillery problem. Koopman was searching for an unknown U-boat and needed to spread the search effort over an area in an optimal way, just as Kolmogorov figured the optimal amount of dispersion in order to destroy a German cannon. Minesweepers, who worked with similar problems, adopted Koopman's techniques.

Three crucial turning points—two of them top secret—occurred in the European war during 1943. First, in what the Russians still call the Great Patriotic War, the Soviets defeated the Germans on the Eastern Front, at a cost of more than 27 million lives. Second, the tide began to turn against the Germans' U-boats; they sank a quarter million tons in May but 41 subs were lost. Third, Bletchley Park became a giant factory employing almost 9,000 people. As more bombes came online, the laborious Banburismus cardboards were phased out. Barring unforeseen changes by German cryptographers, decoding naval Enigma was under control.

Back home safely and free of responsibility for the Enigma and Tunny-

Lorenz codes, Turing, the great theoretician, was free to dream. During long walks in the countryside around Bletchley Park, Turing and Good discussed machines that could think with Donald Michie, who would pioneer artificial intelligence. Michie, who had joined Bletchley Park as an 18-year-old, described the trio as "an intellectual cabal with a shared obsession with thinking machines and particularly with machine learning as the only credible road to achieving such machines." They talked about "various approaches, conjectures, and arguments concerning what today we call AI."[38]

Max Newman, formerly Turing's mathematics instructor at Cambridge, wanted to automate the British attack on Tunny-Lorenz's codes, and he, Michie, and Good were already working on new machines to do it. Michie had refined Turingismus, but it soon became obvious that mechanical switches would be far too slow. The process would have to be electronic; engineer Thomas H. Flowers suggested using glass vacuum tubes because they could switch current on and off much faster. With backing from Newman, Flowers built the first Colossus at the Post Office Research Station, which ran Britain's telephone system. Installed at Bletchley Park, Colossus decrypted its first message on February 5, 1944. Flowers's car broke down that day but not his Colossus.

Flowers had strict orders—no reasons given—to get a second, more advanced Colossus model operational no later than June 1. Working until they thought their eyes were dropping out, Flowers and his team had Colossus II ready on schedule.

Almost as soon as it began operating, Hitler teletyped an encrypted message to his commanding officer in Normandy, Field Marshal Erwin Rommel. He ordered Rommel not to move his troops for five days after any invasion of Normandy. Hitler had decided it would be a diversionary feint to draw German troops away from the ports along the English Channel and that the real invasion would take place five days later. Colossus II decoded the message, and a courier raced a copy from Bletchley Park to Gen. Dwight "Ike" Eisenhower. As Ike and his staff were trying to decide when to launch the invasion of Normandy the courier handed him a sheet of paper containing Hitler's order. Unable to tell his staff about Bletchley Park, Eisenhower simply returned the paper to the courier and announced, "We go tomorrow," the morning of June 6.[39] He later estimated that Bletchley Park's decoders had shortened the war in Europe by at least two years.

The Colossi became the world's first large-scale electronic digital computers, built for a special purpose but capable of making other computations too. Flowers would build ten more models during the war. With the Germans introducing complexities that made manual decrypting methods useless, the Colossi replaced Turing's pencil-and-paper Turingery in August 1944. As Michie reported, Turing's Bayesian scoring system based on bans had started "first as a minor mental aid in a variety of jobs" but then turned into "a major aid in the [Colossi's] wheel pattern breaking."[40] Turing's method also contributed intellectually to the use of the Colossi and produced procedures that made the machines much more effective. Each new Colossus was an improvement over the previous one, and Michie believed the eleventh "nudged the design further in the direction of 'programmability' in the modern sense."[41]

By 1945 Turing had moved on to voice encryption at a nearby military installation at Hanslope Park. Late in the war, others at Bletchley Park, ignorant of Turing's work on Enigma, decided to use Bayesian methods to try to break the Japanese naval codes in the Pacific. Japan's main naval cipher, JN-25, was becoming increasingly complex, and Bletchley Park began working on some particularly difficult versions shortly after September 1943.

A trio of British mathematicians was assigned to work in tandem with Washington. The three were Ian Cassels, later a professor at Cambridge; Jimmy Whitworth; and Edward Simpson, who had joined Bletchley Park in 1942, immediately after earning a mathematics degree at Queen's University, Belfast, at the age of 19. Simpson had been working on Italian codes at Bletchley Park, but after Italy's surrender he was switched to JN-25.

"The unbelievably tight security ethos" at Bletchley Park prevented the group from getting advice from Turing or Good, Simpson explained in 2009 after his wartime work was revealed.[42] As a result, the men adopted and developed Bayes on their own. It was a full year before they were able to speak with Turing's colleague Alexander, who by that time had begun work on Japanese naval codes too.

Japanese coding clerks who used the principal code, JN-25, transmitted their messages in blocks of five digits. The British mathematicians knew that each block was the result of adding a random five-digit group, called an additive, to a five-digit code group taken from the JN-25 codebook. In effect, British cryptanalysts had to perform the reverse operation—but without the JN code and additive books. First, they identified groups that might be additives. Then a team composed of civilians and Wrens who, despite being newcomers to cryptography, had to identify the most probable additives rapidly, objectively, and in a standardized manner. They could judge the plausibility of an additive according to the plausibility or probability of the deciphered code group produced by the additive. As a measure of their belief, team members assigned a Bayesian probability to each speculative code group according to how often it had occurred in already deciphered messages. The most probable blocks, as well as borderline or especially important cases, were studied further.

"For practical purposes, it was not necessary to agonise over the prior odds to be assigned to the hypothesis that an additive was true," Simpson explained. "Instead, the essential judgment to be made was whether the [weight of] collective evidence . . . was sufficiently convincing for it to be accepted as genuine . . . As always in cryptanalysis, the inspired hunch grounded in experience could sometimes make the most important contribution of all."[43]

After October 1944 Alexander, Bletchley Park's finest Banburismus solver, developed an elaborate use of Bayes' theorem and Turing's decibans for the Japanese codes.

By 1945 U.S. cryptanalysts were writing memos to one another about Bayes' theorem. Whether the Americans learned about Bayes from Bletchley Park or discovered its usefulness on their own is not known; 65 years after the war, the British government still refuses to declassify many documents about wartime cryptography. A young American mathematician, Andrew Gleason, who was working on Japanese naval codes and who looked after Turing during his stay in Washington, almost certainly knew about Bayes during the war. He, Good, and Alexander continued to work on top-secret cryptography for decades after the war. Gleason helped establish a postwar curriculum for training cryptanalysts at the U.S. National Security Agency (NSA), taught mathematics at Harvard and NSA, and published a probability textbook that instructed a generation of NSA's cryptanalysts in how to use Bayes' theorem, Turing's decibans and centibans, Bayesian inference, and hypothesis testing. Some 20 of his students became leaders in Soviet code breaking during the 1960s and 1970s. Gleason was deeply knowledgeable but pragmatic about Bayes; his textbook also discussed methods developed by Neyman, the arch anti-Bayesian.

A few days after Germany's surrender in May 1945 Bletchley Park received surprising and shocking orders. The fact that cryptography, Bletchley Park, Turing, Bayes' rule, and the Colossi had contributed to victory was to remain super-secret after the peace and continue into the Cold War. Turing's assistant Good complained later that everything about decryption and the U-boat fight "from Hollerith [punch] cards to sequential statistics, to empirical Bayes, to Markov chains, to decision theory, to electronic computers" was to remain ultraclassified.[44] Most of the Colossi were dismantled and broken into unidentifiable pieces. Those who built the Colossi and broke Tunny were gagged by Britain's Official Secrets Acts and the Cold War; they could not even say that the Colossi had existed. Books by British and U.S. participants in the U-boat war were almost immediately classified, confined to high-level military circles, and not published for years or in some cases decades. Even classified histories of the war excluded the decryption campaign against the U-boats. Only after 1973 did the story of Bayes, Bletchley Park, and Turing's nation-saving efforts begin to emerge.

Why was the story concealed for so long? The answer seems to be that the British did not want the Soviet government to know they could decrypt Tunny-Lorenz codes. The Russians had captured a number of Lorenz machines, and Britain used at least one of the two surviving Colossi to break Soviet codes during the Cold War. Only when the Soviets replaced their Lorenz machines with new cryptosystems was Bletchley Park's story revealed.

The secrecy had tragic consequences. Family and friends of Bletchley Park employees went to their graves without ever knowing the contributions their loved ones had made during the war. Those connected with Colossus, the epitome of the British decryption effort, received little or no credit. Turing was given an Order of the British Empire (OBE), a routine award given to high civil servants. Newman was so angry at the government's "derisory" lack of gratitude to Turing that he refused his own OBE.

Britain's science, technology, and economy were losers, too. The Colossi were built and operational years before the ENIAC in Pennsylvania and before John von Neumann's computer at the Institute for Advance Study in Princeton, but for the next half century the world assumed that U.S. computers had come first.

Obliterating all information about the decryption campaign distorted Cold War attitudes about the value of cryptanalysis and about antisubmarine warfare. The war replaced human spies with machines. Decryption was faster than spying and provided unfiltered knowledge of the enemy's thinking in real time, yet the Cold War glamorized military hardware and the derring-do of spydom.

The secrecy also had a catastrophic effect on Turing. At the end of the war he said he wanted "to build a brain."[45] To do so, he turned down a lecture-ship at Cambridge University and joined the National Physical Laboratory in London. Because of the Official Secrets Act he arrived as a nobody. Had he been knighted or otherwise honored he would surely have found it easier to get more than two engineers as support staff. Ignorant of Turing's achievements, the director of the laboratory, Charles Galton Darwin, a grandson of Charles Darwin, repeatedly reprimanded Turing for morning tardiness after working late the night before. Once an afternoon committee meeting with Darwin and others stretched late in the day. At 5:30 p.m. Turing promptly stood up and announced to Darwin that he was leaving—"punctually."[46]

At the laboratory, Turing designed the first relatively complete electronic stored-program digital computer for code breaking in 1945. Darwin deemed it too ambitious, however, and after several years Turing left in disgust. When the laboratory finally built his design in 1950, it was the fastest computer in the world and, astonishingly, had the memory capacity of an early Macintosh built three decades later.

Turing moved to the University of Manchester, where Newman was building the first electronic, stored-program digital computer for Britain's atomic bomb. Working in Manchester, Turing pioneered the first computer software, gave the first lecture on computer intelligence, and devised his famous Turing Test: a computer is thinking if, after five minutes of questioning, a person cannot distinguish its responses from those of a human in the next room. Later, Turing became interested in physical chemistry and how huge biological molecules construct themselves into symmetrical shapes.

A series of spectacular international events in 1949 and 1950 intruded on these productive years and precipitated a personal crisis for Turing: the Soviets surprised the West by detonating an atomic bomb; Communists gained control of mainland China; Alger Hiss, Klaus Fuchs, and Julius and Ethel Rosenberg were arrested for spying; and Sen. Joseph McCarthy of Wisconsin began brandishing his unsubstantiated list of so-called Communists in the U.S. State Department.

Even worse, two upper-crust English spies—an openly promiscuous and alcoholic homosexual named Guy Burgess and his friend from Cambridge student days Donald Maclean—evaded arrest by fleeing to the USSR in 1950. The United States told British intelligence they had been tipped off by Anthony Blunt, another homosexual graduate of Cambridge, a leading art historian, and the queen's surveyor of paintings. With both the British and American governments panicked by visions of a homosexual spy scandal, the number of men arrested for homosexuality in Britain spiked.

On the first day of Queen Elizabeth II's reign, February 7, 1952, Turing was arrested for homosexual activity conducted in the privacy of his home with a consenting adult. As Good protested later, "Fortunately, the authorities at Bletchley Park had no idea Turing was a homosexual; otherwise we might have lost the war."[47]

In the uproar over Burgess and Maclean, Turing was viewed not as the hero of his country but as yet another Cambridge homosexual privy to the most closely guarded state secrets. He had even worked on the computer involved in Britain's atomic bomb test. As a result of his arrest, Britain's leading cryptanalyst lost his security clearance and any chance to continue work on decoding. In addition, because the U.S. Congress had just banned gays from entering the country, he was unable to get a visa to travel or work in the United States.

As the world lionized the Manhattan Project physicists who engineered the atomic and hydrogen bombs, as Nazi war criminals went free, and as the United States recruited German rocket experts, Turing was found guilty. Less than a decade after England fought a war against Nazis who had conducted medical experiments on their prisoners, an English judge forced Turing to choose between prison and chemical castration. He chose the estrogen injections. Over the next year he grew breasts. And on June 7, 1954, the day after the tenth anniversary of the Normandy invasion he helped make possible, Alan Turing committed suicide. Two years later the British government knighted Anthony Blunt, the spy who later admitted tipping off his friends Burgess and Maclean and precipitating the witch hunt against homosexuals. Even today, it is difficult to write—or read—about Turing's end. In 2009, 55 years after Turing's death, a British prime minister, Gordon Brown, finally apologized.

Turing's Bayesian work lived on in cryptography. Secretly for decades, an American colleague of Turing's taught Bayes to NSA cryptographers. With Turing's blessing, Good developed Bayesian methods and theory and became one of the world's leading cryptanalysts and one of the three leaders in the Bayesian renaissance of the 1950s and 1960s. He wrote roughly 900 articles about Bayes' rule and published most of them.

Outside of cryptography, however, no one knew that some of the most brilliant thinkers of the mid-twentieth century had used Bayes to defend their countries during the Second World War. It emerged from the war as vilified as ever.

dead and buried again

With its wartime successes classified, Bayes' rule emerged from the Second World War even more suspect than before. Statistics books and papers stressed repeatedly and self-righteously that they did not use the rule. When Jack Good discussed the method at the Royal Statistical Society, the next speaker's opening words were, "After that nonsense . . ."[1]

"Bayes" still meant equal priors and did not yet mean making inferences, conclusions, or predictions based on updating observational data. The National Bureau of Standards suppressed a report to Aberdeen Proving Ground, the U.S. Army's weapons-testing center, during the 1950s because the study used subjective Bayesian methods. During Sen. Joseph McCarthy's campaign against Communists, a bureau statistician half-jokingly called a colleague "un-American because [he] was Bayesian, and . . . undermining the United States Government."[2] Professors at Harvard Business School called their Bayesian colleagues "socialists and so-called scientists."[3]

"There still seems to remain in some quarters a lingering idea that there is something 'not quite nice,' something unsound, about the whole concept of inverse probability," a prominent statistician wrote.[4] Unless declared otherwise, a statistician was considered a frequentist.

The Bayesian community was small and isolated, and its publications were well-nigh invisible. Prewar theory by Frank Ramsey, Harold Jeffreys, and Bruno de Finetti lay unread. Nearly all the papers published in the *Annals of Mathematical Statistics* concerned issues framed by Jerzy Neyman's frequentist work from the 1930s. Thanks to Ronald Fisher's genetics research and the powerful anti-Bayesian stance of an Iowa State University statistician named

Oscar Kempthorne, agricultural studies at most land-grant institutions relied on frequentism. When Gertrude Cox, the president of the American Statistical Society in 1956, spoke about the future of statistics, she barely mentioned Bayes. The first practical article telling scientists how to use Bayesian analysis would not appear until 1963.

Not even civilian researchers for the military knew much about Bayes in 1950. When an economist was preparing a research budget for the U.S. Air Force at RAND, a California think tank, he asked visiting statistician David Blackwell how to assess the probability that a major war would occur within five years. Blackwell, who had not yet become a Bayesian, answered, "Oh, that question just doesn't make sense. Probability applies to a long sequence of repeatable events, and this is clearly a unique situation. The probability is either 0 or 1, but we won't know for five years." The economist nodded and said, "I was afraid you were going to say that. I have spoken to several other statisticians, and they have all told me the same thing."[5]

The Bayesian theorist Dennis V. Lindley concluded, "The upstart Bayesian movement is being contained, largely by being ignored."[6] Another statistician recalled, "A lot of us thought [Bayes] was dead and buried."[7]

part III

the glorious revival

arthur bailey

After the Second World War the first public challenge to the anti-Bayesian status quo came not from the military or university mathematicians and statisticians but from a Bible-quoting business executive named Arthur L. Bailey.

Bailey was an insurance actuary whose father had been fired and blackballed by every bank in Boston for telling his employers they should not be lending large sums of money to local politicians. So ostracized was the family that even Arthur's schoolmates stopped inviting him and his sister to parties. Turning his back on the New England establishment, Bailey enrolled at the University of Michigan in Ann Arbor. There he studied statistics in the mathematics department's actuarial program, earned a bachelor of science degree in 1928, and met his wife, Helen, who became an actuary for John Hancock Mutual Life before their children were born.[1]

Bailey's first job was, he liked to say, "in bananas," that is, in the statistics department of the United Fruit Company headquarters in Boston. When the department was eliminated during the Depression, Bailey wound up driving a fruit truck and chasing escaped tarantulas down Boston streets. He was lucky to have the job, and his family never lacked for bananas and oranges.

In 1937, after nine years in bananas, Bailey got a job in an unrelated field in New York City. There he was in charge of setting premium rates to cover risks involving automobiles, aircraft, manufacturing, burglary, and theft for the American Mutual Alliance, a consortium of mutual insurance companies.

Preferring church and community connections to the fair-weather friends of his youth, Bailey hid his growing professional success by living quietly in unpretentious New York suburbs. He relaxed by gardening, hiking with his four children, and annotating a copy of *Gray's Botany* with the locations of his favorite wild orchids. His motto was, "Some people live in the past, some people live in the future, but the wisest ones live in the present."

Settling into his new job, Bailey was horrified to see "hard-shelled underwriters" using the semi-empirical, "sledge hammer" Bayesian techniques developed in 1918 for workers' compensation insurance.[2] University statisticians had long since virtually outlawed those methods, but as practical business people, actuaries refused to discard their prior knowledge and continued to modify their old data with new. Thus they based next year's premiums on this year's rates as refined and modified with new claims information. They did not ask what the new rates should be. Instead, they asked, "How much should the present rates be changed?" A Bayesian estimating how much ice cream someone would eat in the coming year, for example, would combine data about the individual's recent ice cream consumption with other information, such as national dessert trends.

As a modern statistical sophisticate, Bailey was scandalized. His professors, influenced by Ronald Fisher and Jerzy Neyman, had taught him that Bayesian priors were "more horrid than 'spit,'" in the words of a particularly polite actuary.[3] Statisticians should have no prior opinions about their next experiments or observations and should employ only directly relevant observations while rejecting peripheral, nonstatistical information. No standard methods even existed for evaluating the credibility of prior knowledge (about previous rates, for example) or for correlating it with additional statistical information.

Bailey spent his first year in New York trying to prove to himself that "all of the fancy actuarial [Bayesian] procedures of the casualty business were mathematically unsound."[4] After a year of intense mental struggle, however, he realized to his consternation that actuarial sledgehammering worked. He even preferred it to the elegance of frequentism. He positively liked formulae that described "actual data. . . . I realized that the hard-shelled underwriters were recognizing certain facts of life neglected by the statistical theorists."[5] He wanted to give more weight to a large volume of data than to the frequentists' small sample; doing so felt surprisingly "logical and reasonable." He concluded that only a "suicidal" actuary would use Fisher's method of maximum likelihood, which assigned a zero probability to nonevents.[6] Since many businesses file no insurance claims at all, Fisher's method would produce premiums too low to cover future losses.

Abandoning his initial suspicions of Bayes' rule, Bailey spent the Second

World War studying the problem. He worked alone, isolated from academic thinkers and from his actuarial colleagues, who scratched their heads at Bailey's brilliance.

After the war, in 1947, Bailey moved to the New York State Insurance Department as the regulatory agency's chief actuary. An insurance executive called him "the keeper of our consciences." As his colleagues boozed in hotel bars at conferences, Bailey sipped soft drinks and quoted occasionally from the Bible. During slack times, he read it. Some actuaries said "all manner of nasty things about Arthur Bailey," the executive continued, "but we learned to respect his integrity and stature from knowing him in the after-hours."[7]

Bailey began writing an article summarizing his tumultuous change in attitude toward Bayes' rule. Although his old-fashioned notation was difficult to understand, he was building a mathematical foundation to justify the use of current rates as the priors in Bayes' theorem. He started his paper with a biblical justification for using prior beliefs: "If thou canst believe," he wrote, quoting the apostle Mark, "all things are possible to him that believeth." Then, reviewing Albert Whitney's mathematics for workers' compensation, Bailey affirmed the Bayesian roots of the Credibility theory developed for workers' compensation insurance years before. Credibility was central to actuarial thought, and while relative frequencies were relevant, so were other kinds of information. Bailey worked out mathematical methods for melding every scrap of available information into the initial body of data. He particularly tried to understand how to assign partial weights to supplementary evidence according to its credibility, that is, its subjective believability. His mathematical techniques would help actuaries systematically and consistently integrate thousands of old and new rates for different kinds of employers, activities, and locales. His working library included a 1940 reprint of Bayes' articles with a preface by Bell Telephone's Edward Molina. Like Molina, Bailey used Laplace's more complex and precise system instead of Thomas Bayes'.

By 1950 Bailey was a vice president of the Kemper Insurance Group in Chicago and a frequent after-dinner speaker at black-tie banquets of the Casualty Actuarial Society. He read his most famous paper on May 22, 1950. Its title explained a lot: "Credibility Procedures: Laplace's Generalization of Bayes' Rule and the Combination of Collateral [that is, prior] Knowledge with Observed Data."

For actuaries who could concentrate on a long scholarly paper after a heavy (and no doubt alcoholic) meal, Bailey's message must have been thrilling. First, he praised his colleagues for standing almost alone against the statistical establishment and for staging the only organized revolt against the frequentists' sampling philosophy. Insurance statisticians marched "a step ahead" of others. Actuarial practice was an obscure and profound mystery, and it went "beyond anything that has been proven mathematically." But, he declared triumphantly, "it works. . . . They have made this demonstration many times. It does work!"[8]

Then he announced the startling news that their beloved Credibility formula was derived from Bayes' theorem. Practical actuaries had thought of Bayes as an abstract, temporal solution treating time sequences of priors and posteriors. But Bailey reminded his colleagues that Bayes' friend and editor, Richard Price, would be considered today an actuary. And he turned Bayes' imaginary table into a frontal attack on frequentists and the contentious Fisher. He concluded with a rousing call to reinstate prior knowledge in statistical theory. His challenge would occupy academic theorists for years. It was a fighting speech. Reading it later, Professor Richard von Mises of Harvard praised it wholeheartedly. Von Mises wrote Bailey that he hoped his speech would make "the unjustified and unreasonable attacks on the Bayes theory, initiated by R. A. Fisher, fade out."[9]

Unfortunately, Bailey did not live long to campaign for Bayes' rule. Four years after giving his most important speech, he suffered a heart attack at the age of 49 and died on August 12, 1954. His son blamed the fact that Bailey had started smoking in college and been unable to stop.

Still, a few practicing actuaries understood his message. The year of Bailey's death, one of his admirers was sipping a martini at the Insurance Company of North America's Christmas party when INA's chief executive officer, dressed as Santa Claus, asked an unthinkable question: Could anyone predict the probability of two planes colliding in midair?

Santa was asking his chief actuary, L. H. Longley-Cook, to make a prediction based on no experience at all. There had never been a serious midair collision of commercial planes. Without any past experience or repetitive experimentation, any orthodox statistician had to answer Santa's question with a resounding no. But the very British Longley-Cook stalled for time. "I really don't like these things mixed with martinis," he drawled. Nevertheless, the question gnawed at him. Within a year more Americans would be traveling by air than by railroad. Meanwhile, some statisticians were wondering if they could avoid using the ever-controversial subjective priors by making predictions based on no prior information at all.

Longley-Cook spent the holidays mulling over the problem, and on

January 6, 1955, he sent the CEO a prescient warning. Despite the industry's safety record, the available data on airline accidents in general made him expect "anything from 0 to 4 air carrier-to-air carrier collisions over the next ten years." In short, the company should prepare for a costly catastrophe by raising premium rates for air carriers and purchasing reinsurance. Two years later his prediction proved correct. A DC-7 and a Constellation collided over the Grand Canyon, killing 128 people in what was then commercial aviation's worst accident. Four years after that, a DC-8 jet and a Constellation collided over New York City, killing 133 people in the planes and in the apartments below.[10]

Later, Arthur Bailey's son, Robert A. Bailey, used Bayesian techniques to justify offering merit rates to good drivers. Motor vehicle casualty rates soared so high during the 1960s that half the Americans then alive could expect to be injured in a car accident during their lifetimes. Americans were buying more cars and driving more miles, but laws had not kept pace. There was no uniform road signage; most drivers and vehicles were tested or inspected only once in their lifetimes, if at all; penalties for drunk driving were light; and cars were designed without safety in mind. Insurers suffered heavy losses. A direct, up-front system to reward good drivers was needed, but merit rating was regarded as unsound because a single car had inadequate credibility. Using Bayes' rule, Robert Bailey and Leroy J. Simon showed that relevant data from Canada's safe-driving discounts could be used to update existing U.S. statistics.

Robert Bailey also used Bayesian procedures to rate insurance companies themselves by incorporating nonstatistical, subjective information such as opinions about a company's ownership, including the quality and drinking habits of its managers. In time, the insurance industry accumulated such enormous amounts of data that Bayes' rule, like the slide rule, became obsolete.

To the few actuaries who understood Arthur Bailey's work, he was a da Vinci or a Michelangelo: he had led their profession out of its dark ages.[11] News of his achievement percolated slowly and haphazardly to university theorists. During the early 1960s an actuarial professor at the University of Michigan, Allen L. Mayerson, wrote about Bailey's seminal role in Credibility theory. Professor of statistics Jimmie Savage, a new convert to Bayesian methods, was working in Ann Arbor at the time and later visited Bruno de Finetti, the Bayesian actuarial professor, at his vacation home on an island off Italy. The two attended a conference together in Trieste, where the Italian spread the word about Bailey and the Bayesian origin of insurance Credibility. It was the first time most statisticians had heard of him.

Hans Bühlmann, who became a mathematics professor and president of ETH Zurich, remembers the excitement of that conference. He had spent a leave of absence studying in Neyman's statistics department in Berkeley in the 1950s, "when it was kind of dangerous to pronounce the Bayesian point of view." Taking up Bailey's challenge, Bühlmann produced a general Bayesian theory of credibility, which statisticians carried far beyond the world of actuaries and insurance. Carefully renaming the prior a "structural function," Bühlmann believed he helped Continental Europe escape some of the "religious" quarrels over Bayes' rule, quarrels that lay ahead for Anglo-Americans.[12]

7.

from tool
to theology

While Arthur Bailey was transforming the sledgehammer of Credibility into Bayes' rule for the insurance industry, a postwar boom in statistics was elevating the method's lowly status. Gradually, Bayes would shed its reputation as a mere tool for solving practical problems and emerge in glorious Technicolor as an all-encompassing philosophy. Some would even call it a theology.

The Second World War had radically upgraded the stature, financial prospects, and career opportunities of applied mathematicians in the United States. The military was profoundly impressed by its wartime experience with statistics and operations research, and during the late 1940s the government poured money into science and statistics. Military funding officers roamed university hallways trying to persuade often reluctant statisticians to apply for grants. Naval leaders, convinced that postwar science needed a jump-start to prime technology's pump, organized the Office of Naval Research, the first federal agency formed expressly to finance scientific research. Until the National Science Foundation was created in 1950, the U.S. Navy supported much of the nation's mathematical and statistical research, whether classified or unclassified, basic or applied. Other funding came from the U.S. Army, the U.S. Air Force, and the National Institutes of Health.

A generation of pure mathematicians who had made exciting, life-or-death decisions during the war soon switched to applied mathematics and statistics. As the statistics capital of the world moved from Britain to the United States, the field exploded. Amid such spectacular growth, the number of theoretical statisticians increased a hundred-fold. Settling into mathematics departments, they coined new terms like "mathematical statistics" and "theoretical statistics."

In these boom times, even Bayesians could get jobs in elite research institutions. At one end of the Bayesian spectrum was a small band of evangelists intent on making their theories mathematically and academically respectable. At the other end were practitioners who wanted to play key roles in science instead of in formalistic mathematical exercises.

In the face of jangling changes and new attitudes, the wartime marriage of convenience between abstract and applied mathematicians fell apart. Statisticians complained that pure mathematicians regarded useful research as "something for peasants," akin to washing dishes and sweeping streets.[1] Jack Good claimed the mathematicians at Virginia Tech, home of the nation's third largest statistics department in the 1960s, loathed problem solvers.[2]

Frisky with federal funds, statisticians and data analysts divorced themselves from mathematics departments and formed their own enclaves. Yet even there tension sizzled between abstract theorizing and scientific applications, albeit in more decorous privacy. Serial schisms continue to this day, with applied mathematicians occupying—depending on the university—departments of mathematics, applied mathematics, statistics, biostatistics, and computer science.

Jerzy Neyman's laboratory at Berkeley, then the largest and most important statistics center in the world, developed fundamental sampling theories and reigned over this fractious profession for years after the Second World War. But Neyman's laboratory developed fissures of its own. Unable to compete with the soaring demand for statisticians, the department hired and promoted its own students and became ingrown. When a student tried to solve a blackboard problem unconventionally, Neyman grabbed his hand and forced it to write the answer Neyman's way. For 40 years most of his hires were frequentists, and outsiders called the group "Jesus and his disciples."[3] Neyman continued to run his institute into his 80s.

Although both were fervent anti-Bayesians, Neyman and Fisher battled to the end, neither willing to admit that the other might be using the technique that best fit his own needs. For Fisher, the stakes were high: "We are quite in danger of sending highly trained and intelligent young men out into the world with tables of erroneous numbers under their arms, and with a dense fog in the place where their brains ought to be. In this century, of course, they will be working on guided missiles and advising the medical profession on the control of diseases, and there is no limit to the extent to which they could impede every sort of national effort."[4] He described Neyman as "some hundred years out of date, . . . partly incapacitated by the crooked reasoning."[5] Neyman called Fisher's researches "insidious because, in a skillfully hidden manner, they involve unjustified claims of priority."[6] And so it went. At the age of 85, Neyman declared loftily, "[Bayes] does not interest me. I am interested in frequencies."[7]

To Bayesian sympathizers, frequentism began to look like a Rube Goldberg cartoon of loosely connected ad hockeries, tests, and procedures that arose independently instead of growing in a unified, logical manner out of probability. The joke was that if you didn't like the result of your frequentist analysis, you just redid it using a different test. By comparison, Bayes' rule seemed to have an overall rationale. As the number of statisticians, symposia, articles, and journals multiplied, a series of publications issued around 1950 began to attract attention to the heretofore invisible world of Bayes' rule.

Bayes stood poised for another of its periodic rebirths as three mathematicians, Jack Good, Leonard Jimmie Savage, and Dennis V. Lindley, tackled the job of turning Bayes' rule into a respectable form of mathematics and a logical, coherent methodology. The first publication heralding the Bayesian revival was a book by Good, Alan Turing's wartime assistant. As Good explained, "After the war, he [Turing] didn't have time to write about statistics because he was too busy designing computers and computer languages, and speculating about artificial intelligence and the chemical basis of morphogenesis, so with his permission, I developed his idea . . . in considerable detail."[8] Good finished the first draft of *Probability and the Weighing of Evidence* in 1946 but could not get it published until 1950, the same year Arthur Bailey issued his Bayesian manifesto for actuaries. Much of the delay, Good explained, was caused by the continuation of wartime secrecy during the Cold War.

At first, his book fell on deaf ears. He was unaccustomed to teaching or explaining his ideas, and no one knew he had used Bayes to help break the Enigma codes. When he gave a talk about his "neo-Bayesian or neo/Bayes-Laplace philosophy" at a Royal Statistical Society conference, his style was clipped, and he did not waste words.[9] Lindley, who was in the audience, reported, "He did not get his ideas across to us. We should have paid much more respect to what he was saying because he was way in advance of us in many ways."[10]

After the war Good continued doing classified cryptography for the British government and frequently used equal priors to help decide what hypotheses he should follow up. When David Kahn's bestseller *The Codebreak-*

ers was published in 1967, the National Security Agency censored a passage identifying Good as one of Britain's top three cryptanalysts. He was at the time one of the world's most knowledgeable people about the coding industry. Good was quick, smart, original, armed with a fabulous memory, and unconventional enough to think about the paranormal and astrology and to join Mensa, the organization for people with high IQs. He introduced himself with a handshake and the words, "I am Good."[11]

From the Second World War on, everything technical about cryptography was classified, and while Good obeyed the restraints, he chafed against them and looked for ways to evade censorship. To reveal an ultraclassified technique used by Turing to find pairs and triplets of letters indicating the German submariners' code of the day, Good wrote about a favorite British hobby, bird watching. What if, he suggested, an avid bird watcher spotted 180 different bird species? Many of them would be represented by only one bird; logically, the bird watcher must have entirely missed many other species. Counting those missing species as zero (as a frequentist would have done) has the deleterious effect of asserting that missing species can never be found. Turing decided to assign those missing species a tiny possibility, a probability that is not zero. He was trying to learn about rare letter groupings that did not appear in his collection of intercepted German messages. By estimating the frequency of missing species in his sample, he could use Bayes to estimate the probability of those letter groupings appearing in a much larger sample of messages and in the very next Enigma message he received. Decades later, DNA decoders and artificial intelligence analysts would adopt the same technique.

Clever as he was, Good could be difficult to get along with, and he moved from post to post. After he spent a year at a cryptography think tank, the Institute for Defense Analyses (then at Princeton University), many coworkers were relieved to see him go. In 1967 Good moved permanently to Virginia Polytechnic Institute and State University in Blacksburg. At his insistence, his contract stipulated that he would always be paid one dollar more than the football coach. He worked far from the Bayesian mainstream, however; during the 1960s Bayes' rule in the United States was concentrated at the universities of Chicago and Wisconsin and at Harvard and Carnegie Mellon.

Sidelined by geography and silenced by the British government's classification of his work with Turing, Good mailed unsolicited carbons of his typed curriculum vitae—what he called his Private List of more than 800 articles and four books[12]—to startled colleagues. He numbered every work and marked a significant portion of them as classified. Only as the British slowly declassified his cryptanalysis work could he reveal Bayes' success with the Enigma code. When that happened, he bought a vanity license plate emblazoned with his James Bond spy status and his initials, 007 IJG.

Hampered by governmental secrecy, his own personality, and an inability to explain his work, Good remained an independent voice within the Bayesian community as two others became its intellectual leaders.

Unlike Good, Dennis Lindley and Jimmie Savage evolved almost accidentally as Bayesians. When Lindley was a boy during the German bombing of London, a remarkable mathematics teacher named M. P. Meshenberg tutored him in the school's air raid shelter. Meshenberg convinced Dennis's father, a roofer proud to have never read a book, that the boy should not quit school early or be apprenticed to an architect. Because of Meshenberg, Dennis stayed in school and won a mathematics scholarship to attend Cambridge University. Later in the war, when the British government asked mathematicians to learn some statistics, Lindley helped introduce statistical quality control and inspection into armaments production for the Ministry of Supply.

After the war he returned to Cambridge, the British center of probability, where Jeffreys, Fisher, Turing, and Good had either worked or studied. There Lindley became interested in turning the statisticians' collection of miscellaneous tools into a "respectable branch of mathematics," a complete body of thought based on axioms and proven theorems.[13] Andrei Kolmogorov had done the same for probability in general in the 1930s. Since Fisher in particular often arrived at his ideas intuitively and neglected mathematical details, there was plenty of room for another mathematician to straighten things out logically.

In 1954, a year after publishing a lengthy article summarizing his project, Lindley visited the University of Chicago, only to realize that Savage had done an even better job of it. Although Lindley and Savage would soon become leading spokesmen for Bayes' rule, neither realized at this point he was headed down a slippery slope toward Bayes. Each thought he had merely put traditional statistical techniques on a rigorous mathematical footing. Only later did they realize they could not move logically from their rigorous axioms and theorems to the ad hoc methods of frequentism. Lindley said, "We were both fools because we failed completely to recognize the consequences of what we were doing."[14]

Despite being almost blind, Savage was immensely learned on an encyclopedic range of topics. His father, a Jewish East European immigrant with a third-grade education, had changed the family's name from Ogushevitz to Savage and settled in Detroit. Both Jimmie and his brother Richard were born with extreme myopia and involuntary eye movements. As an adult, before crossing a street Jimmie would wait five or ten minutes to make sure there were no oncoming cars, and attending lectures he would approach the blackboard and peer at it through a powerful monocular. The brothers could read quite comfortably, however, and as children called themselves "reading machines";[15] their mother, a high school graduate and nurse, had kept them supplied with library books. Reading was always a privilege to be cherished, and Jimmie read with a rare intensity and developed the embarrassing habit of questioning everything. His wide-ranging studies and insatiable curiosity would alter the history of Bayes' rule.

Because of his eyesight, however, Savage almost missed getting a college education. His teachers considered him feebleminded and unsuited for higher studies. He was finally admitted to Wayne (later Wayne State) University in Detroit. From there he transferred to the chemistry department at the University of Michigan, only to be rejected again, this time as unfit for laboratory work. A kindly mathematics professor, G. Y. Rainich, rescued him by teaching a class of visually impaired students in total darkness. Rainich called it "mental geometry . . . just like in Russia," where many schools could not afford candles.[16] Three students in the class, including Savage, earned doctorates.

During the Second World War Savage worked in the Statistical Research Group at Columbia University with the future Nobel Prize–winning economist Milton Friedman. The experience persuaded Savage to switch from pure mathematics to statistics. After the war he moved to the University of Chicago, a center of scientific excitement, thanks in large part to the dazzling Nobel Prize winner Enrico Fermi, the last physicist to excel at both experimentation and theory. Fermi himself used Bayes. In the autumn of 1953, when Jay Orear, one of Fermi's graduate students, was struggling with a problem involving three unknown quantities, Fermi told him to use a simple analytic method that he called Bayes' theorem and that he had derived from C. F. Gauss. A year later, when Fermi died at the age of 53, Bayes' rule lost a stellar supporter in the physical sciences.

Fermi was not the only important physicist to use Bayesian methods during this period. A few years later, at Cornell University, Richard Feynman suggested using Bayes' rule to compare contending theories in physics.

Feynman would later dramatize a Bayesian study by blaming rigid O-rings for the *Challenger* shuttle explosion.

During this exciting period in 1950s Chicago, Savage and Allen Wallis founded the university's statistics department, and Savage attracted a number of young stars in the field. Reading widely, Savage discovered the work of Émile Borel, Frank Ramsey, and Bruno de Finetti from the 1920s and 1930s legitimizing the subjectivity in Bayesian methods.

Savage's revolutionary book *Foundations of Statistics* was the third in the series of pathbreaking Bayesian publications in the fifties. It appeared in 1954, four years after Bailey's insurance paper and Good's book and one year after Lindley's paper. Because of Ramsey's early death, it fell to Savage to develop the young philosopher's ideas about utility and to turn Bayes' rule for making inferences based on observations into a tool for decision making and action.

Almost defiantly, Savage proclaimed himself a subjectivist and a personalist. Subjective probability was a measure of belief. It was something you were willing to use for a bet, particularly on a horse race, where bettors share the same information about a horse but come to different conclusions about its chances and where the race itself can never be precisely replicated. Subjective opinions and professional expertise about science, medicine, law, engineering, archaeology, and other fields were to be quantified and incorporated into statistical analyses.

More than anyone else Savage forced people to think about combining two concepts: utility (the quantification of reward) and probability (the quantification of uncertainty). He argued that rational people make subjective choices to minimize expected losses.

Savage was confronting the thorniest objection to Bayesian methods: "If prior opinions can differ from one researcher to the next, what happens to scientific objectivity in data analysis?"[17] Elaborating on Jeffreys, Savage answered as follows: as the amount of data increases, subjectivists move into agreement, the way scientists come to a consensus as evidence accumulates about, say, the greenhouse effect or about cigarettes being the leading cause of lung cancer. When they have little data, scientists disagree and are subjectivists; when they have piles of data, they agree and become objectivists. Lindley agreed: "That's the way science is done."[18]

But when Savage trumpeted the mathematical treatment of personal opinion, no one—not even he and Lindley—realized yet that he had written the Bayesian Bible. "Neither of us would have known at the time what was meant by saying we were Bayesians," Lindley said. Savage's book did not use the term "Bayesian" at all and referred to Bayes' rule only once. Savage's views and his book gained popularity slowly, even among those predisposed to Bayes' rule. Many had hoped for a how-to manual like Fisher's *Statistical Methods for Research Workers*. Lacking computational machinery to implement their ideas, Bayesians were limited to a few simple problems involving easily solved integrals and would spend years adapting centuries-old methods for calculating them. Savage, though, said he was "little inclined to high speed machines for help. This is no doubt partly due to my being reactionary . . . but my main interests are in the qualitative. . . . Tables of functions depending on several parameters are almost unprintable and, when printed quite unintelligible."[19] Savage continued instead to prove abstract mathematical theorems and work on building a logical foundation for Bayesian methods.

His applications were too whimsical to be useful: what is the probability that aspirin curls rabbits' ears? what is the most probable speed of neon light through beer? Some thought Savage's failure to tackle serious problems impeded the spread of Bayesian methods. Lindley complained, "Perhaps statistics would have benefited more if he had not been so punctilious in replying to correspondents and so helpful with students, and instead developed more operational methods that the writers and graduates could have used."[20]

Some readers were also troubled by the fact that Savage used aspects of frequentism to argue for Bayes' subjective priors, taboo since the nineteenth century. As Savage explained, when he wrote the book he was "not yet a personalistic Bayesian." He thought he came to Bayesian statistics "seriously only through recognition of the likelihood principle; and it took me a year or two to make the transition."[21]

According to the likelihood principle, all the information in experimental data gets encapsulated in the likelihood portion of Bayes' theorem, the part describing the probability of objective new data; the prior played no role. Practically speaking, the principle greatly streamlined analysis. Scientists could stop running an experiment when they were satisfied with the result or ran out of time, money, and patience; non-Bayesians had to continue until some frequency criterion was met. Bayesians would also be able to concentrate on what happened, not on what *could* have happened according to Neyman-Pearson's sampling plan.

The transition to Bayes took Savage several years, but by the early 1960s he had accepted its logic wholeheartedly, fusing subjective probability with new statistical tools for scientific inference and decision making. As far as Savage was concerned, Bayes' rule filled a need that other statistical procedures could not. Frequentism's origin in genetics and biology meant it was involved with group phenomena, populations, and large aggregations of similar objects. As for using statistical methods in biology or physics, the Nobel Prize–winning physicist Erwin Schrödinger said, "The individual case [is] entirely devoid of interest."[22] Bayesians like Savage, though, could work with isolated one-time events, such as the probability that a chair weighs 20 pounds, that a plane would be late, or that the United States would be at war in five years.

Bayesians could also combine information from different sources, treat observables as random variables, and assign probabilities to all of them, whether they formed a bell-shaped curve or some other shape. Bayesians used all their available data because each fact could change the answer by a small amount. Frequency-based statisticians threw up their hands when Savage inquired whimsically, "Does whiskey do more harm than good in the treatment of snake bite?" Bayesians grinned and retorted, "Whiskey *probably* does more harm than good."[23]

As a movement, Bayes was looking more akin to a philosophy—even a religion or a state of mind—than to a true-or-false scientific law like plate tectonics. According to David Spiegelhalter of Cambridge University, "It's much more basic. . . . A huge sway of scientists says you can't use probability to express your lack of knowledge or one-time events that don't have any frequency to it. Probability came very late into civilization . . . [and many scientists find it] rather disturbing because it's not a process of discovery. It's more a process of interpretation."[24]

"Mathematical scientists often sense a combination of harmony and power in certain formulas," explains Robert E. Kass, a Bayesian at Carnegie Mellon University. "There is at once a deep esthetic experience and a pragmatic recognition of profound consequences, leading to what Einstein called 'the cosmic religious feeling.' Bayes Theorem gives such a feeling. It says there is a simple and elegant way to combine current information with prior experience in order to state how much is known. It implies that sufficiently good data will bring previously disparate observers to agreement. It makes full use of available information, and it produces decisions having the least possible error rate. Bayes' Theorem is awe-inspiring." Unfortunately, Kass continued, "when people are captivated by its spell, they tend to proselytize and become blinded to its fundamental vulnerability. . . . [that] its magical powers depend on the validity of its probabilistic inputs."[25]

With zealots proselytizing Bayes as an all-encompassing panacea, the method inspired both religious devotion and dogmatic opposition. The battle between Bayesians and their equally fervent foes raged for decades and alienated many bystanders. As one onlooker reflected, "It was a huge food fight. It was devastating. They hated each other."[26] A prominent statistician lamented, "Bayesian statisticians do not stick closely enough to the pattern laid down by Bayes himself: if they would only do as he did and publish posthumously we should all be saved a lot of trouble."[27]

Savage became one of the believers. He developed into a full-blown, messianic Bayesian, "the most extreme advocate of a Bayesian . . . ever seen," William Kruskal of the University of Chicago said. Savage recast the controversy over Bayes' rule in its most extreme form as subjectivity versus objectivity. For him, as for Lindley, the rule was the one-and-only, winner-take-all method for reaching conclusions in the face of uncertainty. Bayes' rule was right and rational, they felt, and other views were wrong, and it was neither necessary nor desirable to admit compromise.

"Personal probability . . . became for [Savage] the only sensible approach to probability and statistics," Kruskal recalled sadly. "If one were not in substantial agreement with him, one was inimical, or stupid, or at the least inattentive to an important scientific development. This attitude, no doubt sharpened by personal difficulties and by the mordant rhetoric of some anti-Bayesians, exacerbated relationships between Jimmie Savage and many old professional friends."[28]

Savage's last year at Chicago, 1960, was fraught with turmoil. Although his department colleagues knew nothing about it, the administration was trying to abolish the statistics department, and Savage was fighting to get the decision reversed. His marriage was disintegrating and, hoping to save it, he moved to the University of Michigan. As he departed, he told his colleagues, "I proved the Bayesian argument in 1954. None of you have found a flaw in the proof and yet you still deny it. Why?"[29] When he tried to return to Chicago, members of the department he had formed and chaired voted against rehiring him. At first, no other American or British university would offer him a position. In 1964 he moved to Yale University, remarried, and achieved some level of tranquility.

In 1971, at the age of 53, Savage died suddenly of a heart attack. His death at midcareer deprived American Bayesians of their leading spokesman. The New Haven Register had another perspective. Savage had cowritten a book called How to Gamble If You Must. For Bayesians, all assumptions about the future were risky, and gambling was the paradigm of decision making.

The *Register* headlined his obituary, "Yale Statistician Leonard Savage Dies; Authored Book on Gambling."

In the meantime, Lindley had moved back to Britain, where for many years he was the only Bayesian in a position of authority. In time he built not just Bayesian theory but also strong Bayesian research groups, first at the University College of Wales in Aberystwyth and then at University College London. The latter had England's most important statistical department and was a temple of frequentism. When Lindley arrived, a colleague said it was "as though a Jehovah's Witness had been elected Pope."[30] Lindley complained that he "inherited" several statisticians who "would not change their view of statistics."[31] He said, "The general attitude [was] to turn their heads the other way."[32]

In an era when many sneered at Bayes, it took courage to create Europe's leading Bayesian department. Often the only Bayesian at meetings of the Royal Statistical Society and certainly the only combative one, Lindley defended Bayes' rule like a fearless terrier or a devil's advocate. In return, he was tolerated almost as comic relief. "Bayesian statistics is not a *branch* of statistics," he argued. "It is a way of looking at the *whole* of statistics."

Lindley became known as a modern-age revolutionary. He fought to get Bayesians appointed, professorship by professorship, until the United Kingdom had a core of ten Bayesian departments. Eventually, Britain became more sympathetic to the method than the United States, where Neyman maintained Berkeley as an anti-Bayesian bunker. Still, the process left scars: despite Lindley's landmark contributions he was never named a Fellow of the Royal Society. In 1977, at the age of 54, Lindley forsook the administrative chores he hated and retired early. He celebrated his freedom by growing a beard and becoming what he called "an itinerant scholar" for Bayes' rule.[33]

Thanks to Lindley in Britain and Savage in the United States, Bayesian theory came of age in the 1960s. The philosophical rationale for using Bayesian methods had been largely settled. It was becoming the only mathematics of uncertainty with an explicit, powerful, and secure foundation in logic. How to apply it, though, remained a controversial question.

Lindley's enormous influence as a teacher and organizer bore fruit in the generation to come, while Savage's book spread Bayesian methods to the military and to business, history, game theory, psychology, and beyond. Although Savage wrote about rabbit ears and neon light in beer, he personally encouraged researchers who would apply Bayes' rule to life-and-death problems.

8.

jerome cornfield,
lung cancer,
and heart attacks

Bayes came to medical research through the efforts of a single scientist, Jerome Cornfield, whose only degree was a B.A. in history and who relied on the rule to identify the causes of lung cancer and heart attacks.

Lung cancer, extremely rare before 1900 and still uncommon in 1930, sprang up as if out of nowhere shortly after the Second World War. By 1952 it was killing 321 people per million per year in England and Wales. A year later approximately 30,000 new cases were diagnosed in the United States. No other form of cancer showed such a catastrophic leap. Studies in Europe, Turkey, and Japan confirmed the puzzling plague. There seemed to be something special about the disease.

But what could it be? Its cause was unknown. Pathologists thought the increase in lung cancer might be due to improvements in diagnostic methods or to the natural aging of the population. Others blamed exhaust from factories or the growing number of automobiles, tar particles from modern asphalt pavements, or England's infamous smog from homes heated with open coal-burning fires. Cigarettes, mass-produced since the invention of a cigarette-making machine in 1880, had been patriotically shipped to soldiers during the First World War. Animal studies, though, had failed to demonstrate that tobacco tar was carcinogenic.

As early as 1937 a small-scale study in Germany had pointed ever so tentatively to cigarette smoke. But there were doubts about that too. While 80% of middle-aged men in England and Wales smoked cigarettes, tobacco consumption per capita had dropped slightly. And fumes from cigarettes, which had replaced cigars and pipes, did not seem worse than other smoke.

The world's most famous biostatistician, Austin Bradford "Tony" Hill, was intrigued. He called himself an arithmetician rather than a mathematician or statistician and, in a series of articles in The Lancet, used straightforward logic to persuade the medical community to objectively quantify its research findings. During the late 1940s, two decades after Ronald Fisher had introduced randomization to agricultural experimentation, Hill introduced randomization to medical research. Inaugurating the modern controlled clinical trial, Hill showed that pertussis vaccine reduced children's whooping cough cases by 78% and that streptomycin was effective against pulmonary tuberculosis. Bradford Hill became so famous that a letter addressed to "Lord Hill, Bradford, England" reached him.

To identify the most probable causes of the catastrophic increase in lung cancer, Hill and a young physician and epidemiologist, Richard Doll, organized interviews with patients with and without lung cancer in 20 London-area hospitals. All were questioned about their past activities and exposures. The results, published in 1950, were shockingly clear. Of 649 men with lung cancer, only 2 were nonsmokers; a high proportion of the lung cancer patients were heavy cigarette smokers, and their death rate was 20 times higher than that of nonsmokers. A large American study by Ernst L. Wynder and Evarts A. Graham confirmed the British result the same year.

The startling news that cigarettes and lung cancer were linked caused an instant international uproar. Newspapers, radio, television, and magazines competed with medical journals for the latest scoop. With the exception of the influenza epidemic of 1918, no disease had ever sprung as fast from obscurity to worldwide consciousness. Few have engendered such enormous controversy.

The Hill and Doll study remains one of the crowning glories of medical statistics. It was the first sophisticated case-control study of any noninfectious disease. And it persuaded Hill and Doll to quit smoking. Despite its dramatic results, their study did not show that smoking cigarettes actually causes lung cancer. No one could say that for sure. Jerome Cornfield, an American government bureaucrat at the National Institutes of Health (NIH), took up the challenge. And with Hill organizing clinical studies in Britain and Cornfield developing their mathematical defense in the United States, the two tackled complementary aspects of the same problem from different sides of the Atlantic.

The two men had totally dissimilar backgrounds. Hill's father was a physician with a knighthood, and one of his ancestors had invented the postage stamp. Cornfield was the son of Russian Jewish immigrants and had earned a bachelor's degree from New York University in 1933. The federal government, desperate for economic data during the Depression, had hired "bright guys" to replace the clerks who had traditionally compiled statistics on unemployment, national income, housing, agriculture, and industry.[1] Cornfield qualified as a bright guy, so he signed on as a government statistician for $26.31 a week, $1,368 a year.

Washington, D.C., was still a segregated, southern city. "The rule of thumb was that, if you were Jewish, you could work for the Department of Labor and, if you were Catholic, you could work for the Department of Commerce," explained Marvin Hoffenberg, a friend of Cornfield's and later a UCLA professor.[2] So Cornfield went to Labor. The U.S. Department of Agriculture ran a so-called Graduate School where mathematically inclined government employees could study statistics, and Cornfield took his only mathematics and statistics courses there.

As Cornfield recalled, "Nobody knew how many unemployed there were, and sampling seemed the way to find out. . . . Statistics had me hooked."[3] Although both Fisher and Neyman lectured on sampling methods at the Graduate School, its director, W. Edwards Deming, was open-minded; he published Thomas Bayes' essay with an introduction by Edward Molina of Bell Laboratories.

Friends referred to Cornfield's tenure at the Department of Labor as his serious and exotic phase. He played a major role in revising the Consumer Price Index and in creating one for occupied Japan after the Second World War. But he was "a different kind of a guy," a friend recalled.[4] Unable to think of any good reason for shaving, he grew a little pointed beard, and with his long gaunt frame and an umbrella over his arm he resembled an elegant diplomat strolling jauntily to work. At a time when few others would, he shared his office with a woman statistician and an African American statistical clerk. Next to his mechanical Marchant desktop calculator he installed a Turkish water pipe and could be seen puffing nonchalantly from its two-foot tube.

Cornfield moved to the federal government's new NIH in 1947. Because infectious diseases were in decline in the United States, NIH epidemiologists were attacking chronic diseases, particularly cancer, heart attacks, and diabetes. To assist them, NIH hired a few people with strong quantitative backgrounds. Only one of them had so much as a master's degree. Biostatistics was a professional backwater, and throughout the 1950s and 1960s NIH employed only ten or 20 statisticians at any one time. It was this small band that introduced statistical methods to NIH researchers in biology and medicine.

By 1950 most men in the United States smoked, and smoking rates were increasing, especially among women. The favorite brands were unfiltered Camels, Lucky Strikes, Chesterfields, and Philip Morris. When Lorillard Tobacco Company introduced filtered Kents in 1952, the filters contained asbestos, which was not removed until 1957. When 14 studies conducted in five countries showed that lung cancer patients included an alarming percentage of heavy smokers, both Cornfield and his wife quit their 2½-pack-a-day habits.

Cornfield realized that the Hill and Wynder studies did not directly answer the questions physicians and their frightened patients were asking: what is my risk? The studies showed the percentages of smokers among groups of people with and without lung cancer, but they did not say what proportion of smokers and nonsmokers was likely to develop lung cancer.

The surest and most direct way to answer the public's fears was to follow large groups of smokers and nonsmokers for years, prospectively, to see how many of each group developed lung cancer. Unfortunately, studies about the future of large populations require a great deal of money and time, especially for relatively rare problems like lung cancer. That is why Hill and Doll had organized their study as a retrospective one, choosing people who already had lung cancer and asking them about their health histories. Such studies are a relatively quick, cheap way to identify potential causes of a particular disease. As a statistician, however, Cornfield suspected that retrospective studies like Hill and Doll's could also be used to answer the individual's haunting question, What's the chance that I or my loved ones will get this fatal disease?

In 1951 Cornfield used Bayes' rule to help answer the puzzle. As his prior hypothesis he used the incidence of lung cancer in the general population. Then he combined that with NIH's latest information on the prevalence of smoking among patients with and without lung cancer. Bayes' rule provided a firm theoretical link, a bridge, if you will, between the risk of disease in the population at large and the risk of disease in a subgroup, in this case smokers. Cornfield was using Bayes as a philosophy-free mathematical statement, as a step in calculations that would yield useful results. He had not yet embraced Bayes as an all-encompassing philosophy.

Cornfield's paper stunned research epidemiologists. More than anything else, it helped advance the hypothesis that cigarette smoking was a cause of lung cancer. Out of necessity, but without any theoretical justification, epidemiologists had been using case studies of patients to point to possible causes of problems. Cornfield's paper showed clearly that under certain conditions (that is, when subjects in a study were carefully matched with controls) patients' histories could indeed help measure the strength of the link between a disease and its possible cause. Epidemiologists could estimate disease risk rates by analyzing nonexperimental clinical data gleaned from patient histories. By validating research findings arising from case-control studies, Cornfield made much of modern epidemiology possible. In 1961, for example, case-control studies would help identify the antinausea drug thalidomide as the cause of serious birth defects.

Two massive efforts in England and the United States during the mid-1950s confirmed Cornfield's judgment. Because many people had rejected the findings of their retrospective study, Hill and Doll had decided to take a direct approach and conduct a prospective study. They questioned 40,000 British physicians about their current smoking habits and then followed them for five years to see who got lung cancer. In a parallel U.S. study, E. Cuyler Hammond and Daniel Horn followed 187,783 men aged 50 to 69 in New York State for more than 3½ years. Death rates in both countries were similar: heavy smokers were 22 to 24 times more likely to get lung cancer than nonsmokers and, in another surprise discovery, were 42% and 57% more likely to get, respectively, heart and circulatory diseases. Research also showed that cigarettes were more dangerous than pipes, although the risk declined after smoking stopped.

Surprisingly, neither Fisher nor Neyman could accept research results showing that cigarettes caused lung cancer. Both anti-Bayesians were heavy smokers, and Fisher was a paid consultant to the tobacco industry. But more important, neither found epidemiologic studies convincing. And both were correct in pointing out that tobacco could be associated with cancer without causing it. In 1955 they launched a vigorous counterattack, arguing that only experimental data from strictly controlled laboratory and field experiments could predict future disease rates. The most eminent American medical statistician of the day, Joseph Berkson, of the Mayo Clinic in Rochester, Minnesota, joined the attack; Berkson did not believe cigarettes could cause both cancer *and* heart disease.

Fisher kept up a barrage of angry attacks, including a book and two articles published in highly prestigious journals, *Nature* and the *British Medical Journal*. According to Doll, Fisher even went so far as to accuse Hill of scientific dishonesty. Over the course of three years Fisher developed two remarkable hypotheses. The first, believe it or not, was that lung cancer might cause smoking. The second was that a latent genetic factor might give some people hereditary predilections for both smoking and lung cancer. In neither case would smoking cause lung cancer.

Cornfield maintained a running argument with Fisher through the 1950s. Cornfield was already thinking deeply about the standards of evidence needed before observational data could establish cause and effect. Finally, in 1959, he raked Fisher over the coals about smoking with a common-sense, nonmathematical paper that reads like a legal brief. In that seminal paper he and five coauthors systematically addressed every one of Fisher's alternative explanations for the link between cigarette smoking and lung cancer. They hurled one counterargument after another at Fisher's hypothetical genetic factor. If cigarette smokers were nine times more likely than nonsmokers to get lung cancer, Fisher's latent genetic factor must be even larger—though nothing approaching that had ever been seen.

Cornfield dismissed out of hand Fisher's suggestion that cancer might cause smoking: "Since we know of no evidence to support the view that the bronchogenic carcinoma diagnosed after age 50 began before age 18, the median age at which smokers begin smoking, we shall not discuss it further."[5] Cornfield pointed out that Fisher's genetic factor would have to spread rapidly and occur more among cigarette smokers than nonsmokers; cause tumors on mouse skin but not on human lungs; weaken with age after a smoker quit; and be more likely in men than women, 60 times more prevalent among two-pack-a-day smokers, and different in pipe and cigar smokers. Yet none of these phenomena had ever been observed.

Fisher wound up looking ridiculous. As Cornfield coolly noted, "A point is reached . . . when a continuously modified hypothesis becomes difficult to entertain seriously."[6] Scientists who can find only one viable explanation for associations in their data have probably found its causal agent. The existence of possible alternative explanations indicates that the cause has probably not yet been found. Cornfield was laying out the road map for future smoking and lung cancer research.

By now, Cornfield the history major had become the most influential biomedical statistician in the United States. When the U.S. surgeon general concluded in 1964 that "cigarette smoking is causally related to lung cancer in men," he cited Cornfield's work.[7] Nonexperimental studies had helped identify an association between smoking and lung cancer. With the help of Bayes' rule—what Laplace had called "the probability of causes and future events, derived from past events"—Cornfield provided the theoretical justification for using case-control studies to estimate the strength of links between exposure and disease. Today, thanks to Cornfield, case-control studies are the primary tool epidemiologists use to identify likely causes of chronic diseases.

Over his career, Cornfield would become involved in every major public health problem of the day. Most of them, including smoking, the safety of polio vaccines, and the efficacy of diabetes treatments, were fiercely controversial.

To calm the statistics phobia of physicians and epidemiologists, Cornfield developed an easygoing bedside manner. Abandoning his serious phase, he cultivated an infectious laugh and an irrepressible air of informality. By mixing humor into conversations, telling stories, and laughing heartily he inspired tremendous confidence. Even his gait and prose became sprightly. Soon every biomedical scientist with a committee and a controversy wanted Cornfield on board. By pointing out common elements that everyone shared, he could unify the most disparate group. After one particularly onerous series of meetings and reports, a committee member asked him, "Did you get my last letter about sample size?" There was a pause, and Cornfield grinned and said, "Christ, I hope so." When the committee finally produced its massive procedural manual, Cornfield waved it over his head, declaring, "You know, say what you will about the Ten Commandments, you must come back to the pleasant fact that there were only ten of them."

Cornfield typically rose at 5 a.m. to write and make paper-and-pencil calculations. He came up with clever approximations and computational tricks, much as Laplace had done. He visualized particularly difficult distributional functions by carving them out of a bar of soap. To collaborate with biochemists, he studied basic biology. And although he was a brilliant speaker, he never prepared a talk until the night before. He procrastinated even the day before he was scheduled to speak at 8:30 a.m. at Yale University about the contentious Salk poliomyelitis vaccine tests. "Don't worry, Max," he told a friend. "God will provide."

Cornfield was a voracious reader but did not own a TV and was blissfully unaware of popular culture. Once a biostatistician who dated Hollywood stars begged him to speed up a morning meeting: "I've got to get through because at 12 o'clock I have a luncheon date with Kim Novak." Puzzled, Cornfield asked, "Kim Novak? Who's he?"[8] At the time she was Columbia Pictures' answer to Marilyn Monroe.

Another watershed medical study also occupied Cornfield's attention during the 1950s. Death rates from cardiovascular disease had been rising steadily in the United States since 1900. Heart disease had been the nation's leading cause of death since 1921, and strokes the third leading cause since 1938. Yet researchers at the midpoint of the twentieth century were as ignorant of the causes of heart disease and stroke as they had been of lung cancer.

Understanding the causes of deaths from cardiovascular disease would require following a population for many years. A prospective study, however, was more feasible than with lung cancer because heart problems were far more common. In 1948 Cornfield helped design the Framingham Heart Study, which has since followed the health of three generations of Framingham, Massachusetts, residents.

As one of the first important studies based on Framingham, Cornfield followed 1,329 adult male residents for a decade. Between 1948 and 1958, 92 of the group experienced myocardial infarction or angina pectoris.

Longitudinal studies like Framingham are designed to investigate a large variety of variables, both singly and jointly, on the risk of developing a disease. Traditionally, epidemiologists studied their data by inspecting— "contemplating" was Cornfield's word—the resulting multiple cross-classification arrays. Three risk factors, each considered at low, medium, and high levels, would produce a tidy 3x3 table of cells, but as the number of variables increased and they were considered singly and jointly, the number of cells to be contemplated quickly became impracticable. A cross-classification study with 10 risk factors at low, medium, and high levels would produce 59,049 cells. To get even 10 patients per cell, the study would need a cohort of 600,000 people, more than the population of Framingham.

Cornfield realized he needed a "more searching form of analysis than simple inspection."[9] He would have to develop a mathematical model for summarizing the observations. He chose Bayes' rule and used cardiovascular death rates as the prior. Framingham gave him data about two groups of people, those who had died of heart disease and those who had not. Within each group he had information about seven risk factors. Calculating Bayes' rule, he got a posterior probability in the form of a logistic regression function, which he then used to identify the four most important risk factors for cardiovascular disease. In addition to age itself, they were cholesterol, cigarette smoking, heart abnormalities, and blood pressure.

Bayes allowed Cornfield to reframe Framingham's data in terms of the probability that people with particular characteristics would get heart disease.

There was no critical level for cholesterol or blood pressure below which people were safe and above which they got the disease. And patients cursed with both high cholesterol and high blood pressure were 23% more at risk for heart attacks than those with low cholesterol and blood pressure rates.

Cornfield's identification in 1962 of the most critical risk factors for cardiovascular disease produced one of the most important public health achievements of the twentieth century: a dramatic drop in death rates from cardiovascular disease. Between 1960 and 1996 they fell 60%, preventing 621,000 fatalities. His report also showed researchers how to use Bayes' rule to analyze several risk factors at a time; his multiple logistic risk function has been called one of epidemiology's greatest methodologies.

To measure the efficacy of a particular therapy, Cornfield used an early multicenter trial at NIH to introduce another Bayesian concept—Harold Jeffreys's relative betting odds. Now known as the Bayes' Factor, it is the probability of the observed data under one hypothesis divided by its probability under another hypothesis.

When Cornfield worked with researchers who used mice to screen for anticancer drugs, the rigidity of frequentist methods struck him like a blow from behind. According to their rules, even if their initial test results disproved their hypothesis, they had to take six more observations before stopping the experiment. Frequentist methods also banned switching a patient to better treatment before a clinical trial was finished. Frequentist experimenters could not monitor interim results during the clinical trial, examine treatment effects on subgroups of patients, or follow leads from the data with further unplanned analyses. When Cornfield discovered that Bayesian methods would let him reject some hypotheses after only two solidly adverse observations, he was converted. He had started out using Bayes' theorem as an enabling tool to solve a particular problem, the way it had been used for cryptography, submarine hunting, and artillery fire during the Second World War. But now he was moving gradually toward making Bayes' theorem the foundation of a broad philosophy for handling information and uncertainties. As he began thinking about Bayes as a philosophy rather than just a tool, he became part of the profound conversion that Jeffreys, Savage, Lindley, and others were also going through in the 1950s and 1960s. While Fisher considered a hypothesis significant if it was unlikely to have occurred by chance, Cornfield declared airily, "If maintenance of [Fisher's] significance level interferes with the release of interim results, all I can say is so much the worse for the significance level."[10]

Interestingly, most of NIH's other statisticians failed to follow their leader into Bayesian fields. Cornfield published scientifically important papers about Bayesian inference in mainstream statistics journals. Nevertheless, when he included Bayesian methods in some of the trials he worked on, their main conclusions were based on frequentism. It would be another 30 years before NIH started using Bayes for clinical trials. Savage thought that many researchers were content to reap the benefits of Bayes' theorem without embracing the method.

Cornfield declared cheerily, however, "Bayes' theorem has come back from the cemetery to which it has been consigned."[11]

In 1967 Cornfield retired from NIH and moved later to George Washington University, where he chaired the statistics department and developed Bayes' rule into a full-scale logical mathematical approach. In one paper he proved to the satisfaction of many Bayesians that, according to frequency rules, any statistical procedure that does not stem from a prior can be improved upon.

Despite his Bayesian conversion, Cornfield was in great demand as a consultant. He advised the U.S. Army on experimental design; the investigating committee critiquing the best-selling Kinsey Report on female sexuality; the U.S. Department of Justice on sampling voting records to reveal bias against black voters; and the State of Pennsylvania after the Three Mile Island nuclear power plant incident.

In 1974 the Bayesian biostatistician with a bachelor's degree in history became president of the American Statistical Association. In his presidential address, the man who used humor and good cheer to reassure physicians about randomized trials, who gave epidemiologists some of their most important methodologies, and who established causes of both lung cancer and heart attacks, asked, "Why should any person of spirit, of ambition, of high intellectual standards, take any pride or receive any real stimulation and satisfaction from serving an auxiliary role [as a statistician] on someone else's problem?" Smiling at his own question, Cornfield continued, "No one has ever claimed that statistics was the queen of the sciences. . . . The best alternative that has occurred to me is 'bedfellow.' Statistics—bedfellow of the sciences—may not be the banner under which we would choose to march in the next academic procession, but it is as close to the mark as I can come."[12]

When Cornfield was diagnosed with pancreatic cancer in 1979, he knew as well as anyone at NIH the disease's appalling six-month survival rate. Nonetheless, he was determined to continue living to the fullest. Despite serious postoperative complications, his humor remained intact. A friend told him, "Jerry, I'm *so* glad to see you." Smiling, Cornfield replied, "That's nothing compared to how happy I am to be able to see you."[13] As he was dying he said to his two daughters, "You spend your whole life practicing your humor for the times when you really need it."[14]

there's always
a first time

Bayes' military successes were still Cold War secrets when Jimmie Savage visited the glamorous new RAND Corporation in the summer of 1957 and encouraged two young men to calculate a life-and-death problem: the probability that a thermonuclear bomb might explode by mistake.

RAND was the quintessential Cold War think tank. Gen. Curtis E. LeMay, the commander of the Strategic Air Command (SAC), had helped start it in Santa Monica, California, 10 years earlier as "a gimmick" to cajole top scientists into applying operations research to long-range air warfare.[1] But RAND, an acronym for Research ANd Development, considered itself a "university without students" and its 1,000-odd employees "defense intellectuals." Their mission was to use mathematics, statistics, and computers to solve military problems, pioneer decision making under conditions of uncertainty, and save the United States from Soviet attack. The U.S. Air Force, which funded RAND, gave its researchers carte blanche to choose the problems they wanted to investigate. But since President Eisenhower's "New Look" military policy depended on early nuclear bombing ("massive retaliation") as the cheapest way to respond to a Soviet attack, RAND's top issues were nuclear strategy, surviving nuclear attack, and response options. Because SAC bombers had a monopoly on transporting America's nuclear arsenal and General LeMay sat at the pinnacle of the world's military might, RAND's voice was often influential.

By the time Savage visited Santa Monica that summer RAND reports had already challenged some of SAC's sacred cows. To drop nuclear weapons on Soviet targets, macho air force pilots wanted to fly the new B-52 Stratofortress jets; RAND recommended fleets of cheaper conventional planes. RAND had also described SAC's overseas bases for manned bombers as sitting ducks for Soviet attack. A year after Savage's visit, RAND would challenge Cold War dogma by arguing that victors typically fare better with negotiated settlements than with unconditional surrenders. RAND would even urge counterbalancing LeMay's B-52s with the navy's submarine-based missiles. In retaliation, SAC would almost break off relations with RAND on several occasions between Savage's visit in 1957 and 1961.

Circulating gregariously among RAND researchers that summer, Savage met Fred Charles Iklé, a young Swiss-born demographer who had studied the sociological effects of nuclear bombing on urban populations. At 33, Iklé was seven years younger than Savage and had earned a Ph.D. in 1950 from the University of Chicago, where Savage was teaching. Seeking a wide-open field that no one else at RAND was studying, Iklé chose nuclear catastrophes that an Anglo-American nuclear arsenal would not deter: those caused by accident or by someone mentally deranged. Referring to massive retaliation a few years later, Iklé would declare, "Our method of preventing nuclear war rests on a form of warfare universally condemned since the Dark Ages—the mass killing of hostages."[2] With SAC poised to expand its bomb-bearing flights, Iklé and Savage talked about assessing their impact on nuclear accidents. Eventually the issue came around to the question, what was the probability of an accidental H-bomb explosion?

After a summer of conversations, Savage was preparing to return to academia when Albert Madansky, a 23-year-old Ph.D. graduate of Savage's statistics department, arrived at RAND. Madansky had financed his graduate studies by working part-time for Arthur Bailey, the Bayesian theorist in the insurance industry. Until Bailey's death, he had considered an actuarial career. Savage, who had published his book *Foundations of Statistics* but had not yet embraced Bayes' rule, talked over the H-bomb problem with Madansky without considering it in Bayesian terms. As he left Santa Monica, he handed over the H-bomb study to Madansky but left the young man to do as he saw fit. Madansky would come up with a Bayesian approach on his own.

Because RAND's report on the project would eventually be classified, Madansky could not talk about his work for 41 years. But when Savage returned to Chicago he lectured openly about the fundamental statistical issues involved. Bayes' rule was emerging in fits and starts from the secrecy of the Second World War and the Cold War.

The H-bomb problem facing Madansky was politically and statistically tricky. No atomic or hydrogen weapon had ever exploded accidentally. In the 12 years since the United States had dropped atomic bombs on Japan in August 1945, nuclear bombs had been detonated, but always deliberately, as part of a weapons test. Barring accidents, the nation's leaders believed that their stocks of nuclear weapons deterred all chance of thermonuclear war—and that accidents could not occur in the future because none had occurred in the past. Nonetheless, the question remained: could the impossible happen?

According to more than a century of conventional statistics, the impossible could not be calculated. Jakob Bernoulli had decreed in 1713 that highly improbable events do not happen. David Hume agreed, arguing that because the sun had risen thousands of times in the past it would continue to do so. It was Thomas Bayes' friend and editor, Richard Price, who took the contrary view that the highly improbable could still occur. During the nineteenth and early twentieth centuries, Antoine-Augustin Cournot concluded that the probability of a physically impossible event was infinitely small and thus the event would never occur. Andrei Kolmogorov slightly refined "never" by saying that if the probability of an event is very small, we can be practically certain that the event will not happen in the very next trial.

Fisher was no help either. He argued that probability is a simple relative frequency in an infinitely large population; until a nuclear bomb accident occurred, he had no way of judging its future probability. Mercifully, Madansky did not have an infinitely large population of H-bomb accidents, and further experimentation was out of the question. Fisher's approach left him with the banal observation that zero accidents had occurred and that the probability of a future accident was also zero.

Madansky concluded, "As long as you are set that the probability is going to be zero, then nothing's going to change your mind. If you have decided that the sun rises each morning because it has always done so in the past, nothing is going to change your mind except one morning when the sun fails to appear."[3]

He did not buy the argument that an accident could never occur simply because none had happened in the past. First, the political and military establishment's assumption that a well-stocked American arsenal of nuclear weapons would deter war rested on increasingly shaky grounds. In the six years between 1949 and 1955 the Soviets had exploded their first atomic bomb, the United States had detonated the world's first hydrogen bomb, and Britain had tested both an atomic and a hydrogen bomb. The USSR had launched the first man-made satellite into orbit around Earth in 1957. Meanwhile, the

United States trained countries in the North Atlantic Treaty Organization (NATO) to fire nuclear weapons and supplied Britain, Italy, and Turkey with nuclear missiles. By the time the Anglo-American Agreement for Co-operation on the Uses of Atomic Energy for Mutual Defense Purposes was signed in 1958, all hope of preventing the spread of nuclear weapons had evaporated. France tested its first atomic bomb in 1960.

In addition to the rapid expansion of nuclear weapons, Madansky had 16 top-secret reasons for doubting that the probability of a future accident was zero. A classified list detailed 16 of "the more dramatic incidents" involving nuclear weapons between 1950 and 1958.[4] They included accidental drops, jettisons, aircraft crashes, and testing errors. Incidents had occurred off British Columbia and in California, New Mexico, Ohio, Florida, Georgia, South Carolina, and overseas. RAND's list omitted accidents that had not attracted public attention.

An atomic or hydrogen bomb consists of a small capsule, or "pit," of uranium or plutonium inside a case covered with powerful conventional explosives. Only if these high explosives blow up at the same instant can the uranium or plutonium capsule be sufficiently compressed on all sides to trigger a nuclear blast. In a few cases these conventional explosives had detonated, generally on impact in a plane crash. Because the capsule of nuclear material had not been installed inside the weapons, however, there had been no nuclear accidents. That fact had convinced SAC that its procedures were sound and that no nuclear accident would occur.

Still, scores of people had died when the high explosives in unarmed nuclear weapons blew up. The year 1950 was a banner year for accidents. On April 11, 1950, 13 people died near Kirtland Air Force Base, outside Albuquerque, New Mexico, when a B-29 crashed into a mountain; flames from the high explosives were visible for 15 miles. On July 13, 16 were killed when a B-50 nose-dived near Lebanon, Ohio. Nineteen people, including Gen. Robert F. Travis, died when a B-29 with mechanical problems crash-landed in California on August 5, injuring 60 people in a nearby trailer camp. Also that year, two bombs without nuclear capsules were jettisoned and abandoned in deep water in the Pacific Ocean off British Columbia and in an unnamed ocean outside the United States.

Newspaper reporting was meager until an improperly closed lock in a B-47's bomb bay opened in 1958 and a "relatively harmless" bomb fell into Walter Gregg's garden in Mars Bluff, South Carolina.[5] The conventional explosives detonated on impact, digging a crater 30 feet deep and 50 to 70 feet across, destroying Gregg's house, damaging nearby buildings, and killing several chickens. No people died, but news coverage was extensive; typically reporters said that a "TNT triggering device" had exploded. RAND noted disapprovingly that a *Time* magazine article was "astonishingly accurate."[6] Congress, Britain's Labour Party, and Radio Moscow complained.

The air force paid the Greggs $54,000, and all B-47 and B-52 flights carrying nuclear weapons were suspended until new safety measures could be introduced. SAC also established a new policy: nuclear bombs were to be jettisoned on purpose only into oceans or designated "water masses. . . . Hence only uncontrolled drops may attract public attention in the future."[7]

As the press became increasingly suspicious of the accidents, Iklé and another RAND researcher, Albert Wohlstetter, became concerned. Iklé recommended that the government remain silent about the presence of nuclear weapons in aircraft crashes. As Iklé and Madansky worked on their study, an accident occurred that could have provoked an international scandal. A wheel casting on a B-47 failed at the air force's fueling base in Sidi Slimane, French Morocco. High explosives detonated and a fire raged for seven hours, destroying the nuclear weapon and capsule aboard.

Given all these accidents involving unarmed nuclear weapons, Madansky felt he could no longer assume, as SAC and frequentist statisticians had, that an accident involving an H-bomb could never occur. Instead, he decided, he needed "another theology, . . . another brand of inference," where the possibility of an accident was not necessarily zero.

Frequentism was no help. "But, but, but," Madansky said later, "if you're willing to admit a shred of disbelief, you can let Bayes' theorem work. . . . Bayes is the only other theology that you can go to. It's just sort of natural for this particular problem. At least that's how I felt back then."[8] As Dennis Lindley had argued, if someone attaches a prior probability of zero to the hypothesis that the moon is made of green cheese, "then whole armies of astronauts coming back bearing green cheese cannot convince him." In that connection, Lindley liked to quote Cromwell's Rule from the Puritan leader's letter to the Church of Scotland in 1650: "I beseech you, in the bowels of Christ, think it possible you may be mistaken."[9] In the spirit of Cromwell's Rule, Madansky adopted Bayes as his "alternate theology."

Many Cold War statisticians knew it well. They were using it to deal with one of their biggest problems, estimating the reliability of the new intercontinental ballistic missiles. "We didn't know how reliable the missiles were," Madansky explained, "and we had a limited amount of test data to determine it, and so a number of the people working in reliability applied Bayesian methods. The entire field, from North American Rockwell, Thompson Ramo Wooldridge, Aerospace, and others, was involved. I'm sure Bayesian ideas floated around them too. We all knew about it. . . . It was just a natural thing to do."[10]

Madansky immediately set out to measure how much credence could be given to the belief that no unauthorized detonations would occur in the future. He started with "statistically speaking, this simple commonsense idea based on the notion that there is an *a priori* distribution of the probability of an accident in a given opportunity, which is *not all concentrated at zero.*"[11] The decision to incorporate a whisper of doubt into his prior was important. Once a Bayesian considered the possibility that even one small accident might have occurred in the past 10,000 opportunities, the probability of an accident-free future dropped significantly.

Politically and mathematically, Madansky faced an extremely difficult problem. As a young civilian challenging one of the military's fundamental beliefs about the Cold War, he would have to convince decision makers that, although nothing catastrophic had occurred, something might in future. He would have to explain the Bayesian process to nonspecialists. And because the military was often suspicious of civilians making unwarranted suggestions, he would have to make as few initial assumptions as possible. In addition, there was the statistical problem involving the small number of accidents and, fortunately, no Armageddons.

Not wanting to stick his neck out, Madansky decided to make his prior odds considerably weaker than 50–50. "I tried to figure out how to avoid making any specifications about what the prior should be."[12] To refine his minimally informative prior, he added another common-sense notion: the probability of an accident-free future depends on the length of the accident-free past and the number of future accident opportunities. Madansky had no direct evidence because there had never been a nuclear accident. But the military had plenty of indirect data, and he began using it to modify his prior.

He knew that the military was already developing plans to greatly increase the number of flights carrying nuclear weapons. SAC was planning a system of 1,800 bombers capable of carrying nuclear weapons; approximately 15% of them would be in the air at all times, armed and ready for attack. At that time, SAC's B-52 jet-powered Stratofortress bombers carried up to four nuclear bombs, each with an explosive power between 1 million and 24 million tons of TNT, or up to 1,850 Hiroshima bombs. The United States was also planning to outfit intercontinental ballistic missiles with hydrogen warheads, accelerate the production of intermediate range ballistic missiles, and negotiate with NATO countries for launching rights and military bases. The military would soon be dealing with shorter alarm times, increased alertness, and more decentralized weaponry, all factors that could increase the likelihood of a catastrophe.

Madansky calculated the number of "accident opportunities" based on the number of weapons, their longevity, and the number of times they were aboard planes or handled in storage.[13] Accident opportunities corresponded to flipping coins and throwing dice. Counting them proved to be an important innovation.

"A probability that is very small for a single operation, say one in a million, can become significant if this operation will occur 10,000 times in the next five years," Madansky wrote.[14] The military's own evidence indicated that "a certain number of aircraft crashes" was inevitable. According to the air force, a B-52 jet, the plane carrying SAC's bombs, would average 5 major accidents per 100,000 flying hours. Roughly 3 nuclear bombs were dropped accidentally or jettisoned on purpose per 1,000 flights that carried these weapons. In that 80% of aircraft crashes occurred within 3 miles of an air force base, the likelihood of public exposure was growing. And so it went. None of these studies involved a nuclear explosion, but to a Bayesian they suggested ominous possibilities.

Computationally, Madansky was confident that RAND's two powerful computers, a 700 series IBM and the Johnniac, designed by and named for John von Neumann, could handle the job. But he hoped to avoid using them by solving the problem with pencil and paper.

Given the power and availability of computers in the 1950s, many Bayesians were searching for ways to make calculations manageable. Madansky latched onto the fact that many types of priors and posteriors share the same probability curves. Bailey had used the same technique in the late 1940s, and later it would be known as Howard Raiffa and Robert Schlaifer's conjugate priors. When Madansky read their book describing them, he was pleased to know that his prior had a name and rationale: "I was just doing it ad hoc."[15]

Using his tractable prior, classified military data, and informed guesswork, Madansky came to a startling conclusion. It was highly probable that SAC's expanded airborne alert system would be accompanied by 19 "conspicuous" weapon accidents each year.

Madansky wrote an elementary summary that high-level military decision makers could understand and inserted it into RAND's final report, "On the Risk of an Accidental or Unauthorized Nuclear Detonation. RM-2251 U.S. Air Force Project Rand," dated October 15, 1958. The report listed Iklé as its primary author in collaboration with Madansky and a psychiatrist, Gerald J. Aronson. Until then, many RAND reports had been freely published, but air force censors were clamping down on its think tank, and this report was classified, its initial readership limited to a select few. It was finally released more than 41 years later, on May 9, 2000, with numerous passages blacked out.

In view of what would happen a few years later in Spain, much of the report seems prescient. Madansky could not predict when or where the accident would occur, but he was sure of two things. The likelihood of an accident was rising, and it was in the military's interest to make its nuclear arsenal safer. Given the media's increasingly savvy coverage of accidents involving nuclear weaponry, Iklé foresaw Soviet propaganda, citizens' campaigns to limit the use of nuclear devices, and foreign powers demanding an end to American bases on their soil. Who knew, but the British Labour Party might even win an election or NATO might crumble?

As a result, Iklé and Madansky advocated safety features that included requiring at least two people to arm a nuclear weapon; electrifying arming switches to jolt anyone who touched them; arming weapons only over enemy territory; installing combination locks inside warheads; preventing the release of radioactive material in accidental fires of high-energy missile fuels; and switching the nuclear matter inside weapons from plutonium to uranium because the latter would contaminate a smaller area. The report also recommended placing reassuring articles in scientific journals to publicize research indicating that plutonium emissions were less dangerous to humans than had been thought; the source of the research was to be disguised.

Later, after SAC implemented its airborne alert program and began keeping significant numbers of nuclear-armed planes aloft at all times, Iklé and Madansky followed up with a more pointed and less mathematical summary about accident rates. As of 2010 this internal document, meant only for internal circulation, was still classified. In an appendix to the first report, Iklé and Aronson, the psychiatrist who consulted at RAND, tackled the topic of mental illness among military personnel in charge of nuclear bombs. Such concerns were widespread at the time. Aronson believed that men who worked near the bombs should be psychologically tested by being confined in a chamber by themselves, deprived of sleep and sensory stimuli for several hours, and perhaps dosed with hallucinogens such as LSD. He predicted "only between one-third and one-quarter of 'normal' volunteer-subjects would be able to withstand [it] for more than several hours."[16] It was later learned that the Central Intelligence Agency financed a variety of LSD (lysergic acid diethylamide) experiments on people without their knowledge or consent during the 1950s, although the practice violated ethical standards even then.

After the report was issued, Iklé, his knees shaking, went to brief "a sizable audience of Air Force generals."[17] RAND researchers assumed that General LeMay would scorn their conclusions. LeMay had directed the fire-bombing of Japanese cities during the Second World War, his ghostwritten autobiography would propose bombing the Vietnamese "back into the Stone Age" in the mid-1960s, and he was the model for "Buck" Turgidson, the insanely warlike, cigar-chomping general played by George C. Scott in the film Dr. Strangelove. Iklé described LeMay's attitude about nuclear weapons as one of "injudicious pugnacity."[18]

But the general surprised him. The day after RAND presented its report in Washington, LeMay asked for a copy. Iklé said LeMay later issued a "bliz-zard" of orders, among others calling for the two-man rule and coded locks. The army and navy followed suit. Iklé put LeMay's response "in the 'success' column of my life's ledger."[19]

According to most reports, however, few of the coded locks were actu-ally installed on nuclear weapons until John F. Kennedy became president. Four days after Kennedy's inauguration, a SAC B-52 disintegrated in midflight. One of its two 24-megaton hydrogen bombs smashed into a swamp near Goldsboro, North Carolina, and a large chunk of enriched uranium sank more than 50 feet, where it presumably remains to this day. Analysis showed that only one of the bomb's six safety devices had functioned properly. JFK was told of scores of nuclear weapon accidents—according to Newsweek, more than 60 since the Second World War. From then on, the Kennedy administration vigorously pursued nuclear weapon safety and added coded locks to nuclear weapons.

Iklé went on to become a leading hard-line specialist in military and foreign policy and would be awarded two Distinguished Public Service Med-als, the Department of Defense's highest civilian award. Madansky became a professor at the University of Chicago, where he developed a reputation as a neutral pragmatist in the battles between Bayesians and frequentists. RAND gradually weaned itself from air force funding by diversifying into social welfare research.

The world can be thankful that Madansky's Bayesian statistics forced the military to tighten safety measures. A number of false alerts suggestive of Soviet nuclear attacks were identified correctly before SAC could launch a counterattack. Phenomena causing false alerts included the aurora borealis, a rising moon, space debris, false U.S. radar signals, errors in the use of computers (such as a mistaken Pentagon warning of incoming Soviet missiles in 1980), routine Soviet maintenance procedures after the Chernobyl accident, a Norwegian weather research missile, and "more hidden problems of unauthorized acts."

10.

46,656 varieties

In sharp contrast to the super secrecy of Madansky's H-bomb report, the schism between entrenched frequentists and upstart Bayesians was getting downright noisy. As usual, the bone of contention was the subjectivity of Thomas Bayes' pesky prior. The idea of importing knowledge that did not originate in the statistical data at hand was anathema to the anti-Bayesian duo Fisher and Neyman. Since they were making conclusions and predictions about data without using prior odds, Bayesian theoreticians on the defensive struggled to avoid priors altogether.

Bayesian theories mushroomed in glorious profusion during the 1960s, and Jack Good claimed he counted "at least 46656 different interpretations," far more than the world had statisticians.[1] Versions included subjective, personalist, objective, empirical Bayes (EB for short), semi-EB, semi-Bayes, epistemic, intuitionist, logical, fuzzy, hierarchical, pseudo, quasi, compound, parametric, nonparametric, hyperparametric, and non-hyperparametric Bayes. Many of these varieties attracted only their own creators, and some modern statisticians contend that hairsplitting produced little pathbreaking Bayesian theory. When asked how to differentiate one Bayesian from another, a biostatistician cracked, "Ye shall know them by their posteriors."

Almost unnoticed in the hoopla, the old term "inverse probability" was disappearing, and a modern term, "Bayesian inference," was taking its place. As the English language swept postwar statistical circles, articles by British theorists began to look more important than Laplace's French ones. "Much that's been written about the history of probability has been distorted by this English-centric point of view," says Glenn Shafer of Rutgers University.[2]

129

More than language may have been involved. In 2008, when the Englishman Dennis Lindley was 85 years old, he said he was now almost convinced that Laplace was more important than Thomas Bayes: "Bayes solved one narrow problem; Laplace solved many, even in probability. . . . My ignorance of the Frenchman's work may be cultural, since he did not figure prominently in my mathematical education." Then he added with characteristic honesty, "But I am biased: the French let us down during WW2 and then there was the ghastly de Gaulle."[3]

In England the ferment over Bayes' rule extended even into Fisher's family. His son-in-law George E. P. Box was the young chemist who had to say he was transporting a horse in order to consult Fisher during the Second World War. Like Fisher, Box came to believe that statistics should be more closely entwined with science than with mathematics. This view was reinforced by Box's later work for the chemical giant ICI in Britain and for the quality control movement with W. Edwards Deming and the Japanese auto industry.

When Box organized a statistics department at the University of Wisconsin in 1960, he taught for the first time, a class called Foundations of Statistics. "Week after week," he said, "I prepared my notes very carefully. But the more I did, the more I became convinced that the standard stuff I'd studied under Egon Pearson was wrong. So gradually my course became more and more Bayesian. . . . People used to make fun of it and say it was all nonsense."[4]

Helping scientists with scanty data, Box found that traditional statistics produced messy, unsatisfactory solutions. Still, frequentism worked well for special cases where data fell into bell-shaped probability curves and middle-values were assumed to be averages. So, as Box said, "Comparing averages looked right to me until Stein."[5]

Stein's Paradox called those averages into question. Charles Stein, a theoretical statistician, had been thinking about something that looked quite simple: estimating a mean. Statisticians do not concern themselves with individuals; their bread and butter is a midvalue summarizing large amounts of information. The centuries-old question was which midvalue works best for a particular problem. In the course of his investigation, Stein discovered a method that, ironically, produced more accurate predictions than simple averages did. Statisticians called it Stein's Paradox. Stein called it shrinkage. As a frequency-based theorist, he studiously avoided discussing its relationship with Bayes.

Stein's Paradox, however, works for comparisons between related statistics: the egg production of different chicken breeds, the batting averages of various baseball players, or the workers' compensation exposure of roofing companies. Traditionally, for example, farmers comparing the egg production of five chicken breeds would average the egg yields of each breed separately. But what if a traveling salesman advertised a breed of hens, and each one supposedly laid a million eggs? Because of their prior knowledge of poultry, farmers would laugh him out of town. Bayesians decided that Stein, like the farmers, had weighted his average with a sort of super- or hyperdistribution about chicken-ness, information about egg laying inherent in each breed but never before considered. And intrinsic to poultry farming is the fact that one hen never lays a million eggs.

In like manner, Stein's system used prior information to explain the hitherto mediocre batter who begins a new season hitting a spectacular .400. Stein's Paradox tells fans not to forget their previous knowledge of the sport and the batting averages of other players.

When Stein and Willard D. James simplified the method in 1961, they produced another surprise—the same Bayes-based formula that actuaries had used earlier in the century to price workers' compensation insurance premiums. Whitney's actuarial Credibility theorem used $x = P + z(p-P)$, and Stein and James used $z = \overline{y} + c(y-\overline{y})$: identical formulas but with different symbols and names. In both cases, data about related quantities were being concentrated, made more credible or shrunk, until they clustered more tightly around the mean. With it, actuaries made more accurate predictions about the future well-being of workers in broad industrial categories. Only Arthur Bailey had recognized the formula's Bayesian roots and realized it would be equally valid for noninsurance situations.

Delighted Bayesians claimed Stein was using the prior context of his numbers to shrink the range of possible answers and make for better predictions. Stein, however, continued to regard Bayes' philosophical framework as a "negative" and subjective priors as "completely inappropriate."[6]

Box, who believed Stein should have admitted his method was Bayesian, immediately thought of other relationships that work the same way. Daily egg yields produced on Monday are related to yields produced on Tuesday, Wednesday, and Thursday. In this case, the link between different items is a time series, and successive observations tend to be correlated with each other just as tomorrow tends to be rather like today. But Box discovered gleefully that analyzing time series with Bayesian methods improved predictions and that without those methods Stein's Paradox did not work for time series. As Box explained, "If someone comes into your office with some numbers and says, 'Analyze them,' it's reasonable to ask where they come from and how they're linked. The quality that makes numbers comparable has to be taken into account. You can't take numbers out of context."[7]

Frequentists and Bayesians battled for years over Stein's Paradox, in part because neither side seemed to be all right or all wrong. Box, however, was a convinced Bayesian, and he wrote a snappy Christmas party song to the tune of "There's No Business like Show Business." One verse went:

> There's no Theorem like Bayes theorem
> Like no theorem we know
> Everything about it is appealing
> Everything about it is a wow
> Let out all that a priori feeling
> You've been concealing right up to now.
> . . . . There's no theorem like Bayes theorem
> Like no theorem we know.[8]

As Bayesian interpretations multiplied like bunnies and cropped up in unlikely places like Stein's Paradox, cracks formed in Fisher's favorite theory, fiducial probability. He had introduced it as an alternative to Bayes' rule during an argument with Karl Pearson in 1930. But in 1958 Lindley showed that when uniform priors are used, Fisher's fiducial probability and Bayesian inference produce identical solutions.

Still another fissure occurred when Allan Birnbaum derived George A. Barnard's likelihood principle from generally accepted frequentist principles and showed that he needed to take into account only observed data, not information that might have arisen from the experiment but had not. Another frequentist complained that Birnbaum was "propos[ing] to push the clock back 45 years, but at least this puts him ahead of the Bayesians, who would like to turn the clock back 150 years."[9] Jimmie Savage, however, praised Birnbaum's work as a "historic occasion."[10]

Savage also damned Fisher's fiducial method for using parts of Bayes while avoiding the opprobrium attached to priors. Savage considered Fisher's theory "a bold attempt to make the Bayesian omelet without breaking the Bayesian eggs."[11] Box thought his father-in-law's fiducial probability was beginning to look like "a sneaky way of doing Bayes."[12]

Yet another disagreement between Bayesians and anti-Bayesians surfaced in 1957, when Lindley, elaborating on a point made by Jeffreys, highlighted a theoretical situation when the two approaches produce diametrically opposite results. Lindley's Paradox occurs when a precise hypothesis is tested with vast amounts of data. In 1987, Princeton University aeronautics engineering professor Robert G. Jahn conducted a large study which he concluded supported the existence of psychokinetic powers. He reported that a random event generator had produced 104,490,000 trials testing the hypothesis that someone on a couch eight feet away *cannot* influence their results any more than random chance would. Jahn reported that the random event generator produced 18,471 more examples (0.018%) of human influence on his sensitive microelectronic equipment than could be expected with chance alone. Even with a p-value as small as 0.00015, the frequentist would reject the hypothesis (and conclude in favor of psychokinetic powers) while the same evidence convinces a Bayesian that the hypothesis against spiritualism is almost certainly true.

Six years later, Jimmie Savage, Harold Lindman, and Ward Edwards at the University of Michigan showed that results using Bayes and the frequentist's p-values could differ by significant amounts even with everyday-sized data samples; for instance, a Bayesian with any sensible prior and a sample of only 20 would get an answer ten times or more larger than the p-value.

Lindley ran afoul of Fisher's temper when he reviewed Fisher's third book and found "what I thought was a very basic, serious error in it: Namely, that [Fisher's] fiducial probability doesn't obey the rules of probability. He said it did. He was wrong; it didn't, and I produced an example. He was furious with me." A sympathetic colleague warned Lindley that Fisher was livid, but "it was not until the book of Fisher's letters was published that I realized the full force of his fury. He was unreasonable; he should have admitted his mistake. Still, I was a bumptious young man and he was entitled to be a bit angry." Lindley compounded his tactlessness by turning his discovery into a paper. The journal editor agreed that the article had to be published because it was correct, but asked Lindley if he knew what he was doing: "We would have the wrath of Fisher on our heads."[13] For the next eight months Fisher's letters to friends included complaints about Lindley's "rather abusive review."[14]

Bayes frayed Neyman's nerves too. When Neyman organized a symposium in Berkeley in 1960, Lindley read a paper about prior distributions that caused "the only serious, public row that I can recall having had in statistics. Neyman was furious with me in public. I was very worried, but Savage leapt to my defense and dealt with the situation, I think very well."[15]

One day in the mid-1960s Box dared to tackle his irascible father-in-law about equal priors. Fisher had come to see his new granddaughter, and friends had warned Box he could get only so far with Fisher before he exploded. Walking up the hill to the University of Wisconsin in Madison, Box told the old man, "I'll give them equal probabilities, so if I have five hypotheses, each has a probability of one-fifth."[16]

Fisher responded in a rather annoyed way, as if to say, "This is what I'm going to say and then you're going to shut up." He declared, "Thinking that you don't know and thinking that the probabilities of all the possibilities are equal is not the same thing."[17] That distinction, which Box later agreed with, prevented Fisher from accepting Bayes' rule. Like Neyman, Fisher agreed that if he ever saw a prior he could believe, he would use Bayes-Laplace. And in fact he did. Because he knew the pedigrees of his laboratory animals genera-tions back, he could confidently specify the initial probabilities of particular crosses. For those experiments Fisher did use Bayes' rule.

A compromise between Bayesian and anti-Bayesian methods began to look attractive. The idea was to estimate the initial probabilities according to their relative frequency and then proceed with the rest of Bayes' rule. Empirical Bayes, as it was called, seemed like a breakthrough. Egon Pearson had made an early stab at it in 1925, Turing had used a variant during the Second World War, Herbert Robbins proposed it in 1955, and when Neyman recommended it, a flurry of publications appeared. Empirical Bayesians had little impact on mainstream statistical theory, however, and almost none on applications until the late 1970s.

At the same time, others tackled one of the practical drawbacks to Bayes: its computational difficulty. Laplace's continuous form of Bayes' rule called for integration of functions. This could be complicated, and as the number of unknowns rose, integration problems became hopelessly difficult given the computational capabilities of the time. Jeffreys, Lindley, and David Wallace were among those who worked on developing asymptotic approximations to make the calculations more manageable.

Amid this mathematical fervor, a few practical types sat down in the 1960s to build the kind of institutional support that frequentists had long enjoyed: annual seminars, journals, funding sources, and textbooks. Morris H. DeGroot wrote the first internationally known text on Bayesian decision theory, the mathematical analysis of decision making. Arnold Zellner at the University of Chicago raised money, founded a conference series, and began testing standard economics problems one by one, solving them from both Bayesian and non-Bayesian points of view. Thanks to Zellner's influence, Savage's subjective probability would have one of its biggest impacts in economics. The building process took decades. The International Society for Bayesian Analysis and the Bayesian section of the American Statistical Association were not formed until the early 1990s.

The excitement over Bayesian theory extended far beyond statistics and mathematics though. During the 1960s and 1970s, physicians, National Security Agency cryptanalysts, Central Intelligence Agency analysts, and lawyers also began to consider Bayesian applications in their fields.

Physicians began talking about applying Bayes to medical diagnosis in 1959, when Robert S. Ledley of the National Bureau of Standards and Lee B. Lusted of the University of Rochester School of Medicine suggested the idea. They published their article in *Science* because medical journals were uninterested. After reading it, Homer Warner, a pediatric heart surgeon at the Latter-day Saints Hospital and the University of Utah in Salt Lake City, developed in 1961 the first computerized program for diagnosing disease. Basing it on 1,000 children with various congenital heart diseases, Warner showed that Bayes could identify their underlying problems quite accurately. "Old cardiologists just couldn't believe that a computer could do something better than a human," Warner recalled.[18] A few years after Warner introduced his battery of 54 tests, Anthony Gorry and Octo Barnett showed that only seven or eight were needed, as long as they were relevant to the patient's symptoms and given one at a time in proper sequence. Few physicians used the systems, however, and efforts to computerize diagnostics lapsed.

Between 1960 and 1972 the National Security Agency educated cryptanalysts about advanced Bayesian methods with at least six articles in its in-house *NSA Technical Journal*. Originally classified as "Top Secret Umbra," an overall code word for high-level intelligence, they were partially declassified at my request in 2009, although the authorship of three of the six was suppressed. (At least one of those three has the earmarks of a Jack Good article.) In another article, an agency employee named F. T. Leahy quoted the assertion in Van Nostrand's *Scientific Encyclopedia* that Bayes' theorem "has been found to be unscientific, to give rise to various inconsistencies, and to be unnecessary." In fact, Leahy declared in 1960, Bayes is "one of the more important mathematical techniques used by cryptanalysts . . . [and] was used in almost all the successful cryptanalysis at the National Security Agency. . . . It leads to the *only* correct formulas for solving a large number of our cryptanalytic problems," including those involving comparisons among a multiplicity of hypotheses.[19] Still, "only a handful of mathematicians at N.S.A. know about all the ways" Bayes can be used. The articles were presumably intended to remedy the situation.

At the CIA, analysts conducted dozens of experiments with Bayes. The CIA, which must infer information from incomplete or uncertain evidence, had failed at least a dozen times to predict disastrous events during the 1960s and 1970s. Among them were North Vietnam's intervention in South Vietnam and the petroleum price hike instituted by the Organization of Petroleum Exporting Countries (OPEC) in 1973. Agency analysts typically made one prediction and left it at that; they ignored unlikely but potentially catastrophic possibilities and failed to update their initial predictions with new evidence. Although the CIA concluded that Bayes-based analyses were more insightful, they were judged too slow. The experiments were abandoned for lack of computer power.

The legal profession reacted differently. After several suggestions that Bayes might be useful in assessing legal evidence, Professor Laurence H. Tribe of Harvard Law School published a blistering and influential article about the method in 1971. Drawing on his bachelor's degree in mathematics, Tribe condemned Bayes and other "mathematical or pseudo-mathematical devices" because they are able to "distort—and, in some instances, to destroy—important values" by "enshrouding the [legal] process in mathematical obscurity."[20] After that, many courtroom doors slammed shut on Bayes.

The extraordinary fact about the glorious Bayesian revival of the 1950s and 1960s is how few people in any field publicly applied Bayesian theory to real-world problems. As a result, much of the speculation about Bayes' rule was moot. Until they could prove in public that their method was superior, Bayesians were stymied.

part IV

to prove its worth

11.

business decisions

With new statistical theories cropping up almost daily during the 1960s, the paltry number of practical applications in the public arena was becoming a professional embarrassment.

Harvard's John W. Pratt complained that Bayesians and frequentists alike were publishing "too many minor advances extracted from real or mythical problems, sanitized, rigorized, polished, and presented in pristine mathematical form and multiple forums."[1]

Bayesians in particular seemed unwilling to apply their theories to real problems. Savage's rabbit ears and 20-pound chairs were textbook showpieces, even less substantial than Egon Pearson's chestnut foals and pipe-smoking males of 30 years before. They were "dumb," a Bayesian at Harvard Business School protested later;[2] they lacked the practical world's ring of truth. When analyzing substantial amounts of data even diehard Bayesians preferred frequency. Lindley, already Britain's leading Bayesian, presented a paper about grading examinations to the Harvard Business School in 1961 without mentioning Bayes; only later did he analyze a similar problem using wine statistics and Bayesian methods.

Mathematically and philosophically, the rule was simplicity itself. "You can't have an opinion afterwards," said Pratt, "without having had an opinion before and then updating it using the information."[3] But making a belief quantitative and precise—there was the rub.

Until their logically appealing system could prove itself as a powerful problem solver in the workaday world of statistics, Bayesians were doomed to minority status. But who would undertake a cumbersome, complex, technically fearsome set of calculations to explore a method that was professionally almost taboo? At the dawn of the electronic age, powerful computers were few and far between, and packaged software nonexistent. Bayesian techniques for dealing with practical problems and computers barely existed. Brawn—lots of it—would have to substitute for computer time. The faint of heart need not apply.

Nevertheless, a few hardy, energetic, and supremely inventive investigators tried to put Bayes to work for business decision makers, social scientists, and newscasters. Their exploits dramatize the formidable obstacles that faced anyone trying to use Bayes' rule.

The first to try their luck with Bayes was an unlikely duo at Harvard Business School, Robert Osher Schlaifer and Howard Raiffa. They were polar opposites. Schlaifer was the school's statistical expert, but, in a sign of the times, he had taken only one mathematics course in his life and was an authority on ancient Greek slavery and modern airplane engines. Raiffa was a mathematical sophisticate who became the legendary Mr. Decision Tree, an advisor to presidents and a builder of East–West rapprochement. Together they tackled the problem of turning Bayes into a tool for business decision makers.

Fortunately, Schlaifer enjoyed using his laserlike mind, hyperlogic, and outsider status to bludgeon through convention and orthodoxy. Years later, when Raiffa was asked to describe his colleague, he did so in two words, "imperious and hierarchical."

Schlaifer was an opinionated perfectionist. Plunging into a topic, he saw nothing else. When his passion turned to bicycle racks, he persuaded a Harvard dean to install his new design on campus. Because he loved old engines, an MIT physicist volunteered to come each week to maintain his Model A Ford and hi-fi to his exacting standards. And when he studied consumer behavior, his faculty colleagues pondered instant coffee as soberly as they would have nuclear fusion. Like an autocrat atop an empire, Schlaifer bestowed nicknames on his colleagues: "Uncle Howard" for Raiffa, "Great Man Pratt" for John W. Pratt, and, most memorably, "Arturchick" for his graduate student and near namesake, Arthur Schleifer Jr. All three survived to become chaired professors at Harvard.

Whatever else, the man had panache and fun. Few Harvard professors invited research assistants for Sunday dinners and bottles of 1938 Clos Vougeot, but Schlaifer did. And few took off an entire month or sabbatical year to relax in Greece or France, as he did with his French-born wife, Geneviève, whom he publicly referred to as Snuggle Buggle.

Schlaifer was not to the manor born. He was born in Vermillion, South Dakota, in 1914 and grew up near Chicago, in a small town where his father was superintendent of schools. At Amherst College he majored in classics and ancient history and took courses in economics and physics. He enrolled in calculus, his only formal math class, solely to capture a large cash prize for the best student. After graduating Phi Beta Kappa at the age of 19, he studied in Athens at the American School of Classical Studies between 1937 and 1939 and earned a Ph.D. in ancient history at Harvard in 1940. Over the next few years he published several articles about religious cults and slavery in ancient Greece. Schlaifer was a quick study, and he filled in for Harvard historians, economists, and physicists as they left for defense work during the Second World War.

Eventually, Schlaifer was assigned to the university's Underwater Sound Laboratory, where sonar was developed. He and theoretical physicist Edwin Kemble tried to silence submarine torpedo propellers the better to attack German U-boats. Schlaifer understood the scientific issues well enough to solve equations using Marchant or Frieden electromechanical calculators and to turn technical reports into prose. The war gave him a voracious appetite for practical, real-world problems, and he abandoned ancient history.

After the war Schlaifer's physics impressed the Harvard Business School enough to hire him to fulfill a departmental obligation: a study of the aircraft engine industry. He turned the low-status assignment into a triumph, a 600-page classic on aviation history, *Development of Aircraft Engines, Development of Aircraft Fuel*.

Between his war work and the book, Schlaifer acquired a usefully intimidating campus reputation as a physicist, later cemented by his obituary in the *New York Times*. He was teaching accounting and production when the business school, despite his spectacularly unsuitable training, assigned him to teach statistical quality control. Knowing nothing about statistics, Schlaifer crammed by reading the dominant theoreticians of the day, Fisher, Neyman, and Egon Pearson. Wartime operations research had mathematized problem solving for two common business problems, inventory control and transportation scheduling. But frequentism offered businesses no help when it came to issues like launching a new product or changing a price.

Schlaifer's publications later described modest pleas for help from a newsstand owner unsure of how many copies of the *Daily Racing Form* to stock, and from a wholesaler worried about allocating his ten delivery trucks between two warehouses. With luck, business owners might make the right decisions. But given the uncertainties involved in even these simple problems, Schlaifer wondered how they could ever hope to *systematically* make the best possible choices. Even if they could collect additional information by sampling or experimentation, would it be worth the cost?

According to frequentists, objective statistics were synonymous with long-run relative frequency, and probabilities were invalid unless based on repeatable observations. Frequentists dealt with a rich abundance of directly relevant data and took samples to test hypotheses and draw inferences about unknowns. The method worked for repetitive, standardized phenomena such as successive grain crops, genetics, gambling, insurance, and statistical mechanics.

But business executives often had to make decisions under conditions of extreme uncertainty, without sample data. As Schlaifer realized, "Under uncertainty, the businessman is forced, in effect, to gamble, . . . hoping that he will win but knowing that he may lose."[4] Executives needed a way to assess probabilities without the repeated trials required by frequentist methods. Schlaifer said teaching frequentism made him feel like a jackass. It simply did not address the main problem in business: decision making under uncertainty.

Thinking through the problem, Schlaifer wondered how executives could make decisions based on no data. Whatever prior information they had about the demand for their product was obviously better than none. From there, Schlaifer got to the problem of how sample data should be used and how much money should be spent getting it. Updating prior information with sample data got him to Bayes' rule because it could combine subjectively assessed prior probabilities with objectively attained data. It was a fundamental insight that changed his life.

Schlaifer did not have much mathematics. Innocent of the raging philosophical divide between objectivists and subjectivists, Schlaifer threw away his books and reinvented Bayesian decision theory from scratch. As a self-made statistician working in a business school, he owed nothing to the statistical establishment. And given his iconoclastic zeal, he brazenly challenged giants in the field. Like Savage, Schlaifer was combining uncertainty with economics in order to make decisions. Savage thought Schlaifer was "hot as a pistol, sharp as a knife, clear as a bell, quick as a whip, and as exhausting as a marathon run."[5]

Schlaifer realized that his mathematical competence was almost nil, "of order of magnitude epsilon."[6] To compensate, he worked 75 to 80 hours a week, puffed through four packs of unfiltered cigarettes a day, and traced his thoughts with different colored chalk across the blackboard in his smoke-filled office. Single-mindedly pursuing what he thought was relevant to the real world, he would lurch into one theory, back off, try it another way, and race on to yet another. His voice boomed down the hall almost hourly—"Oh, my God!" or "How could I have been so dumb?"—as he overturned one firmly held opinion with another. Ever curious, he demanded the absolute, best possible analyses. He realized he needed mathematical help.

Hearing about a young closet Bayesian named Howard Raiffa at Columbia University, Schlaifer scouted him out and persuaded Harvard to hire him. For the next seven years Raiffa and Schlaifer collaborated closely. Raiffa went on to become an international negotiator who wielded enormous influence on education, business, the law, and public policy in the United States and abroad. But he always regarded Schlaifer as "the great man. . . . I revered him; I was in awe of him. . . . He was so positive, so certain, so opinionated, but so smart, so smart. . . . [He was] "a man—a real man—who independently discovered Bayesianism, mocked those who didn't agree with him, and not only theorized and philosophized but applied the approach to real problems." Raiffa called him "the—not a but the—most important person in my intellectual development."[7]

Both Schlaifer and Raiffa were elite intellectuals, but Schlaifer's style was arrogant while Raiffa was, in the words of a collaborator, "a sweetheart, a very warm, open, embracing person."[8] Schlaifer was Phi Beta Kappa in the Ivy League; Raiffa attended City College of New York, "the college of choice for poor and middle-income students in New York—I was on the poor side."[9] During the Second World War, a misplaced template for grading the air force's arithmetic and elementary algebra test flunked Raiffa and doomed the future advisor to U.S. presidents—not to a prestigious research laboratory like the one where Schlaiffer spent the war but to three rounds of basic training and to grunt assignments in cooks-and-bakers school, meteorology, and a radar blind-landing system.

Anti-Semitism finally determined Raiffa's career choice. One day he overheard his army sergeants saying they wanted to line up America's Jews on a beach and use them for target practice. Later, real estate agents in Fort Lauderdale, Florida, refused to find housing for Raiffa and his wife because they were Jewish. When a friend told him that engineering and science also discriminated against Jews, Raiffa was prepared to believe it. Then he learned that insurance actuaries were graded on objective, competitive examinations. Seeking a field where competence counted more than religion, Raiffa enrolled in the University of Michigan's actuarial program, where Arthur Bailey had studied.

To Raiffa's great surprise, he became a superb and "deliriously happy" student who raced through a bachelor's degree in mathematics, a master's degree in statistics, and a doctorate in mathematics in six years between 1946 and 1952. "In the year I studied statistics, I don't think I heard the word 'Bayes.' As a way of inference, it was nonexistent. It was all strictly Neyman-Pearson, classical, objectivistic (frequency-based) statistics."[10]

Although Schlaifer had embraced Bayes in one fell swoop, Raiffa inched grudgingly toward its subjectivity. But reading John von Neumann and Oskar Morgenstern's book *Game Theory* (1944), he instinctively assessed how others would play in order to determine how he himself should compete: "In my naiveté, without any theory or anything like that. . . . [I began] assessing judgmental probability distributions. I slipped into being a subjectivist without realizing how radically I was behaving. That was the natural thing to do. No big deal."[11]

When Raiffa gave a series of seminars on Abraham Wald's new book *Statistical Decision Functions*, he discovered it was full of Bayesian decision-making rules for use in a frequentist framework. Independently of Turing and Barnard, Wald had discovered sequential analysis for testing ammunition while working with the Statistical Research Group in the Fire Control Division of the National Defense Research Committee. Though a confirmed frequentist, he sometimes solved problems in a curiously roundabout manner. After inventing a Bayesian prior, he would solve the Bayesian version of his problem and then analyze its frequentist properties. He also said that every good decision procedure is Bayesian and confided to the statistician Hilda von Mises that he was a Bayesian but did not dare say so publicly. His work would have a large impact on many mathematical statisticians and decision theorists, including Raiffa.

Until Wald's book appeared, the word "Bayesian" had referred only to Thomas Bayes' controversial suggestion about equal priors, not to his theorem for solving inverse probability problems. After Wald died in a plane crash in India in 1950, the Columbia University statistics department hired Raiffa to teach Wald's course. Raiffa kept a day ahead of his students by reading the textbook every night. He was moving gradually toward a viewpoint opposed by almost every statistics department in the country, including Columbia's. Giving up "scientific" objectivity and embracing subjectivity would not be easy.

At first, like Schlaifer, Raiffa taught straight frequentism, using the then-canonical Neyman-Pearson theory, tests of hypotheses, confidence intervals, and unbiased estimation. But by 1955, Raiffa, like Schlaifer, no longer believed these concepts were central. Columbia faculty members were auditing Raiffa's lectures, and his transformation into a closet Bayesian left him a nervous wreck. He did not "come out" because colleagues whom he admired greatly were vociferously opposed to Bayesianism. "Look, Howard, what are you trying to do?" they asked. "Are you trying to introduce squishy judgmental, psychological stuff into something which we think is science?"[12]

He and his Columbia colleagues were working on totally different kinds of problems. Empowered by the Second World War, statisticians like Raiffa and Schlaifer were increasingly interested in using statistics not just to analyze data but to make decisions. In contrast, Neyman and Pearson considered the errors associated with various strategies or hypotheses and then decided whether to accept or reject them; they could not infer what to do based on an observed sample outcome without thinking about all the potential sample outcomes that could have occurred but had not. This was Jeffreys's objection to using frequentism for scientific inference. Raiffa felt the same way for different reasons; he wanted to make decisions tied to "real economic problems; not phony ones."[13]

Raiffa was interested in concrete, one-of-a-kind decisions requiring quick judgments about how much of a product to stock or how to price it. Like Schlaifer at Harvard, he wanted to help businesses resolve uncertainties and use indirectly relevant information. As Raiffa put it, anti-Bayesians "would never—and I mean never—assign probabilities to some such statement as 'the probability that p falls in the interval from .20 to .30.'"

Bayesian subjectivists, on the other hand, actually wanted answers expressed in terms of probabilities. They did not want to merely accept or reject a hypothesis. As Raiffa realized, a business owner wanted to be able to say that "on the basis of my formerly held beliefs . . . and of the specific sample outcomes, I now believe there is a .92 probability that $\tilde{p}$ is greater than .25."[14]

This was *verboten* for frequentists, who recognized only those sample outcomes that were "significant at the .05 level." Raiffa regarded their focus as "a very, very cursory description of the distribution. I wanted my students to think probabilistically about [the entire distribution of p, about] where the uncertain p could be, and then figure out from the decision point of view where the correct action would be. So the whole question of a test of hypotheses seemed to me to be leading students in a wrong direction."[15]

The chasm between the two schools of statistics crystallized for Raiffa when Columbia professors discussed a sociology student named James Coleman. During his oral examination Coleman seemed "confused and fuzzy . . . clearly not of Ph.D. quality."[16] But his professors were adamant that he was otherwise dazzling. Using his new Bayesian perspective, Raiffa argued that the department's prior opinion of the candidate's qualities was so positive that a one-hour exam should not substantially alter their views. Pass him, Raiffa urged. Coleman became such an influential sociologist that he appeared on both the cover of *Newsweek* and page one of the *New York Times*.

So far, Raiffa regarded his transformation from a Neyman-Pearsonite to a Bayesian as an intellectual conversion; his emotional conversion was still to come.

In a campaign to raise the intellectual standards of business schools, the Ford Foundation donated money to Harvard to hire a mathematical statistician in 1957. The university made Raiffa an attractive offer: joint posts in its new department of statistics and in the business school. When the statistics department learned about Raiffa's conversion to Bayesian ideas, the chair, Frederick Mosteller, seemed tolerant but lukewarm, and another prominent professor, William Cochran, said, "Well, you'll grow up."[17] At the business school, though, Schlaifer welcomed Raiffa with open arms.

Schlaifer was "the most opinionated person I ever met," Raiffa recalled. At first, he did not realize "how *wonderful* Schlaifer was. . . . I didn't realize that Schlaifer was as great a man as he was. He was already focused on the real business decision problems, not on testing hypotheses. He said they were getting it wrong." Then Raiffa corrected himself: "No, no, he didn't say that. He said they're getting it wrong for business decision-making under uncertainty."[18]

Each morning Raiffa tutored Schlaifer in calculus and linear algebra, vectors, transformations, and the like. The next morning Schlaifer would conjecture new theorems, and the day after that apply what he had learned to a concrete problem. "His mind was receptive, razor-sharp, tenacious, persistent, creative," Raiffa discovered.[19] Both men were workaholics, but Schlaifer worked longer hours than anyone else. As Raiffa recalled, "He was really a *fabulous* student. . . . He had raw mathematical abilities provided he could see how it might be put to use."[20] The two never referred to journals or books. Everything they did together was self-generated.

Schlaifer did not know nearly as much statistics as Raiffa, but he was much better read. Raiffa had not studied the great prewar theorists Jeffreys, Fisher, and Egon Pearson. Later, when he discovered Savage's work, he was amazed at its clarity. Raiffa took the advice of his colleagues and named his Harvard chair for Frank Ramsey without ever having read the young man's work.

Writing articles with Schlaifer, Raiffa always produced the first draft. Then Schlaifer "analyzed everything seven ways to Sunday and changed the text endlessly, putting in commas, and taking them out again," recalled John Pratt, who wrote several important works with them.[21] Schlaifer almost refused to let Harvard Business School Press publish one of his books because its editors put quotation marks outside the punctuation instead of inside. (Thus the battle raged over "over." and "over".)

"The unsettling part of him," Raiffa said, ". . . was that he made so much sense—as long as he stayed away from politics!"[22] So Raiffa argued statistics with his collaborator but closed his ears when Schlaifer inveighed against the income tax and advocated solving Haiti's problems by turning all the Haitians in the United States into soldiers and sending them back home.

To Schlaifer, Bayes' rule was not just something to use, it was something to believe in—fervently. A true believer, he refused to accept that there might be several ways to approach a problem. He used "sheer brutal insistence and intellectual discourse by finding holes in every argument that Howard used," recalled his student Arthur Schleifer Jr. "He would show that these alternative ways led to untenable paradoxes. . . . Robert's approach was that there was this one way to do it, you have to do it this way, and if you do it any other way, I'll show you you're wrong."[23]

Raiffa wanted to expose his students to both frequentist and Bayesian methods so that if some read the establishment's point of view they would not get confused. But Schlaifer regarded this as teaching falsehood. Anyway, he pronounced loftily, "businessmen don't read the literature."[24]

By 1958 Raiffa had also converted passionately to subjectivism. It seemed obvious that four businesses serving four different markets could use the same information to produce four different nonstatistical priors and four different conclusions. Although this still bothers some scientists and statisticians, others were content to overwhelm their initial nonstatistical prior opinions with large amounts of new statistical information. Just as many Americans remember where they were when they heard about Kennedy's assassination or the 9/11 attack, so many Bayesians of Raiffa's generation remember the exact moment when Bayes' overarching logic suddenly hit them like an ir-resistible epiphany. Critics began to call Harvard Business School "a Bayesian hothouse."

Betting—assigning different probabilities to the same phenomenon—became the tangible expression of Bayesian beliefs. "Every day there'd be a half dozen bets around [Schlaifer's group] about anything—elections and sports. Dollar bills were changing hands all the time. It was part of the in-grained way of life. You really believed this stuff," Schleifer said.[25] Schlaifer and Raiffa were developing reputations as zealots with a cause.

Schlaifer sent 700 pages of his first textbook to McGraw-Hill for publica-tion as *Probability and Statistics for Business Decisions: An Introduction to Managerial Economics under Uncertainty.* Then he discovered his usual host of errors and infelicities and insisted that McGraw-Hill withdraw the first printing and replace it with a second. It was a classic case of placing intellectual rigor above business economics, and Schlaifer won. The book sold for $11.50 in 1959, and Harvard promoted its former tutor to a professorship in business administration.

Probability and Statistics for Business Decisions was the first textbook written entirely and wholeheartedly from the Bayesian point of view. Students could solve inventory, marketing, and queuing problems using simple arithmetic, slide rules, or, at most, a desk calculator. The book acknowledged few previ-ous authorities. Schlaifer had come to his subjectivist position independently of Ramsey, de Finetti, and Savage. In turn, Savage recognized that Schlaifer had developed his ideas "wholly independently" and was more "down to earth and less spellbound by tradition."[26]

Mulling over the state of Bayes' rule, Schlaifer and Raiffa realized that Bayesians, unlike frequentists, had no bookshelves of mathematical tools ready for use. As a result, Bayesian methods were regarded as impractically complicated, particularly by business students, who were often mathemati-cally unprepared. While theoreticians like Savage and Lindley tried to make Bayes mathematically respectable, Raiffa and Schlaifer set out in 1958 to make it fully operational and easy to use for bread-and-butter problems. Like George Box, they parodied a popular song, this one from *Annie Get Your Gun,* claiming that anything a frequentist could do, they could do better.

To make calculations easier, they introduced decision trees, tree-flipping, and conjugate priors. "I began using decision tree diagrams that depicted the sequential nature of decision problems faced by business managers," Raiffa said. "Should I, as decision maker, act now or wait to collect further marketing information (by sampling or further engineering)? . . . I never made any claim to being the inventor of the decision tree but . . . I became known as Mr. Decision Tree."[27] Soon the diagrams of Bayes' decision-making process were, like many-branched trees, rooted in undergraduate business curricula. The trees are probably the best-known practical application of Bayes' rule.

Tree-flipping began as a simplification to help one of Raiffa's graduate students who was interested in wildcat drilling for oil. Normally, a wildcatter decided whether to test a particular site *before* deciding to drill or not drill. To avoid some messy algebra, Raiffa flipped the order of the wildcatter's decision. He dealt with the probability that test results would be positive or negative *before* he considered whether or not to conduct the test. Working through the diagram produced information about *x*'s followed by *y*'s. Tree-flipping put the *y*'s first. It amounted to using Bayes' rule because the probability of *x* given *y* and the probability of *y* given *x* are the two critical elements of its formula.

"So you flip trees," Raiffa said. "We didn't call it Bayes. The worst thing you can do is to use Bayes' theorem. It's too complicated. Just use common sense and play around with these things, then it was pretty easy. We had people doing complicated things that could have been done by Bayes, but we didn't do it by Bayes. We did it by tree-flipping."[28]

Raiffa also developed a handy shortcut for updating priors and posteriors. Called conjugate prior distributions, it used the fact that in many cases the shape or curve of a probability's distribution is the same in both prior and posterior. Thus, if you start with normal Gaussians, you'll end up with normal Gaussians. Conjugate priors paid dividends with the repeated updating called for by Bayes' method. Albert Madansky used a similar concept for his H-bomb study. The shortcut would later become unnecessary with the adoption of Monte Carlo Markov Chain methods.

In a further simplification, some business Bayesians even dropped the prior odds called for by Bayes' rule. Schleifer said, "My take on it was to forget the priors unless there was overwhelming prior evidence that you really know a lot about the parameter you're interested in."[29]

Today, when TV and radio are filled with talking heads, it is hard to imagine that the use of expert opinion was terra incognita in the early 1960s. No one knew whether business executives would be willing to offer their opinions for incorporation into a mathematical formula. And no one was sure whether an expert's subjective judgment would be valid. John Pratt asked his wife, Joy, whose job was promoting films in local theaters, to estimate their daily attendance. At first, her estimates fell into too narrow a range. By comparing them with actual attendance figures—hundreds of data points taken night after night at two local theaters—Joy Pratt learned to make such accurate predictions that her husband became convinced that expert opinion could be useful. Bayesians objected that Pratt and Schlaifer analyzed the data using frequentist techniques. Bayesian methods—comparing different kinds of movies, the length of time they played, the popularity of their stars, and so on—would have been too complex. The use of expert opinion for decision making later became a major field of study.

It turned out that Joy and John Pratt were right: marketing executives risked a lot of money on the basis of very little information and loved being asked for their professional judgment. Accustomed to waiting until the end of a frequentist study to voice their opinions, they actually liked having their "managerial intuition" or "feel for a situation" folded into preliminary assessments.

Raiffa and Schlaifer began exploring such nitty-gritty questions as how to interview experts and measure their expertise. DuPont, trying to decide how big a factory to build for artificial shoe leather in 1962, was delighted to assess the prior odds of the demand for its new product. Design engineers at Ford Motor Company were equally pleased that incorporating their opinions into Bayes' prior let Ford use smaller opinion samples. Their work opened up just about any business problem to mathematical analysis. An engineering problem might have 20 sources of uncertainty; of these, perhaps 12 could be handled by single guesses; 5 needed more testing; and 2 might be so critical that experts had to be interviewed. Bayes' rule was solving far more complex problems than Savage's mental exercises about curling rabbit ears.

Between 1961 and 1965 an exciting weekly seminar, generally followed by drinks in Schlaifer's office, focused on decision making under uncertainty (DUU for short). The seminar explored utility analysis, portfolio analysis, group decision processes, theory of syndicates, behavioral anomalies, and ways to ask about uncertainties and values. Said Raiffa, "We helped shape a field."[30] The seminar and two books Raiffa and Schlaifer cowrote during this period spurred the Bayesian revival of the 1960s. Raiffa was later surprised to realize that the most fertile period of his collaboration with Schlaifer had lasted only four years.

Raiffa's and Schlaifer's classic book for advanced statisticians, *Applied Statistical Decision Theory*, was published in 1961. Its careful, detailed analytical methods set the direction of Bayesian statistics for the next two decades. Today it sits on almost every decision analyst's bookshelf.

When Pratt joined Raiffa and Schlaifer to write *Introduction to Statistical Decision Theory*, he soon realized that what was easy for him to do mathematically was quite difficult for Schlaifer, who could understand the mathematics but not produce it himself. By the time the book was ready for editing, Schlaifer and Raiffa had moved on to other interests. They received so many requests for their preliminary manuscript, however, that McGraw-Hill published it as a typescript in 1965. Thirty years later, Pratt and Raiffa finished it, and MIT published it as an 875-page book.

To introduce business school professors to mathematical methods, Raiffa ran an 11-month-long Ford Foundation program in 1960 and 1961. As a result, the next generation of business school deans at Harvard, Stanford, Northwestern, and elsewhere had received a heavy dose of Bayesian subjectivism for decision making, and the gospel radiated outward to schools of management. Raiffa even gave his students an 84-page handout, "An Introduction to Markov Chains," more than 30 years before their widespread adoption by the statistical profession. By 2000 Bayesian methods were often centered in university business schools rather than statistics departments.

Raiffa and Schlaifer drifted apart after 1965. Raiffa still called himself a Bayesian who, "roughly speaking. . . . wish[es] to introduce intuitive judgments and feelings directly into the formal analysis of a decision problem."[31] Broadening what he knew about subjective probability, game theory, and Bayes' rule, he left Harvard's statistics department to take a joint chair in the business school and the economics department. There he pursued societal, rather than primarily statistical, issues in medicine, law, engineering, international relations, and public policy.

By any measure Raiffa's move was a success. As a pioneer in decision analysis, he was one of four organizers of the Kennedy School of Government at Harvard; the founder and director of a joint East–West think tank to reduce Cold War tensions long before perestroika; a founder of Harvard Law School's widely replicated role-playing course in negotiations; and scientific adviser to McGeorge Bundy, the national security assistant under Presidents Kennedy and Johnson. Raiffa also supervised more than 90 Ph.D. dissertations at Harvard in business and economics and wrote 11 books—no articles, only books—one of which has been in print for more than fifty years. As a Bayesian, Raiffa would cast a long shadow.

Ultimately, however, Raiffa and Schlaifer failed in their bold attempt to permeate business curricula, statistical theory, and American business life with Bayes' rule. Schlaifer built managerial economics into a strong program at the Harvard Business School, but Bayesian decision analysis faded from its curriculum, and Bayes' rule never supplanted "the old stuff" in American classrooms. Since the 1970s, when all the top business schools emphasized Bayesian decision theory, it has been compressed into a few weeks' study. Business students no longer do their own calculations; presumably they can hire a consultant or buy a computer program.

Many theoretical statisticians also ignored Raiffa's and Schlaifer's contributions; they were, after all, outsiders working in a business school. From his vantage point in Britain, Lindley was astonished that the statistical community paid so little attention to Schlaifer. "I was bowled over by him. The book with Raiffa is wonderful," and Schlaifer's 1971 book had computer methods "in advance of their time." Lindley considered Schlaifer "one of the most original minds that I have ever met [with] extraordinarily wide knowledge."[32]

Part of their failure lay in the fact that Schlaifer remained a confirmed university theoretician. When confronted with a problem he could not solve, he set it aside and worked on something else, something business managers cannot afford to do. Nor did he consider long-term solutions and their consequences over time; he dealt in short-term results. He did little consulting work, and his lack of experience selling complicated ideas to busy executives limited the impact of Bayes' rule on working business people. He spent weeks exploring a marketing case about cottage cheese packaging in all its abstract complexities but stripped off all the textural surroundings that most caseworkers would have brought back from a field trip to a dairy. He turned his only graduate student's thesis about all the messy glory of IBM's quality control problems into a dry theoretical paper about two-stage sampling. The student's thesis ended up piled so high with abstract issues that it was not until Schlaifer went on sabbatical that Raiffa could intervene and secure the young man's Ph.D. Schlaifer was a passionate intellectual with a deep interest in narrow topics and all the time in the world for disputation.

After Raiffa moved on to other projects, Schlaifer threw himself into designing a new introductory course for Harvard's first-year students in managerial economics. Naturally, it would be based on Bayesian methods, a first in any business school. He wrote a text and titled it *Managerial Economics Reporting Control*, which he nicknamed MERC. Students hated it, called it Murk, and burned their copies on the front steps of Baker Library. When a reporter for the Harvard newspaper asked for a comment, Schlaifer replied, "Well, I'd rather be among those whose books are burned than those who burn books."

Then Schlaifer leaned intently forward: "Tell me. There is one thing that really interests me. This book is printed on very good, very glossy paper. It must have burned very poorly. How do you burn them?"

"Well, sir," the student answered respectfully, "we burn them page by page."[33]

Schlaifer, farseeing to the last, spent the remaining years of his life trying to write computer software for practitioners, even though teams of mathematically sophisticated programmers were already taking over the field. In 1994, at the age of 79, Schlaifer died of lung cancer. After his death, Raiffa and Pratt finished the trio's 30-year-old opus, *Introduction to Statistical Decision Theory*. Dedicating it to their former colleague, Pratt and Raiffa hailed Schlaifer as "an original, deep, creative, indefatigable, persistent, versatile, demanding, sometimes irascible scholar, who was an inspiration to us both."[34]

12.

who wrote
the federalist?

Alfred C. Kinsey's explosive bestseller *Sexual Behavior in the Human Male* was published in 1948, the same year pollsters failed to predict Harry Truman's victory over Thomas Dewey in the presidential election. With the public crying foul, fraud, and debauchery, social scientists feared for the future of their profession. Opinion polling was one of their basic tools, so the Social Science Research Council, representing seven professional societies, appointed statistician Frederick Mosteller of Harvard University to investigate the scandals.

Mosteller's forthright report on Truman's election blamed the nation's pollsters for rejecting randomized sampling and for clinging to outdated sampling designs that underrepresented blacks, women, and the poor—all of whom voted more heavily Democratic than the population reached by the pollsters.

In the case of Kinsey's research, powerful men—including John Foster Dulles, secretary of state under Eisenhower; Arthur Sulzberger, publisher of the *New York Times*; Harold W. Dodds, president of Princeton University; and Henry P. Van Dusen, president of the liberal Union Theological Seminary—were demanding an end to funding for research on human sexuality. But Mosteller underwent Kinsey's standard interview about his sexual history and emerged impressed. Kinsey's lack of randomized sampling was statistically damning, but his work was far better than anyone else's in the field, and the country did not have 20 statisticians who could have done better. It was quietly arranged that when Kinsey wrote his next study, on the sexuality of women, Jerome Cornfield of the National Institutes of Health would help with the statistics.

Both scandals involved discrimination problems, also called classification problems, which struck at the heart of polling, science, social science, and statistics. Researchers tended to assign people or things to categories without being totally sure that the assignments were accurate or that the categories were well defined. Pollsters classified people as Republicans or Democrats; marketers divided consumers into users of one detergent or another; scientists classified plants in biology and skulls in anthropology; and social scientists categorized individuals according to personality.

Finishing up with the Kinsey committee, Mosteller looked around for a research topic involving classification issues. He had a feet-on-the-ground attitude, perhaps the result of having been raised by his divorced mother, who never graduated from high school but who had insisted, over the objections of her ex-husband, that Fred get an education. Mosteller had earned his bachelor's and master's degrees in mathematics from Carnegie Institute of Technology (now Carnegie Mellon University) in Pittsburgh and enrolled in graduate statistics at Princeton University's highly abstract mathematics department. As the primary liaison between Princeton and Columbia statisticians working on military research, he learned that he loved working on deadline on real-world problems. After the war, in 1946, Mosteller finished his Ph.D. at Princeton and, driven by his interest in health, education, and baseball, moved to Harvard. The investigations of campaign polling and sexual research left him ripe for a problem of his own choosing.

Mosteller began looking around for a large database to use for developing ways to discriminate between two cases. He began thinking—not about Bayes' rule—but about a minor historical puzzle: *The Federalist* papers. Between 1787 and 1788, three founding fathers of the United States, Alexander Hamilton, John Jay, and James Madison, anonymously wrote 85 newspaper articles to persuade New York State voters to ratify the American Constitution. Historians could attribute most of the essays, but no one agreed whether Madison or Hamilton had written 12 others.

Mosteller had learned about the problem during a summer job he held as a graduate student in 1941. Counting the number of words in each sentence of *The Federalist* papers with psychologist Frederick Williams, he discovered "an important empirical principle—people cannot count, at least not very high." He also learned that, stylistically, Hamilton and Madison were practically twins, skilled in a complicated oratorical style popularized in 1700s England. Mosteller agreed to "leave general style as a poor bet and pay attention to words."[1] The job was daunting because he would need a lot of single words to supply a pool of thousands of variables. When the summer job ended and the Second World War intervened, Mosteller forgot about The Federalist.

After the war, he decided that The Federalist might fill the bill for his classification project. By 1955 he was far enough along to rope in David L. Wallace, a young statistician at the University of Chicago. In his disarmingly casual way, Mosteller asked Wallace, "Why don't you come up and spend some time in New England this summer, and work on this little project I've sort of started?"[2] The two wound up spending more time studying The Federalist papers than Hamilton and Madison did writing them—"a horrible thought," Wallace said later.

Wallace urged Mosteller to use Bayes' rule for the project. Wallace had earned a Ph.D. in mathematics from Princeton in 1953 and would become a professor at the University of Chicago. But in 1955, his first year at Chicago, Savage was teaching from his recently published book on Bayes' rule. Despite the hostility of most American statisticians, Wallace had been receptive to Bayesian ideas.

Wallace thought The Federalist might be an application where Bayes could be very helpful. "If you stick to relatively simple problems," he explained, "like the ones taught in elementary statistics books, you can do it by Bayesian or non-Bayesian methods, and the answers are not appreciably different." Laplace had investigated such problems in the early 1800s and discovered the same thing. "I'm not really a Bayesian," Wallace said. "I haven't done much more than The Federalist, but . . . when you have large numbers of parameters, of unknowns to deal with, the difference between Bayes and non-Bayes grows immense."

Mosteller was open to suggestion. Unlike Savage, Lindley, Raiffa, and Schlaifer, he was no fervent Bayesian. He was an eclectic problem solver who liked any technique that worked. He accepted the validity of both kinds of probability: probability as degrees of belief and probability as relative frequency. The problem, as he saw it, was that treating a unique event like "Hamilton wrote paper No. 52" was difficult with sampling theory. Bayes' degrees of belief would be harder to specify but more widely applicable.

In addition, Mosteller liked grappling with critical social issues, not avoiding controversy by taking refuge in textbook examples. Reality added a certain frisson to a problem. As he put it, difficulties found "in the armchair" seldom resembled those in the field or scientific laboratory. In later years, when asked why he spent so much time on The Federalist papers, Mosteller would point to "that Bayesian hothouse" in the Harvard Business School and to the fact that Raiffa and Schlaifer did not deal with difficult problems or complex data. The discrepancy between too many Bayesian theories and too few practical applications disturbed Mosteller.

With Savage's encouragement, Mosteller and Wallace started their quest to "apply a 200-year-old mathematical theorem to a 175-year-old historical problem." In the process, Mosteller would organize the largest civilian application of Bayes' rule since Laplace studied babies and Jeffreys analyzed earthquakes. Not coincidentally, Wallace and Mosteller would resort to so-called high-speed computers.

They had seemingly vast amounts of data: 94,000 words definitely written by Hamilton and 114,000 by Madison. Of these, they would ignore substantive words like "war," "executive," and "legislature" because their use varied by essay topic. They would keep "in," "an," "of," "upon," and other context-free articles, prepositions, and conjunctions. As work progressed, however, they became dissatisfied with their "rather catch-as-catch-can methods." In a critical decision, they decided to turn their bagatelle of a historical problem into a serious empirical comparison between Bayesian and frequentist methods of data analysis. *The Federalist* papers would become a way to test Bayes' rule as a discrimination method.

By 1960 Wallace was working full-time on developing a Bayesian analysis of *The Federalist*, working out the details for a mathematical model of their data. Given so many variables, Wallace and Mosteller would be mining the papers like low-grade ore, successively sifting, processing text in waves, and discarding words of no help. They would use numerical probabilities to express degrees of belief about propositions like "Hamilton wrote paper No. 52" and then use Bayes' theorem to adjust these probabilities with more evidence.

Initially, they assumed 50–50 odds for the authorship of each paper. Then they used the frequencies of 30 words—one at a time—to improve their first probability estimate. In a two-stage analysis, they first looked at the 57 papers of known authorship and then used that information to analyze the 12 papers of unknown parentage. As their calculations became increasingly complicated, Wallace developed new algebraic methods for dealing with difficult integrals; his asymptotic approximations would form much of the project's statistical meat.

Mosteller and Wallace adopted another important simplification. Instead of using the mathematical vocabulary of probabilities, they adopted the everyday language of odds. They were expert mathematicians, but they found odds easier computationally and intuitively.

During the decade it took to analyze *The Federalist*, Mosteller kept busy on a number of fronts. His back-to-back investigations of the Truman election and the Kinsey report had turned him into the person to call when something went wrong. Over the years Harvard would ask Mosteller to chair four departments: social relations (as acting chair); statistics (as its founder); and, in the university's School of Public Health, first biostatistics and then health policy and management.

He persuaded Harvard ("VERY SLOWLY," he wrote a friend)[3] to establish the statistics department. Joining the 1950s and 1960s push to apply mathematical models to social problems, he researched game theory, gambling, and learning, where Bayes' theorem itself was not used but served as a metaphor for thinking and accommodating new ideas. Eventually, Mosteller's interest in those fields waned, and he moved on to other pastures. Education remained a major interest. As part of the government's post-Sputnik push to teach students at every level about probability, Mosteller wrote two textbooks about frequentism and Bayes' rule for high school students. In 1961 he taught probability and statistics on NBC's early-morning Continental Classroom series; his lectures were viewed by more than a million people and taken for credit by 75,000. In medical research Mosteller pioneered meta-analysis and strongly advocated randomized clinical trials, fair tests of medical treatments, and data-based medicine. He was one of the first to conduct large-scale studies of placebo effects, evaluations of many medical centers, collaborations between physicians and statisticians, and the use of large, mainframe computers.

How did Mosteller juggle a massive Bayesian analysis on top of his other work? He looked tubby and rumpled, but he was a superb organizer and utterly unfazed by controversy. He was genial; he engaged critics with a touch of humor, and he seemed to believe they were entitled to opinions he disagreed with. He was also patience itself and, with a "Gee, golly, shucks smile," explained things over and over again.[4] Doctrinaire only about grammar and punctuation, he once wrote a student about his paper, "I am in a lonely hotel room, surrounded by whiches."[5]

Mosteller was also very hard working. He once posted a sign in his office, "What have I done for statistics in the past hour?"[6] For a short time he recorded what he did every 15 minutes in the day. He was also, as he himself pointed out, the beneficiary of a bygone era. His wife cared for everything in his life except his professional work, and several women at Harvard devoted their careers to his success, including his longtime secretary and Cleo Youtz, his statistical assistant for more than 50 years.

Mosteller was also said to involve in his research any student who stepped within 50 feet of his office. Persi Diaconis, a professional magician for several years after he ran away from home at the age of 14, met Mosteller on his first day as a graduate student. Mosteller interviewed him in his low-key, friendly way: "I see you're interested in number theory. I'm interested in number theory. Could you help me do this problem?"[7] They published the result together, and Diaconis went on to have a stellar career at Stanford University. It was said that Mosteller's last collaborator on Earth was a mountaintop hermit and that Mosteller climbed the peak and persuaded him to cowrite a book. In truth, Mosteller collaborated only with people he considered worth his time, including Diaconis, John Tukey, the future U.S. senator Daniel Patrick Moynihan, economist Milton Friedman, and statisticians like Savage. Finally, Mosteller credited his success to the fact that "somewhere along the road I got this new way of doing scholarly work in little groups of people."[8] Colleagues and research assistants would divide up an interesting topic, meet every week or two, pass memos back and forth, and in four or five years publish a book. Working four or five such groups at a time, Mosteller wrote or cowrote 57 books, 36 reports, and more than 360 papers, including one with each of his children.

Four years into *The Federalist* project, he and Wallace had a breakthrough. A historian, Douglass Adair, tipped them off to a study from 1916 showing that Hamilton used "while" whereas Madison wrote "whilst." Adair's news told them that screening for words in the anonymous 12 papers might pan out.

The problem was that "while" and "whilst" were not used often enough to identify all 12 papers, and a printer's typographical error or an editor's revision could have tainted the evidence. Single words here and there would not suffice. Wallace and Mosteller would have to accumulate a large number of marker words like "while" and "whilst" and determine their frequency in each and every *Federalist* paper.

Mosteller started *The Federalist* project armed with a slide rule, a Royal manual typewriter, an electric 10-key adding machine, and a 10-bank electric Monroe calculator that multiplied and divided automatically. He greatly missed a device he had enjoyed using at MIT: an overhead projector. He and Wallace soon realized they would have to use computers. Harvard had no computer facilities of its own and relied on a cooperative arrangement with MIT. Mosteller and Wallace wound up using a substantial chunk of Harvard's allocation. Today, a desktop computer would be faster. Also slowing them down was the fact that Fortran was only two years old, awkward, and hard to program for words. They were "straddling the introduction of the computer for linguistic counts and the old hand-counting methods, with the disadvantages of both."

Substituting student brawn for computer power, Mosteller organized an army of 100 helpers—80 Harvard students plus 20 others. For several years, his soldiers punched cards for the supposedly high-speed computer.

Programming proceeded so slowly that Youtz speeded up the search for marker words by organizing a makeshift concordance by hand. Threading electric typewriters with rolls of adding machine tape, students transcribed one word to a line, cut the tape into strips with one word per strip, sorted them alphabetically, and counted the strips. Once someone, forever unnamed, exhaled deeply and blew a cloud of statistical confetti. Within a few days, however, Youtz's typists discovered that Hamilton used "upon" twice per paper while Madison seldom used it at all.

Soon thereafter, students punching computer cards found a fourth marker word, "enough," also used by Hamilton but never by Madison. By now, Mosteller had four words pointing to Madison as the author of the disputed papers. But here again, by editing one another's papers, Madison and Hamilton could have melded their styles. As Mosteller and Wallace concluded, "We are not faced with a black-or-white situation, and we are not going to provide an absolutely conclusive settlement. . . . Strong confidence in conclusions is the most an investigation can be expected to offer." They needed to extend their evidence and measure its strength.

Venturing beyond the simplest Bayesian applications, they found themselves knee-deep "in a welter of makeshifts and approximations." Bayesian methods for data analysis were in their infancy, so they had to develop new theory, computer programs, and simple techniques, like those frequentists had developed before 1935. Discarding whatever did not work, they wound up tackling 25 difficult technical problems and publishing four meaty, parallel studies comparing Bayes, frequentism, and two simplified Bayesian approaches. The calculations became so complicated that they checked their work by hand, largely on slide rules. "Upon" was their single best discriminator by a factor of four. Other good makers were "whilst," "there," and "on." "Even a motherly eye," Mosteller wrote, could see disparities between "may" and "his."

Their biggest surprise was that prior distributions—the bête noir of

Bayes' rule—were not of major importance. "This was awesome," Mosteller said. "We went in thinking that everything depended on what sort of prior information you used. Our finding says that statisticians should remove some attention from the choice of prior information and pay more attention to their choice of models for their data."[9]

In the end, they included prior odds only because Bayes' theorem called for them, and they arbitrarily assigned equal odds to Madison and Hamilton. The prior odds turned out to be so unimportant that Mosteller and Wallace could have let their readers name them. In a timely analogy, Mosteller noted that a single measurement by an astronaut on the moon would be enough to overwhelm any expert's prior opinion about the depth of the dust on the lunar surface. So why include the prior at all? Because with controversial or scanty data, enormous amounts of observational material might be needed to decide the issue.

When Mosteller and Wallace published their report in 1964 they announced happily that "we tracked the problems of Bayesian analysis to their lair and solved the problem of the disputed *Federalist* papers." The odds were "satisfyingly fat" that Madison wrote all 12 of them. Their weakest case was No. 55, where the odds were 240 to 1 in Madison's favor.

Their publication was the fourth about Bayes' rule to appear in a three-year period. It followed works by Jeffreys, Savage, and Raiffa and Schlaifer. Of all these works, only Mosteller and Wallace's had dared treat real issues with Bayesian statistics and modern computers. Mosteller had thought about *The Federalist* papers for 23 years and worked on them for ten. It would long remain the largest problem publicly attacked by Bayesian methods.

The work is still admired as a profound analysis of a difficult problem. Reviewers used words like "ideal," "impressive," "impeccable," and "Herculean." As late as 1990 it was considered the greatest single Bayesian case study.

Despite the razzle-dazzle, no one followed it up. No one—not even Mosteller and Wallace—tried to confirm its results by reanalyzing the material using non-Bayesian methods. Who else could organize committees of collaborators and armies of students to empower a 1960s computer to solve large and complex problems?

As for Mosteller himself, what was his reaction? Satisfaction, of course. But also the feeling that, as a friend said, "Hey, here's a good application, a new technique, let's try it, and then find others too." A number of his books described Bayesian techniques, and to this day Diaconis regards Mosteller as a committed Bayesian who tried hard to get social scientists to accept Bayesian methods. Yet Mosteller never again devoted an entire project to Bayes. While his famous study of poverty in collaboration with Senator Moynihan influenced public policy, it did not make major use of Bayes' rule. When a student won a prize for a Bayesian Ph.D. thesis, Mosteller wrote a congratulatory note: "I think Bayesian methods are about to take off. But then I've been saying that for twenty-five years."[10]

13.

the cold warrior

The Federalist project impressed the still small world of professional statisticians, but John Tukey, a star from the world of Cold War spying, would give Bayes' rule the opportunity to demonstrate its prowess before 20 million American television viewers. But would the statistical community learn from Tukey's example that Bayes had come of age? That was the question.

Bayes' big chance at fame commenced in 1960 with the race between Senator Kennedy and Vice President Richard M. Nixon to succeed Eisenhower as president. The election was far too close to call, but the nation's three major television networks competed fiercely to be the first to declare the victor. Winning the race would translate into prestige and advertising dollars. For the National Broadcasting Corporation (NBC), there was a bonus: the opportunity to show off the latest computers made by its corporate owner, Radio Corporation of America (RCA).

NBC's Huntley–Brinkley Report, the nation's top-rated TV news program, reached 20 million viewers each weeknight. Co-anchors Chet Huntley, broadcasting from New York, and David Brinkley, from Washington, were celebrities; more people could recognize them than Cary Grant or James Stewart. NBC's fast-paced format and informal nightly sign-off—"Good night, Chet," "Good night, David"—transformed TV news.

Despite the program's popularity, memories of the polling industry's spectacularly poor performances in the 1936 and 1948 elections as well as the extraordinarily close Nixon–Kennedy race made network executives nervous. In preparation for Election Day, NBC went looking for someone to help it predict the winner. In the first of a series of surprises, the network approached a Princeton University professor, John W. Tukey.

Today Tukey is best known for the terms "bit" and "software," and few outside statistics and engineering recognize his name. But he was a man of staggering accomplishments in the cloak-and-dagger world of military research, especially in code breaking and high-tech weaponry. He worked two jobs 30 miles apart: at Princeton University, where he was a professor of statistics, and at AT&T's Bell Laboratories, then widely considered the finest industrial research laboratory in the world. From these vantage points, he advised five successive U.S. presidents, the National Security Agency, and the Central Intelligence Agency.

To appreciate the audacity of NBC's job offer to Tukey, one needs to understand how deeply he was embedded in Cold War secrets. He had done research on topology in the late 1930s and military analysis in the 1940s. As a young man during the Second World War, Tukey worked in Princeton with the Operations Research Group that computed how a B-29 bomber speeding over Europe should aim its machine-gun fire. With pencil and paper during the Cold War, he broad-brushed the aerodynamics, trajectory, and warhead for the Nike, the first antiaircraft surface-to-air missile system. He also helped persuade Eisenhower to build the U-2, the spy plane that flew from 1956 until 1960, the year a U-2 pilot Francis Gary Powers was shot down over the USSR.

When NBC News approached Tukey in 1960, he had been a member of the CIA's Science and Technology Advisory Panel and the National Security Agency's Science Advisory Board for eight years. His most famous advisory role had occurred the year before, when, as a delegate to the US–USSR Conference on the Discontinuance of Nuclear Weapon Tests, he surprised the Soviet delegation by showing that seismogram data could distinguish underground nuclear explosions from earthquakes. Once both sides knew they could police each other's compliance, they signed the Partial Test Ban Treaty in 1963 to ban nuclear tests in the atmosphere, space, and sea.

Tukey also helped establish a super-secret cryptography think tank at Princeton University. The communications research division of the Institute for Defense Analyses (IDA) moved into Von Neumann Hall, a new campus building surrounded by an eight-foot-high brick wall. IDA, which had "the most intimate ties" to the National Security Agency, was created to solve advanced cryptographic problems.[1] Although the position did not appear on his curriculum vitae, Tukey served on IDA's board of trustees for decades. Student protests against secret research in universities forced IDA off campus in 1970, despite Tukey's personal appeal to Princeton's president, Robert F. Goheen.

Many university faculty did classified work as part of their regular duties during the 1950s and 1960s. John Pratt and Stephen Fienberg, for example, were cleared for such work at the University of Chicago. Said Fienberg, "When I joined the faculty in the Department of Statistics in 1968, it had a dual contract with the Navy office of research. One part supported basic statistical research, and the other was for statistical consulting. We had a safe in the basement where they kept the classified consulting work, although I do not know which faculty had been working on it."[2]

Tukey also worked closely with members of Princeton's physics department, which was "highly involved in the design of the atomic and later hydrogen bombs."[3] After the United States dropped atomic bombs on Japan in 1945, the director of the Manhattan Project, Gen. Leslie R. Groves, asked the Princeton physics chair Henry Smyth to write the official explanation of the bomb, *Atomic Energy for Military Purposes*. In 1951 Princeton launched a secret undertaking, Project Matterhorn, to design thermonuclear weapons at its nearby Forrestal Research Center. Tukey evaluated Edward Teller's and Stanislaw Ulam's design for the first H-bomb early that year. According to his curriculum vitae, Tukey served Forrestal Research Center as "supervisor, Military Systems Analyst" from 1951 until 1956.[4] Physics professor John A. Wheeler, who headed the weapons program, stated, "I believe that the whole country—scientifically, industrially, financially—is better off because of him and bears evidence of his influence."[5]

In addition to his military research at Princeton, Tukey taught classes and supervised more than 50 graduate students. In appearance he could be "a bouncy and beefy extrovert" and "sort of cherubic looking with a pleasant manner." But his lecture style was oblique at best. Invited to speak at Imperial College in London in 1977, Tukey looked like a great bear of a man in old baggy pants. Sitting in a cross-legged Buddha pose on the podium, he began his lecture by asking slowly and deliberately, "Comments, queries, suggestions?"[6] During the long wait that ensued, he ate prunes—12 of them, one by one—until someone in the audience finally asked if he could explain something or other. Only then did Tukey begin speaking. When a graduate student asked one January for an appointment to discuss his Ph.D. thesis, Tukey checked his diary and said he was going to a meeting in two months and if the student drove him there they could talk about it in the car.

Tukey also advised the federal government on a wide range of civilian problems: air quality, chemical pollution, ozone layer depletion, acid rain, census methodology, and educational testing.

How did he manage all this? Stories are legion about Tukey sitting in the back row at a seminar, dozing, reading mail, scanning newspapers, or editing articles, but then rising at the end of the talk to critique it. Tukey drafted articles in pencil while listening to baroque brass recordings, topped the article with the words "By _____ and John W. Tukey," gave the manuscript to one of his two longtime secretaries, and then searched for a collaborator to finish the piece. He put his name to about 800 publications and worked with more than 105 coauthors, including Jerome Cornfield at NIH, but most frequently with his friend Fred Mosteller at Harvard.

As a result of Tukey's heavy military and teaching workloads, his future father-in-law fully expected him to whip out pad and pencil while waiting at the altar to be married. His bride, Elizabeth R. Rapp, was personnel director of the three-year-old Educational Testing Service. Later, she confided that "as the wife of [a] dedicated workaholic, I understand the selfless love and devotion, accommodation and deprivation required to 'keep them on the road.'" After Elizabeth's death in 1998, Tukey said, "One is so much less than two."[7]

According to Elizabeth, Tukey organized and simplified his personal life like "a New Englander through and through."[8] His conversation was quiet and measured and excluded personal comments and idle chatter. His nephew Frank R. Anscombe contended that Tukey had few wants, although they included a house near the sea, a convertible, a small catamaran, classical music recordings, and mince or apple pie. Tukey traveled with his personal table tennis paddle; collected some 14,000 mystery, sci-fi, and adventure paperbacks; lunched on fistfuls of cheese and six glasses of skim milk; and drove a 1936 wood-paneled station wagon until the passenger door fell off and his papers flew onto Nassau Street in Princeton. For 40 years he wore the same style of black polo shirt, so wrinkled that students sometimes mistook him for a janitor. But he always seemed able to squeeze in one more project, provided it was sufficiently intriguing.

So how, given Tukey's eminence and his time commitments, could NBC convince him that the Huntley–Brinkley news program warranted his attention? First, the reputation of opinion surveys, the mainstay of social science, was abysmal. Although sampling forms the foundation of statistics, commercial pollsters were painfully slow to adopt probabilistic random sampling. Serving on the Kinsey Report study committee with Mosteller, Tukey said he would prefer a random sample of three to a Kinsey sample of 300; Kinsey's wife said she wanted to poison him. If Tukey aimed to improve statistical practices in the polling industry, NBC News was a high-profile place to start.

Second, NBC's RCA computers may have been a draw. If Tukey accepted NBC's offer, he would not need an army of students to snip pieces of adding machine paper. RCA was a major military contractor as well as a giant in communications; it manufactured highly regarded mainframe computers for the military and big business. During the 1940s the company's large research laboratory had designed and built the Selectron memory tube for early computers, including von Neumann's Johnniac.

The opportunity to use RCA's computers to analyze election data must have been tempting. Tukey had foreseen the intimate connection between computers and statistics years earlier. When von Neumann designed an electronic computer for the Institute for Advanced Study at Princeton in late 1945, Tukey was the only Princeton University representative on the committee and helped design the computer's architecture and electronic adding circuit. Still, Tukey's "most striking relationship with the computer was that he didn't use it"; his hardware consisted of pencil and paper.[9]

Polling reform and powerful computers, however, may have paled in importance next to the allure of NBC's vast amounts of voting data. As an undergraduate at Brown University, Tukey had majored in chemistry with doses of physics and geology, and his Ph.D. from Princeton, earned in 1939, was in topology, among the purest branches of abstract mathematics. Military research during the Second World War turned him into a "data analyst" committed to fighting the "mental rigidity" and "ossifications" of pure mathematics and abstract statistics and to bridging the gap between mathematics and science.[10] The war moved him far beyond the statistician's early role as a passive observer.

After the war Tukey decided he wanted to drive "the rocky road of real problems in preference to the smooth road of unreal assumptions, arbitrary criteria, and abstract results without real attachments."[11] To do so, he accepted joint positions a half hour apart at Princeton University and Bell Labs. Later, whenever he was offered professorships at other universities he would ask, "Where could I ever find another Bell Labs?"[12] Like Mosteller, he preferred exploring reality, and NBC News had plenty of that.

But of all the enticements NBC could offer—restoring polling's reputation, fast computers, and real data—the most important must have been the thrill of the chase. To beat other networks to the draw, he would have to work at top speed under international scrutiny to make sense of vast amounts of incomplete, uncertain information. It would be, as he put it later, "the best education in real-time statistics that anybody could have."[13] So the military consultant to presidents joined NBC's Huntley–Brinkley Report.

Tukey's first evening on the job, November 8, 1960, started smoothly. The race between Kennedy and Nixon was the tightest since 1916, and Kennedy would win by 120,000 votes out of 70 million cast. By 2:30 a.m., though, Tukey and his colleagues were ready to call him the winner. The pressure was too much for NBC. Network executives hustled the statisticians into a room without telephones, locked them in and refused to let them out until 8 a.m. Tukey and his team twiddled their thumbs all night, unable to release their results until morning, when it was clear Kennedy had won. Still, Tukey had prevented NBC from mistakenly declaring Nixon the winner. Relieved and impressed, the network asked him to assemble a team for the congressional election in 1962. He would work for NBC News for 18 years.

Tukey's handpicked group eventually included ENIAC coinventor John Mauchly; Cornfield of NIH; Richard F. Link, Tukey's first graduate student and pollster Louis Harris's chief statistician; Yale psychology professor Robert Abelson; and David Brillinger, later a professor of statistics at Berkeley. When David Wallace finished Mosteller's analysis of The Federalist papers, he too joined the group.

Wallace arrived expecting a vacation from Bayes' rule because Tukey was thought to look down on it. Tukey is not known to have ever published anything using Bayes' rule, and in an often-quoted remark, he said, "There are many classes of problems where Bayesian analyses are reasonable, mainly classes with which I have little acquaintance."[14] Among those were business decision making, the bailiwick of Howard Raiffa and Robert Schlaifer. The lack of methodology for quantifying Bayes' initial prior especially irritated Tukey. Publicly, he was a data analyst who was anti-Bayesian and even antiprobability.

Thus when Wallace joined Tukey's NBC team in 1964 he was surprised to find Bayes' rule securely ensconced in the computing program: "I immediately thought, this is all very Bayesian. Also, I did a lot of the coding for a lot of the models over the next decade and a half, and, as far as I'm concerned, I was using Bayesian things."[15] Those who agree include Brillinger, who later became Tukey's biographer and the editor of his papers, Pratt from Harvard, and Fienberg from Carnegie Mellon. Said Fienberg, NBC polling used "a form of empirical Bayes, where the past results were used to construct the prior distribution."[16]

Nevertheless, in almost two decades of election forecasting Tukey never admitted to using Bayes' rule. Why would someone who publicly disdained Bayes' rule and seemed to look down on it use it for something as important as announcing the next president of the United States?

Many colleagues stress that, despite appearances, Tukey was "a very private man." His nephew called him an "elliptical and enigmatic Delphic Oracle in a black polo shirt." Wallace agreed: "Tukey could be close-mouthed. . . . He was a man of extreme power and brilliance and in some ways enigmatic about himself. . . . He didn't always let everyone know what his left hand was doing. He'd deny anything Bayesian in the NBC polling."[17]

His dominating personality could be intimidating. While George Box was giving a seminar at Princeton University, Tukey thought he knew what Box was going to say and kept chiming in with his own commentary. Box finally asked for a show of hands. Who wanted Tukey to continue interrupting, and who wanted him to stop? When Box won, Tukey looked surprised. "In some ways, he was a very clever eight-year-old," Box recalled. "He didn't seem to understand very much about interpersonal relations." Some colleagues pointed to his early years as a child prodigy home-schooled by his mother and said that his wife, Elizabeth, helped "warm him up." Edgar Gilbert from Bell Labs concluded, "He was a very likeable personality but he was hard to understand." Peter McCullagh, an Irish statistician at the University of Chicago, called him a "constructive scientific anarchist, . . . a cultural phenomenon, revered by some, feared by others, understood by few." Part of the problem, Pratt said, was that "Tukey could argue on both sides of anything, and you didn't know where he stood."[18]

Adding to the confusion was Tukey's acceptance of one of Savage's most controversial tenets: subjectivity. Tukey called objectivity "an heirloom" and "a fallacy. . . . Economists are not expected to give identical advice in congressional committees. Engineers are not expected to design identical bridges—or aircraft. Why should statisticians be expected to reach identical results from examinations of the same set of data?"[19]

If Tukey could be hard on Bayes' rule, he was even tougher on Fisher. Tukey believed that Fisher's frequency-based ideas dated from "the world of infancy . . . the childhood of experimental statistics, a childhood spent in the school of agronomy. . . . Almost invariably, when closely inspected, data are found to violate [the] standard assumptions" required by frequency. "Far better an approximate answer to the right question, which is often vague, than an exact answer to the wrong question, which can always be made precise." Tukey publicized how even slight deviations from the normal model could muddle the methods of Fisher, Neyman, and Egon Pearson. He particularly scorned frequentist "techniques for assessing significance and asserting confidence. . . . By and large, the great innovations in [frequency-based] statistics have not had correspondingly great effects upon data analysis." Hard words indeed.[20]

So where did Tukey stand? Anti-Bayes *and* anti-frequentism? Friends contend that, like Mosteller, he opposed any monolithic philosophy. Brillinger thought Tukey was annoyed, "not with Bayesian arguments per se; . . . [but] with some of the Bayesians." Tukey said, "Discarding Bayesian techniques would be a real mistake; trying to use them everywhere, however, would in my judgment, be a considerably greater mistake." The issue was knowing when and where. He often complained about "a natural, *but dangerous* desire for a unified approach," explaining that "the greatest danger I see from Bayesian analysis stems from the belief that everything that is important can be stuffed into a single quantitative framework."[21]

Tukey used almost the same language about Fisher's fiducial alternative to Bayes. While courting his wife, Tukey confided that his mission in life was to emulate Fisher by developing methods for analyzing experimental science. But after writing 64 pages in a search for the logical foundations of Fisher's fiducial probability, Tukey decided that "the belief in a unified structure for inference is a dangerous form of hubris." When Tukey visited Fisher at his home in England and began asking questions about his methods, Fisher stalked angrily away, leaving the Tukeys to find their way out of his house alone. In another version of the story, Fisher threw Tukey out of his office after telling the young man that his paper was a "long screed" and that he would understand probability statements only "if you could ever get your bullheaded mind to stop and think." In both stories, an unstoppable force met an immovable object.[22]

For Tukey the only thing that mattered was the data—shorn of computerization, mathematization, probability, and theory. He named his approach exploratory data analysis (EDA). Like Bayesians, many of its proponents were ridiculed and had trouble finding jobs.

So how did Tukey resolve his paradoxical use of Bayes' rule without admitting it? He called it something else. While Brillinger and Wallace called their NBC polling Bayesian, Tukey said it was "borrowing strength."[23]

"Anything he did, he'd call something else," Wallace said, even if it already had a straightforward, well-established name. New names drew attention to ideas, and a colleague counted 50 terms coined by Tukey. Among those that stuck are linear programming, ANOVA, and data analysis. In one article,

Mosteller had difficulty talking him out of using musical notation—sharps, flats, and naturals. Another colleague threatened to call him J. W. Cutie for terms such as "saphe cracking," "quefrency," and "alanysis." As Wallace said, "It was not always the best way to win friends and influence people. . . . But when I talked to Tukey, I essentially tried to use his terminology."

Still, Bayes' rule by any other name is Bayes' rule. And both Tukey and Mosteller were willing to use whatever statistical tool was needed, even if it was Bayesian. Beginning work long before Election Day, Wallace built a base of initial information by combining data from preelection polls; non-statistical, expert opinion from political scientists; and the voting histories of precincts, counties, cities, and states. Preelection opinion polls did not always ask the right questions, so they often failed to elicit all the information needed. The work of sampling people, surveying them, analyzing their answers, and summarizing the results was complex.

On election night, as partial returns from counties and complete returns from selected precincts flowed in, Tukey and his colleagues watched for swings and deviations from past voting behavior and from political scientists' opinions. Then they modified their initial odds with the new information.

As Wallace relived the moment, he said, "Say we're working at a county level with data coming in. Suppose you had no returns from one county. A strict non-Bayesian would say, 'I can't tell you anything there,' but a slightly Bayesian person would say, 'I don't know what's happening in county A, but county B is very similar and it's showing a swing 5% toward Republicans.' You might say that county A might be going the same way, but not give it great weight, because you do have to come up with a number. . . . 'Okay,' says Tukey, 'go down low, take a group of counties that are similar, weight the data you get in these counties, give zero weight to non-data counties, and upgrade, update it all the time.'" Like Schlaifer at the Harvard Business School, Tukey had concluded that since he had to make a decision with in-adequate information, whatever knowledge existed was better than nothing.

Wallace continued: "You take information where you have it and com-pute it with a lot of error bounds on it into places where you don't have data. . . . You first of all work at rural area counties, then urban areas, north, south areas, or whatever, do it separately and play this game of upward regions and across states. It's 'borrowing strength,' but I'd say it's Bayesian. . . . You're using historical data, from previous elections, to show variability between counties and that's the source of your priors, so it's very Bayesian, with a hierarchical model and historically based prior variances."

Despite weeks of planning and rehearsals, election nights did not always proceed as hoped. Studio 8H in Rockefeller Center, where Huntley sat, was sacrosanct and off-limits to anyone without a special ID badge. But when Brillinger saw that the ID tag resembled the sugar packets in NBC's canteen, he paperclipped a packet to his shirt and wandered happily around. For the election of 1964 between Lyndon Johnson and Barry Goldwater, NBC show-cased seven of its mainframe computers on Studio 8H's stage floor: several of RCA's early 301 models and two spanking new 3301's. All evening view-ers could see their imposing large black boxes on screen behind Huntley. Unfortunately, the computers did not work that night, either because their operating system was unfinished or because the heat of the recording studio's lights fried them. So there they sat all night long, like so many limp rags, Link thought, impressive but useless. With voting results pouring in from around the country, Tukey's team punched furiously away at old-fashioned hand calculators and adding machines. Fortunately, the work was simple that night because LBJ's victory was a foregone conclusion, and Johnson won with a record 61% of the popular vote.

In another election, the team called the winners early for California and New York. Then, late-arriving figures came in contradicting their an-nouncement. Two tense hours passed before voting patterns moved back in line with their predictions. Another extremely tight election kept them at work from Tuesday afternoon straight through to Thursday afternoon. Tukey and Wallace realized they needed to improve their technique.

"It turned out that the problem of projecting turnout was more difficult than that of projecting candidate percentage," Wallace discovered. "The quality of data you get is dubious, and you have clear biases of reporting coming in from one part of the county, and if you're lucky, they might do it randomly but machine votes come in faster than non-machine votes. Some-times there's chicanery as well. The turnout is very hard to predict and that has a startling effect, so Bayes was not a total serving. I had a conversation with a student who was consulting with one of the other networks and he said to me, 'This is just so wonderful because it's the first place in statistics where all your assumptions are totally valid.' I was appalled. . . . It's a highly biased sample. And you're going to get yourself into serious trouble if you don't realize that."[24]

Tukey continued to work for NBC through the election in 1980. After that, NBC switched to exit polls based on interviews of voters emerging from their precincts. Exit polls were cheaper, more photogenic, personal, and chatty. They were the polar opposite of Tukey's secret, highly complex, mathematized approach.

Then came the biggest surprise of all. Like Churchill's muzzle on postwar Bletchley Park, Tukey refused to let any of his colleagues write or even give talks about their polling methods. He never wrote about them either. He said they were proprietary to RCA.

Why the secrecy? Why did Tukey scorn Bayes' rule in public but use it privately for two decades? Toward the end of his life he conceded that "probably the best excuse for Bayesian analysis is the need . . . to combine information from other bodies of data, the general views of experts, etc., with the information provided by the data before us."[25] He even defended Savage's subjectivist gospel, that people could look at the same information yet reach different conclusions. Reject "the fetish of objectivity," Tukey declared.[26] He taught Bayes in 1954 and 1955. And he used a Bayesian argument when testifying before Congress that the U.S. Census should adjust for its undercounts of minorities in some areas by incorporating information from other, similar regions. As of 2010, the Census Bureau had not done so.

So why didn't—or couldn't—Tukey use the B-word? As Brillinger noted, "Bayes is an inflammatory word."[27] Certainly, Tukey's term "borrowing strength" allowed him to avoid it. Perhaps sidestepping it carved out a neutral workspace. Perhaps he felt the need to put his own stamp on another person's work. Or perhaps there was another reason. Given Tukey's personality, it's difficult to know. Halfway through his stint with NBC, RCA gave up trying to compete with IBM and sold off its computer division to Sperry Rand. After that, why would RCA care whether Tukey's system went public? Could RCA's military sponsors have classified Tukey's methods, and was he using Bayes' rule for his classified cryptographic research?

Many details of Tukey's national security career remain "murky, deliberately so on his part," his nephew Anscombe concluded.[28] But as Wallace says, "If you go to the secret coding agencies, you'd find that Bayes had a larger history. I'm not in a position to speak of that but I. J. Good is the principal contributor to the Bayesian group and he was taking that position."[29] Good was Alan Turing's cryptographic assistant during the Second World War. So did Tukey use Bayes' rule for decoding for the National Security Agency? And could he have been distancing himself from Bayesian methods in order to protect work there?

The ties between Tukey, Bayes, and top-secret decoding are many and close. Bayes' rule is a natural for decoders who have some initial guesses and must minimize the time or cost to reach a solution; it has been widely used for decoding ever since Bletchley Park. Tukey's ties to American cryptography were particularly tight. According to William O. Baker, then head of Bell Labs, Tukey was part of the force that helped decrypt Germany's Enigma system during the Second World War and Soviet codes during the Cold War. Tukey served on NSA's Science Advisory Board, which was devoted to cryptography. It was a ten-member panel of scientists from universities, corporate research laboratories, and think tanks; they met twice yearly at Fort Meade, Maryland, to discuss the application of science and technology to code breaking, cryptography, and eavesdropping. Baker, Tukey's close friend, was probably the committee's most important member. Baker chaired a long study of America's coding and decoding resources for the NSA and called for a Manhattan Project–like effort to focus, not on publishable and freely available research, but on top-secret studies. Whether Tukey actually did hands-on cryptography is not known, but as a professional visiting-committee consultant, he was certainly aware of all the statistical methods being used.

Tukey's relationship with Good, one of the leading Bayesians and cryptographers of the 1950s and 1960s, is also suggestive. Tukey visited Good in Britain and invited him to lecture at Bell Labs in October 1955. The day after Good's talk he was surprised to find that Tukey, lying on the floor to relax, had obviously understood it all. Tukey was also sympathetic enough with Good's Bayesian methods to introduce him to Cornfield at NIH and suggest that Good might help with statistical methods there; Cornfield became a prominent Bayesian.

Claude Shannon was also in the audience during Good's talk. Shannon had used Bayes' rule at Bell Labs for his pathbreaking cryptographic and communications studies during the Second World War. Tukey was close to Shannon; in 1946 Tukey coined the word "bit" for Shannon's "binary digit." Tukey, Shannon, and John R. Pierce applied together for a patent for a cathode ray device in 1948.

The evidence is substantial enough to convince some of Tukey's colleagues, including Judith Tanur and Richard Link, that he probably did use Bayes' rule for decoding at Bell Labs. Brillinger, Tukey's biographer and NBC polling colleague, concluded, "I have no problem thinking that he might have."[30]

Whatever the motivation, Tukey's secrecy edict played a major role in the history of Bayes' rule. As Wallace observed, "It's important to the development of Bayesian statistics that a lot was under wraps."[31] Tukey's censorship of his polling methods for NBC News, like the highly classified status of Bayesian cryptography during and after the Second World War, is one reason so few realized how much Bayes' rule was being used.

Tukey's Bayesian polling—conducted in the glare of international publicity for two of the most popular TV anchors of the day—could have spread the news of Bayes' power and effectiveness and reinforced it at regular intervals. But his ban on speaking or writing about it meant that Bayes' rule played a starring role on TV for almost two decades—without most statisticians knowing about it.

As a result, the only large computerized Bayesian study of a practical problem in the public domain during the Bayesian revival of the 1960s was the Mosteller–Wallace study of *The Federalist* in 1964. It would be 11 years before the next major Bayesian application appeared in public. And after Tukey stopped consulting for NBC in 1980, it would be 28 years before a presidential election poll utilized Bayesian techniques again.

When Nate Silver at FiveThirtyEight.com used hierarchical Bayes during the presidential race in November 2008, he combined information from outside areas to strengthen small samples from low-population areas and from exit polls with low response rates. He weighted the results of other pollsters according to their track records and sample size and how up to date their data were. He also combined them with historical polling data. That month Silver correctly predicted the winner in 49 states, a record unmatched by any other pollster. Had Tukey publicized the Bayesian methods used for NBC, the history of political polling and even American politics might have been different.

14.

three mile island

After years of working together, the two old friends Fred Mosteller and John Tukey reminisced in 1967 about how "the battle of Bayes has raged for more than two centuries, sometimes violently, sometimes almost placidly, . . . a combination of doubt and vigor." Thomas Bayes had turned his back on his own creation; a quarter century later, Laplace glorified it. During the 1800s it was both employed and undermined. Derided during the early 1900s, it was used in desperate secrecy during the Second World War and afterward employed with both astonishing vigor and condescension.[1] But by the 1970s Bayes' rule was sliding into the doldrums.

A loss of leadership, a series of career changes, and geographical moves contributed to the gloom. Jimmie Savage, chief U.S. spokesman for Bayes as a logical and comprehensive system, died of a heart attack in 1971. After Fermi's death, Harold Jeffreys and American physicist Edwin T. Jaynes campaigned in vain for Bayes in the physical sciences; Jaynes, who said he always checked to see what Laplace had done before tackling an applied problem, turned off many colleagues with his Bayesian fervor. Dennis Lindley was slowly building Bayesian statistics departments in the United Kingdom but quit administration in 1977 to do solo research. Jack Good moved from the super-secret coding and decoding agencies of Britain to academia at Virginia Tech. Albert Madansky, who liked any technique that worked, switched from RAND to private business and later to the University of Chicago Business School, where he claimed to find more applications than in statistics departments. George Box became interested in quality control in manufacturing and, with W. Edwards Deming and others, advised Japan's automotive industry. Howard

Raiffa also shifted gears to negotiate public policy, while Robert Schlaifer, the nonmathematical Bayesian, tried to program computers.

When James O. Berger became a Bayesian in the 1970s, the community was still so small he could track virtually all of its activity. The first international conference on Bayes' rule was held in 1979, in Valencia, Spain, and almost every well-known Bayesian showed up—perhaps 100 in all.

Gone was the messianic dream that Bayes' rule could replace frequentism. Ecumenical pragmatists spoke of synthesizing Bayesian and non-Bayesian methods. The least controversial ideal, Mosteller and Tukey agreed, was either a frequency-based prior or a "gentle" prior based on beliefs but ready to be overwhelmed by new information.

When Box, J. Stuart Hunter, and William G. Hunter wrote *Statistics for Experimenters* in 1978, they intentionally omitted any discussion of Bayes' rule: too controversial to sell. Shorn of the big bad word, the book was a bestseller. Ironically, an Oxford philosopher, Richard Swinburne, felt no such compunctions a year later: he inserted personal opinions into both the prior hunch *and* the supposedly objective data of Bayes' theorem to conclude that God was more than 50% likely to exist; later Swinburne would figure the probability of Jesus' resurrection at "something like 97 percent." These were calculations that neither the Reverend Thomas Bayes nor the Reverend Richard Price had cared to make, and even many nonstatisticians regarded Swinburne's lack of careful measurement as a black mark against Bayes itself.

Throughout this period Jerzy Neyman's bastion of frequentism at Berkeley remained the premiere statistical center of the United States. Stanford's large statistics department, bolstered by Charles Stein and other University of California professors who had refused to sign a McCarthy-era loyalty oath, was also enthusiastically frequentist, and anti-Bayesian signs adorned professors' office doors.

Bayesians were treading water. Almost without knowing it they were waiting until computers could catch up. In the absence of powerful and accessible computers and software, many Bayesians and anti-Bayesians alike had given up attempts at realistic applications and retreated into theoretical mathematics. Herman Chernoff, whose statistical work often grew out of Office of Naval Research problems, got so impatient with theoreticians spinning their wheels on increasingly elaborate generalizations that he moved from Stanford to MIT in 1974 and then on to Harvard. "We had reached a period," he wrote, "where we had to confront the computer much more intensively and we also had to do much more applied work . . . I thought, for the future, the field needed a lot more contact with real applications in order to provide insights into which way we should go, rather than concentrating on further elaborations on theory." Chernoff was no Bayesian, but he told statistician Susan Holmes, then beginning her career, how to face difficult problems: "Start out as a Bayesian thinking about it, and you'll get the right answer. Then you can justify it whichever way you like."[2]

Within Bayesian circles, opinions were still defended passionately. Attending his first Bayesian conference in 1976, Jim Berger was shocked to see half the room yelling at the other half. Everyone seemed to be good friends, but their priors were split between the personally subjective, like Savage's, and the objective, like Jeffreys's—with no definitive experiment to decide the issue. Good moved eclectically between the two camps.

In a frustrated circle of blame, Persi Diaconis was shocked and angry when John Pratt used frequentist methods to analyze his wife's movie theater attendance data, because there was too much for the era's computers to handle. But one of the low moments in Diaconis's life occurred in a Berkeley coffee shop, where he was correcting galley proofs of an article and Lindley blamed him for using frequency methods in the article. "And you're our leading Bayesian," Lindley complained.[3] Lindley, in turn, upset Mosteller by passing up a chance to do a big project using Bayes instead of frequency. Every opportunity lost for Bayes was a blow to the cause and a reason for recrimination. By 1978 the Neyman–Pearson frequentists held "an uneasy upper hand" over the Bayesians, while a third, smaller party of Fisherians "snipe[d] away at both sides."[4]

Few theorems could boast such a history. Bayesians had developed a broad system of theory and methods, but the outlook for proving their effectiveness seemed bleak. De Finetti predicted a paradigm shift to Bayesian methods—in 50 years, post-2020. The frequentist Bradley Efron of Stanford estimated the probability of a Bayesian twenty-first century at a mere .15.

Politicking for Bayes in Britain, Lindley said, "The change is happening much more slowly than I expected. . . . It is a slow job. . . . I assumed in a naïve way that if I spent an hour talking to a mature statistician about the Bayesian argument, he would accept my reasoning and would change. That does not happen; people don't work that way. . . . I think that the shift will take place through applied statisticians rather than through the theoreticians." Asked how to encourage Bayesian theory, he answered tartly, "Attend funerals."[5]

With Bayesian theory in limbo, its public appearances were few and far between. Consequently, when the U.S. Congress commissioned the first comprehensive study of nuclear power plant safety, the question arose: would anyone dare mention Bayes by name, much less actually use Bayes' rule?

President Eisenhower had launched the nuclear power industry with his Atoms for Peace speech in 1953. Twenty years later, although no comprehensive study of safety risks to the public or the environment had been made, private corporations owned and operated 50 nuclear power plants in the United States. When Congress began debating whether to absolve plant owners and operators of all liability for accidents, the U.S. Atomic Energy Commission finally ordered a safety study.

Significantly, as it turned out, the man appointed to lead the study was not a statistician but a physicist and engineer. Born in Harrisburg, Pennsylvania, in 1927, Norman Carl Rasmussen had served a year in the navy after the Second World War, graduated from Gettysburg College in 1950, and earned a Ph.D. in experimental low-energy nuclear physics at MIT in 1956. He taught physics there until MIT formed one of the first departments of nuclear engineering, in 1958.

When Rasmussen was appointed to assess the safety of the nuclear power industry, there had never been a nuclear plant accident. Believing that any such accident would be catastrophic, engineers designed the plants conservatively, and the government regulated them tightly.

Lacking any data about core melts, Rasmussen decided to do as Madansky had done at RAND when studying H-bomb accidents. He and his co-authors would deal with the failure rates of pumps, valves, and other equipment. When these failure rates did not produce enough statistics either, the Rasmussen group turned to a politically incendiary source of information: expert opinion and Bayesian analysis.

Engineers had long relied on professional judgment, but frequentists considered it subjective and not reproducible and banned its use. Furthermore, the Vietnam War had ended America's enchantment with expert oracles and think tanks. Confidence in leaders plummeted, and a "radical presumption of institutional failure" took its place. Faith in technology dropped too; in 1971 Congress canceled its participation in the supersonic passenger plane, the SST, one of the few times the United States has rejected a major new technology. "No Nukes" activists were demonstrating across the country.

Lacking direct evidence of nuclear plant accidents, Rasmussen's team felt it had no choice but to solicit expert opinion. But how could they combine that with equipment failure rates? Normally, Bayes' theorem provided the way. But Rasmussen's panel already had enough controversy on its hands dealing with nuclear power. The last thing they needed was an argument over methods.

To avoid using Bayes' equation, they employed Raiffa's decision trees. Raiffa was a Bayesian missionary, and his trees had Bayesian roots, but that did not matter. Panel members avoided even the words "Bayes' rule"; they called it a subjectivistic approach. They thought that keeping clear of Bayes' theorem would absolve them of being Bayesians.

The committee's final report, issued in 1974, was loaded with Bayesian uncertainties and probability distributions about equipment failure rates and human mistakes. Frequentists did not assign probability distributions to unknowns. The only reference to Bayes' rule, however, was tucked into an inconspicuous little corner of appendix III: "Treating data as random variables is sometimes associated with the Bayesian approach . . . the Bayesian interpretation can also be used."[6]

But avoiding the use of the word "Bayes" did not acquit the report of blame. Although several later studies approved of its use of "subjective probabilities," some of the report's statistics were roundly damned. Five years later, in January 1979, the U.S. Nuclear Regulatory Commission withdrew its support for the study. The Rasmussen Report seemed doomed to oblivion.

Doomed, that is, until two months later when the core of the Three Mile Island-2 nuclear generating unit was damaged in a severe accident. At almost the same time, Jane Fonda debuted a blockbuster movie, *The China Syndrome*, about the coverup of an accident at a nuclear power plant. The civilian nuclear power industry collapsed in one of the most remarkable reversals in American capitalism. Although approximately 20% of U.S. electric power came from 104 nuclear power plants in 2003, at this writing no new facility has been ordered since 1978.

Three Mile Island revived the Rasmussen Report and its use of subjectivistic analysis. After the accident the committee's insights seemed prescient. Previous experts had thought that the odds of severe core damage were extremely low and that the effects would be catastrophic. The Rasmussen Report had concluded the reverse: the probability of core damage was higher than expected, but the consequences would not always be catastrophic. The report had also identified two significant problems that played roles in the Three Mile Island accident: human error and the release of radioactivity outside the building. The study had even identified the sequence of events that ultimately caused the accident.

Not until 1981 did two industry-supported studies finally employ Bayes' theorem—and admit it. Analysts used it to combine the probabilities of equipment failures with specific information from two particular power plants: Zion Nuclear Power Station north of Chicago and Indian Point reactor on the Hudson River, 24 miles north of New York City. Since then, quantitative risk analysis methods and probabilistic safety studies have used both frequentist and Bayesian methods to analyze safety in the chemical industry, nuclear power plants, hazardous waste repositories, the release of radioactive material from nuclear power plants, the contamination of Mars by terrestrial microorganisms, the destruction of bridges, and exploration for mineral deposits. To industry's relief, risk analysis is also now identifying so-called unuseful safety regulations that can presumably be abandoned.

Subjective judgment still bothers many physical scientists and engineers who dislike mixing objective and subjective information in science. Avoiding the word "Bayes," however, is no longer necessary—or an option.

the navy searches

Surprisingly, given Bayes' success in fighting U-boats during the Second World War, the U.S. Navy embraced the method slowly and grudgingly during the Cold War. High-ranking officers turned to Bayes almost accidentally, hoping at first to garner only the trappings of statistics. Later, the navy would move with increasing confidence and growing computer power to fine-tune the method for antisubmarine warfare. Meanwhile, the Coast Guard eyed the method for rescuing people lost at sea. As was often the case with Bayes' rule, a series of spectacular emergencies forced the issue.

The navy's flirtation with the approach began at dusk on January 16, 1966, when a B-52 jet armed with four hydrogen bombs took off from Seymour Air Force Base near Raleigh, North Carolina. Each bomb was about ten feet long and as fat as a garbage can and had the destructive power of roughly one million tons of TNT. The jet's captain, known for smoking a corncob pipe in the cockpit, and his six-man crew were scheduled to fly continuously for 24 hours and refuel several times in midair.

In a controversial program called Operation Chrome Dome, SAC under Gen. Curtis LeMay kept jets equipped with nuclear weapons flying at all times to protect against Soviet attack. In a costly and hazardous process, tankers refueled the jets in midair.

The jet made the scheduled rendezvous for its third refueling with a SAC KC-135 tanker jet on the morning of January 17. Bomber and tanker maneuvered in tandem over Spain's southeastern coast, six miles above the isolated hamlet of Palomares, Spanish for Place of the Doves. They used a telescoping boom that required the two planes to fly three or four meters apart at 600 miles per hour for up to half an hour. In a split-second miscalculation, the tanker's fuel nozzle struck the metal spine of the bomber, and at 10:22 a.m. local time 40,000 gallons of fuel burst into flame. Seven of the planes' 11 crew members perished.

Airmen, the four bombs, and 250 tons of aircraft debris rained down from the sky. Fortunately, it was a holiday, and most of the area's 1,500 residents were taking time off from working their fields, so no one was hit. Even more important, no nuclear explosion occurred; the bombs had not been "cocked," or activated. However, the parachutes on two of them failed to open, and when the bombs hit the ground their conventional explosives detonated, contaminating the area with an aerosol of radioactive plutonium. Three of the bombs were located within 24 hours, but the fourth was nowhere to be found.

Adding to the crisis was the fact that, unknown to the public, the incident at Palomares was at least the twenty-ninth serious accident involving the air force and nuclear weapons. Ten nuclear weapons involved in eight of these accidents had been jettisoned and abandoned at sea or in swamps, where they presumably remain to this day. The missing weapons, none of which involved a nuclear detonation, included two lost over water in 1950; two nuclear capsules in carrying cases in a plane that disappeared over the Mediterranean in 1956; two jettisoned into the Atlantic Ocean off New Jersey near Atlantic City in 1957; one left at the mouth of the Savannah River off Tybee Beach in Georgia in 1958; the one that fell in Walter Gregg's garden near Florence, South Carolina, in 1958; one in Puget Sound in Washington State in 1959; uranium buried in Goldsboro, North Carolina, in 1961; and a bomb from a plane that rolled off an aircraft carrier into the Pacific in 1965. It was an unenviable record that was only slowly attracting media attention.

When it became obvious that the latest H-bomb to fall from a SAC jet must have landed in the Mediterranean Sea, the Defense Department phoned John Piña Craven, the civilian chief scientist in the U.S. Navy's Special Projects Office.

Craven had a bachelor's degree from Cornell University's naval science training program and a master's in physics from Caltech. While working on a Ph.D. in applied physics at the University of Iowa he spent his spare time taking advanced courses of every kind, from journalism and philosophy of science to partial differential equations. Notably, in view of what was to come, he took statistics and got a C. In 1951 Craven graduated "sort of educated in everything."[1] These were the years when the military was developing crash programs for using navigational satellites and for building ballistic missiles and guidance systems to counter the Soviets. In such an atmosphere, the Pentagon regarded any Caltech grad as a technological whiz kid.

At 31, Craven became what he called the navy's "Oracle at Delphi, . . . an applied physicist advising the Navy whenever they have mission or equipment problems that they cannot handle." His first job was inventing technology to locate Soviet mines blocking Wonson Harbor during the Korean War. Three years later he became chief scientist of the Special Projects Office developing the Polaris Fleet Ballistic Missile Submarine System. When the nuclear submarine USS *Thresher* burst and sank off Cape Cod in 1963 with 129 men on board, he was ordered to develop ways to find objects lost or sunk in deep water. To the military looking for an H-bomb in the Mediterranean Sea, Craven sounded like the man for the job.

"We've just lost a hydrogen bomb," W. M. "Jack" Howard, assistant to the secretary of defense for atomic energy, said when he telephoned Craven.

"Oh, we've just lost a hydrogen bomb," Craven recalls saying. "That's your problem, not mine."

Howard persisted: "But one of the bombs fell in the ocean, so we don't know how to find it; three others are on land."

Craven shot back, "You called the Navy all right but you called the wrong guy. The Supervisor of Salvage is the guy responsible for that." Within hours, though, Craven and the salvage chief, Capt. William F. Searle Jr., formed a joint committee of rejects: Craven had failed twice to get into the naval academy, and Searle graduated from Annapolis before poor vision shunted him into underwater salvage work, where, wits said, everyone is more or less blind.

"Craven, I want a search doctrine," Searle barked. He needed the doctrine—naval-ese for a plan—so he could start work the next morning to send ships and other materiel to Spain. That night Craven kept telling himself, "Jesus, I've got to come up with a search doctrine."

Craven already knew something about Bayesian principles. His minesweeping mentor during the Korean War in 1950–52 was the navy physicist and applied mathematician Rufus K. Reber, who had translated Bernard Koopman's Bayesian antisubmarine studies into practical but classified tables for sea captains planning mine-searching sweeps. Craven had also learned about Bayes while visiting MIT professors doing classified research for the government. Most important, he had heard about Howard Raiffa, who was pioneering the use of subjective probabilistic analyses for business decision making, operational analysis, and game theory at the Harvard Business School.[2]

As Craven understood it, Raiffa used Bayesian probability to discover that horse race bettors accurately predict the odds on horses winning first, second, and third place. For Craven, the key to Raiffa's racetrack culture was its reliance on combining the opinions of people "who really know what's going on and who can't verbalize it but can have hunches and make bets on them." Later, Raiffa commented that he was pleased if he had influenced Craven to assess subjective probabilities and pool them across experts. But he emphasized that Bayes does not get into the act until those subjective views are updated with new information. Furthermore, he remembered talking about weather prediction, not horse races.

"I'm very good at grasping concepts," Craven explained later. "I'm lousy on detail. I got the betting on probabilities and also got the connection with Bayes's conditional probabilities. But I also understand the politics of getting things done in the Navy, and I say I've got to get a search doctrine."

Craven had experts galore at his disposal. Some knew about B-52s while others were familiar with the characteristics of H-bombs; bomb storage on planes; bombs dropping from planes; whether the bomb would stay with the plane wreckage; the probability that one or both of a bomb's two parachutes deploys; wind currents and velocity; whether the bomb will be buried in sand; how big it would look wrapped in its chute, and so on. Craven figured that his experts could work out hypotheses as to where the bomb would fall and then determine the probability of each hypothesis.

Most academic statisticians would have thrown in the towel. They would have believed, with Fisher and Neyman, that sources of information should be confined to verifiable sample data. Craven, of course, had no wish to repeat the experiment. He needed to find the bomb. "At that point, I wasn't looking at the mathematics, I was just remembering what I got from Raiffa."

Then reality intervened. With only a few hours and the assistance of one technician, Craven was forced to be "the guy who interviewed each one of these experts to make the bets. I'm the guy who decides who the bettors are, and I'm also the guy—let's be honest about it—who imagines what I'd say if I were the guy I can't get in touch with. So I'm doing a lot of imagineering. . . . I didn't have time to call together these people." Craven's use of expert guessing would be spectacularly subjective.

Blending hurried phone calls to experts, reports of on-site witnesses, and his own "imagineering," Craven came up with seven hypotheses he called scenarios:

1. The missing H-bomb had remained in the bomber and would be found in the bomber's debris.
2. The bomb would be found in bomb debris along the path of the collision.
3. The bomb had fallen free and was not in the plane's debris.
4. One of the bomb's two parachutes deployed and carried it out to sea.
5. Both of the bomb's parachutes deployed and carried it farther out to sea.
6. None of above.
7. A Spanish fisherman had seen the bomb enter the water. (This hypothesis came later, after naval commanders talked with one Francisco Simo Orts.)

Ideally, at this point, Craven would have gotten "all these scenarios and all these cats [his experts] in a room and have them make bets on them." But with only one night before the search doctrine was needed, Craven realized, "I'm going to invent the scenarios myself *and* guess what an expert on that scenario would bet."

The emergency forced Craven to cut through years of theoretical doubts about building a Bayesian prior and estimating the probability of its success: "As I did this, I knew immediately that I wouldn't be able to sell this concept to any significant operator in the field. So I thought what the hell am I going to do? I'm going to tell them that this is based on Bayes's subjective probability. And second, I'm going to hire a bunch of mathematicians, tell them I want you to put the cloak of authenticity on using Bayes's theorem. . . . . So I hired Daniel H. Wagner, Associates to do this."

Daniel H. Wagner was a mathematician so absent-minded that his car once ran out of gas three times in one day. He had earned a Ph.D. in pure mathematics—none of it applied—from Brown University in 1957. Several years of working for defense contractors convinced him that rigorous mathematics could be applied to antisubmarine warfare and to search and detection work. The fact that both involved innumerable uncertainties made Bayes' rule appealing. As Wagner put it, "Bayes' rule is sensitive to information of all kinds . . . but *every clue has some error attached* because, were there no error, there would be no search problem: You would just go to the target and find it immediately. The problem is that . . . you will rarely be given the value of the expected error and so you will have to deduce the location error from other information."[3]

Operations research was new, but Wagner came recommended by two authorities: Capt. Frank A. Andrews (ret.), the officer who had commanded the Thresher search, and Koopman, by then an influential division head of the Institute for Defense Analyses, the campus-based organization for academics doing secret military research.

Going to Craven's office to learn more about the missing H-bomb, Wagner took along the youngest and greenest of his three-man staff, Henry R. ("Tony") Richardson, who had earned a Ph.D. in probability theory from Brown all of seven months earlier. He would be Bayes' point man at Palomares.

As Wagner reconstructed the scene, Craven showed the mathematicians an interesting chart of the waters off Palomares. The seabed had been divided into discrete rectangular cells, and after interrogating air force experts Craven postulated the first six of his seven scenarios. Then he drew on statistical theory to weight each scenario as to its relative likelihood. His ideas were not quantitative; he had drawn a contour map with mountains of high probabilities and valleys of unlikely regions. Nor was he forthcoming about the reasons for each hypothesis. Richardson realized that, as far as Craven was concerned, he and Wagner were just number crunchers.

For Richardson, the fascinating feature of Craven's probability map was that it was all based on initial information before any searching began. Craven had constructed a rule-of-thumb prior, the first component of Bayes' rule. Richardson was familiar with Koopman's search theory, but Craven's multiple-scenario priors and the promise of Bayesian updating looked intriguing. By assuming that their probabilities would fall into bell shapes, Craven made it possible to use slide rules and desktop electromechanical calculators to develop a map of the bomb's possible locations based on the prior information available to him. Like Laplace, he assigned different probability weights to each scenario.

Wagner and Richardson went to work in the company's headquarters in Paoli, Pennsylvania, verifying and refining Craven's rough calculations. A coworker, Ed P. Loane, constructed a more precise probability distribution for the H-bomb's location by punching data into paper tape and feeding it over public telephone lines into an electronic computer in the nearby office of the Burroughs Corporation. Turning typewriter characters into graphical displays on a teletype machine was challenging. A probability map might wind up looking like this:

```
##$#&
&$&&#
#$##$
```

where # meant a probability between 0 and .05; $ was a probability between .06 and .10, and so forth. Loane worked for Wagner, Associates full time while he was a part-time graduate student in applied mathematics at the University of Pennsylvania, and he wanted desperately to go to Palomares in Richardson's place. Meanwhile, Craven gathered data from the Pentagon for Richardson to take to Spain. The young man was amazed to see Craven and other senior officers milling around, opening doors for him.

Almost daily planning sessions with the military soon convinced the mathematicians that their goal—using Bayes' rule and updating to find the H-bomb—was not the reason they were hired. Bayes was window dressing. If the H-bomb was not found, the navy wanted to be able to prove statistically that it was not there. "The general thrust seemed to be to come up with a credible certification to the President that the H-bomb could not be found, rather than proceeding with an expectation that it could be found. Indeed, the former purpose," Wagner concluded, "is the main reason we were brought into the act."[4]

Richardson agreed: "My recollection of my marching orders was to statistically document the search that was being carried out and, in the event that the bomb was not found, to be able to certify to the President and Congress that everything possible was done and that it was done in a scientifically accurate and careful way. So that was pretty much what I was sent out to do. Having read Koopman's work, and knowing that there was such a thing as optimal search based on Bayesian ideas, I was hoping to do more."[5]

Richardson was not interested in using Bayes as a mathematical excuse for a failed expedition. He wanted to find the bomb. He flew to Spain with Captain Andrews, who had a physics Ph.D. from Yale and, after the *Thresher* search, had retired from the navy to join Catholic University's faculty. Andrews knew the Pentagon had big doubts that the navy search team would ever find the bomb. In addition, he had been warned that, if the bomb was not located, the entire world would know that the search team "had failed professionally." In short, the navy was on the hot seat, and careers were on the line. "The implication was, of course, that if we did not find the weapon, nobody else could," Andrews recalled later.[6]

During the flight, Richardson briefed Andrews on Bayesian search theory. Andrews exclaimed, "Oh, if we'd only had that during the *Thresher* search."[7] Once a large search area was divided into small cells, Bayes' rule said that the failure to find something in one cell enhances the probability of finding it in the others. Bayes described in mathematical terms an everyday hunt for a missing sock: an exhaustive but fruitless search of the bedroom and a cursory look in the bath would suggest the sock is more likely to be found in the laundry. Thus, Bayes could provide useful information even if the search was unsuccessful.

Arriving in Palomares, the men found an impoverished village so small it had no telephone and did not appear on maps or in Spain's census. Beginning in 3500 BC, mining and smelting for lead and silver had pockmarked the desert area with open shafts and, cursed with less than eight inches of rain annually and a saline water supply, agriculture was limited to winter tomatoes grown for export. The B-52's aerial explosion and wind had dusted 558 acres of the town and its fields with radioactive plutonium.

In addition to these problems, the village was under siege from a military camp of 750 Americans, complete with field laundries, bakeries, and a movie theater; an offshore fleet of up to 18 ships at a time; a Soviet trawler snooping in international waters; and scores of international reporters incensed by a blackout on news. Applying Bayes' rule would not be a textbook exercise in abstractions; it would be a high-wire operation conducted under intense scrutiny.

For four days, the U.S. and Spanish governments refused to admit that the bomber might have carried nuclear weapons of any kind. News of nuclear bombs and radioactivity leaked out only after an American sergeant shouted to the first reporter on the scene, "Hey, buddy, can you speak Spanish?"

"Sure."

"Well, tell that peasant over there to get out of that field, for God's sake. I can't make him understand a damned thing. There's radioactivity there and we've got to keep the people cleared out."

A public relations catastrophe was in the making. This was the first accident involving the widespread dispersal of radioactive material and the first that attracted widespread, highly critical scrutiny from the world's media. Within three days, reporters in Palomares knew that a nuclear bomb was missing, but six weeks passed before the U.S. Department of Defense would confirm it. Censorship by Spain's dictator, Francisco Franco, kept news of the radioactivity off local radio stations, while broadcasts from Communist Eastern Europe spread the word. Radio Moscow announced that "the bomb is still in the sea, irradiating the water and the fish," and the Soviet government complained that the United States had broken the nuclear test-ban treaty of 1963. The on-site press corps fumed at being scooped by Radio Moscow, Pentagon-based reporters, and even *Stars and Stripes*.

Locals were understandably terrified. Tourism and exports of Spanish fruit and tomatoes collapsed. Demonstrators in Mexico City, Frankfurt, and the Philippines updated a popular song from *My Fair Lady*, "The bomb in Spain lies mainly in the drain."[8] Adding to the pressure, the Vietnam War was escalating and U.S. military bases around the world were at stake. President Johnson phoned the Department of Defense every day demanding news on the search.

Arriving in this hotbed of tension, Captain Andrews immediately introduced Richardson to RAdm. William S. Guest, commander of the navy task force looking for the bomb. Guest was celebrated as the first U.S. aircraft carrier pilot to sink an enemy ship during the Second World War. He was notoriously stubborn and, behind his back, people called him Bull Dog. Guest understood airplanes and budgets, but not Bayes. However, he did understand Washington's message: "You will listen to Dr. Richardson and we will listen to him, so basically . . . for that reason you should pay attention yourself." Expecting an august authority, Guest had assigned the mathematician a captain's stateroom and steward. When he met Richardson, who looked even younger than his 26 years, Bull Dog harrumphed, "I didn't think we were getting a teenager."

The first thing Guest told Richardson was tongue-in-cheek—but not really. The mathematician was to prove that the missing bomb was on land because Guest's job was to look in the sea, and if it was on land, finding it would be someone else's job. Right off, Richardson declared, "I don't think I have the ability to do that."

Guest commanded 125 swimmers and scuba divers eyeballing the shallow coastline, minesweepers cruising deeper water in heavy surf, 3,000 navy personnel, 25 navy ships, 4 research submersibles, and a host of civilian researchers and contractors. The entire search, called Aircraft Salvops Med, would cost $12 million in 1966 dollars.

Guest wanted to save money by using the equipment where it was best suited and then returning it as soon as possible. This meant he wanted to search some areas that were actually unlikely sites for the H-bomb.

Craven's initial hypotheses were based on prevailing winds, so the early hunt focused on a large rectangular area called Alpha II off the beach of Palomares. Guest ordered his swimmers, divers, and minesweepers to search there over and over again.

Richardson began work immediately by combing the charts of the search to date. The first weakness he saw was that, although there were tracks of where the ships had gone back and forth, there was no mention of effectiveness. "Just going back and forth wouldn't do you much good if you couldn't see the bottom of the ocean," he said. "And that in fact was the case. Some of their sensors couldn't penetrate deep water, so they were basically out there running around but not contributing anything to the effectiveness of the search. . . . None of this is criticism. It was just a horrible situation to be in. With the whole world looking at you, you can't tie boats up in a dock and say they're useless." So, inspired by a conversation with Andrews, Richardson coined the term "Search Effectiveness Probability" (SEP).

Looking at the map of the ocean bottom broken into a grid of little squares, Richardson computed for each of the squares the probability that, if the bomb were there, it would have been found by the amount of search effort applied to that area. "If the Search Effectiveness Probability came in at 95%, you could say to the Admiral, 'This area has been searched pretty thoroughly, and maybe you want to go somewhere else,'" Richardson said.

By then he probably knew as much as anyone about the ongoing search. Quarantined from the curious reporters on land, he worked nights in the ship's accounting office. His luggage, filled with reference books and Reber's declassified tables for minesweepers, had been lost in Madrid, so he painstakingly recreated some of the tables, overlaying bits of paper to superimpose curves. He had no alternative. Portable computers did not exist, and even IBM mainframes had only 32K words (not gigabytes or even megabytes) of memory. Armed with his paper cutouts, his slide rule, and an adding machine that could also multiply, he computed the effectiveness of each day's operations. Each morning he greeted Bull Dog Guest with new probabilities. The admiral enjoyed joking about Richardson's boyish appearance, but the probabilities unsettled him.

"I began computing SEPs—the probability you'd have found the bomb if it were there—and a lot of zeros showed up indicating that, even if it had been there, you probably wouldn't have seen it because your capabilities weren't up to the task." At the other end of the SEP scale, a "one" would have signified that the bomb would have been found, had it been there.

Richardson was calculating very few ones: "All those zeroes. When Guest saw them—remember this is some young teenager talking to him—the minute he saw zeroes, he was quite outspoken in his questions. 'Why are you giving me zeroes when we've been out there for two weeks?'"

Guest began using the search effectiveness evaluations as quantitative guides for moving equipment. He wanted to document that his equipment had conducted their investigations thoroughly; he was not interested in using Bayesian updating to find new, more probable places to inspect. Even when a more likely site for the H-bomb appeared, Admiral Guest stuck to his "plan of squares."

Years later, Craven complained that "the least informed and knowledgeable was Admiral Guest, the on-scene commanding officer." The admiral was furious "because he thinks we're out of our noggin." Richardson is more forgiving. Guest "had other concerns. Bayes was a little bit high-falutin'. SEP was understandable. But you start getting into Bayesian updating and these funny words like priors and posteriors, admirals tend not to be patient with this stuff." As a result, evaluating the search's effectiveness became the focus of the H-bomb search. The idea of using effectiveness data to update the first Bayesian component—Craven's presearch scenarios—faded into the background.

Meanwhile, the eyewitness testimony of the veteran fisherman Francisco Simo Orts was gaining credibility fast. The morning of the crash Orts had watched a large parachute pass over his boat and splash down 100 yards away. He called it "half a man, with the insides trailing." Despite the odd description, his report sounded authentic. Strangely stiff in air, the object sank fast, within 30 seconds, parachute and all. Moreover, Orts said the chute was grayish; air force chutes were orange and white for personnel but gray-white for bombs. Navy personnel had interviewed Orts shortly after the crash but had discounted him because he did not use standard procedures to triangulate the spot. Having fished those waters all his life, he could make a seaman's eye calculation of familiar mountains and villages along the shore and identify the location.

Lt. Cdr. J. Brad Mooney, assistant operations officer for deep submersibles, thought Orts might know what he was talking about. Mooney, later promoted to commander and chief of naval research, came from New Hampshire, where lobstermen used similar methods to find their submerged pots. He and Jon Lindberg, a commercial diving consultant, commandeered a jeep, found Orts in a bar, and took him to sea. When Orts twice pointed minesweepers to the same spot in the Mediterranean, Mooney believed him.

Soon Orts's testimony formed the basis for a high-likelihood hypothesis: with one parachute deployed, the bomb had plunged into a steep, deep-water canyon filled with tailings from an old lead mine. Mooney drew a one-mile radius around Orts's spot and named it Alpha I.

As Craven recalled, "We didn't find the bomb for a long period of time because the place of highest probability is a place we can't get to. It's in a narrow crevasse, too deep." Much of the military's equipment needed for a deep-sea search was inadequate: navigational charts dated from the early 1900s; detectors were "grossly inaccurate" with errors up to 1,000 yards; and many of the most useful devices were available only from commercial or research sources. They included three small submarines: the mini *Alvin* from Woods Hole Oceanographic Institution, the *Aluminaut* from the Reynolds Aluminum Company, and a little yellow sub called the *Perry Cub.*

Of Guest's entire squadron, only the two- or three-man sub *Alvin* could penetrate the rugged depths of the highest probability site. But *Alvin's* battery was fading, and to rejuvenate its power the submersible had to be lifted out of the water and docked for long periods.

Six weeks after the plane crash, Captain Andrews hitched a ride down the crevasse with the *Alvin* crew. Peering through its five-inch portholes, they suddenly spied a strange track leading down a slope, "totally different from anything there," Andrews recalled, "basically like somebody dragging a heavy log or barrel down the slope." *Alvin's* battery was running low again, so they had to abandon the skid marks and surface. Then, for two weeks a large storm circled in, grounded, and thwarted the *Alvin.*

During all this time President Johnson was telephoning the Department of Defense every day, only to be told, "We cannot tell you when we'll recover the bomb. We can only tell you the *probability* of when we'll recover it." Fuming, LBJ replied that he did not want a probability; he wanted a date. Privately, Craven added, "I'm sure his response was profane."

Finally, Johnson's volcanic temper exploded: "I want you to get a series of top-level academics to look at this search plan and tell me what's wrong with it. I don't want this probability stuff. I want a plan that tells me exactly when we're going to find this bomb."

Craven convened a committee of Cornell, Harvard, and MIT professors to come to the Pentagon on the morning of March 15, 1966. Number crunchers from Wagner, Associates presented "a mathematical model whose complexity defied understanding by mere mortals."[9] The professors endorsed the Bayesian plan and adjourned for lunch.

On their return they learned that *Alvin*'s crew, during its nineteenth dive off Palomares, had just spotted the bomb with its enormous parachute strewn over the seafloor rocks. *Alvin* had telephoned the surface that the H-bomb looked "like a ghost down there . . . like a big body in a shroud."[10] It had hit ground in 1,300 feet of water and been dragged by currents down a steep slope almost 2,850 feet deep. It lay within a mile of where Orts had pointed.

After the bomb was safely retrieved, the fisherman sued for $5 million in salvage prize money. At the government's request, Richardson again used optimal search theory based on Bayes' rule to estimate the value of Orts's testimony: he had saved the government at least a year's hard work. In 1971 an admiralty court in New York awarded Orts $10,000. The United States had already settled $600,000 on Palomares residents and given the town a $200,000 desalting plant.

Just as the RAND Corporation's Bayesian study had warned eight years earlier, SAC's crash over Palomares diminished the authority of the U.S. Air Force. Military flights over Spain were forbidden, the number of SAC air-alert missions was halved, and responsibility for American air bases in Spain was transferred from SAC to the U.S. Tactical Air Command in Germany. In return for allowing the United States to keep its bases, Franco demanded American help in getting Spain into NATO and the Common Market.

SAC's next accident involving nuclear weapons, two years after Palomares, was the last straw for Operation Chrome Dome. The accident occurred when a B-52 loaded with four nuclear bombs crashed onto sea ice outside a U.S. air base at Thule, Greenland. The weapons were destroyed by fire, but, as at Palomares, radioactivity contaminated the area. As a result of the two accidents, the rising cost of keeping SAC's planes in the air, and the advent of intercontinental ballistic missiles, Secretary of Defense Robert McNamara ended SAC's airborne alert program in 1968.

In 2002, almost four decades after the Palomares accident, Spanish authorities said they had found no danger in the area from surface radiation. Spanish and U.S. health officials reported that no radiation-related cancers had been detected in Palomares's residents. They also said the 1,600 air force personnel who shipped 1,000 cubic meters of Palomares soil in 4,810 metal drums for burial in South Carolina had been exposed to insignificant amounts of radiation, 1/10 the current limit for radiation workers. Even though the public perceives plutonium as being extremely hazardous, government studies showed that its alpha rays are so weak they do not penetrate skin or clothing and, if ingested, pass out of the body in feces. The greatest danger posed by plutonium occurs when it is inhaled. Despite more than 30 years of living and working in a plutonium-contaminated environment, official reports say, the residents of Palomares have inhaled far less than the maximum safe dose identified by the International Committee on Radiological Protection.

Radioactive snails discovered in 2006, however, prompted fears of dangerous levels of plutonium below ground. A joint Spanish–U.S. study was announced, and children were warned not to play in fields near the explosion sites or to eat the snails, a local delicacy.

But what about Bayes? What did it contribute to the H-bomb search? Richardson concluded that "the numbers I computed were coverage numbers so that [Guest] could say we'd covered these areas. . . . Scientifically, the big thing in my mind was that Bayes was a sidelight to the H-bomb search."[11]

The H-bomb hunt could have been a full-blown Bayesian exercise. The prior probabilities of Craven's prehunt scenarios could have been updated with Richardson's shipboard data to guide the search. However, they were never combined in time to be of any use in locating the lost bomb. And without updating, there was no Bayes. Instead of Bayes, the heroes were Orts and the Alvin. The H-bomb search did develop the methodology for computing SEPs (later called LEPs, for "local effectiveness probability"), but Richardson could not get an article about using probability to find the H-bomb published in an academic journal. The H-bomb hunt was a striking demonstration of how difficult it would be to win operational support for Bayes' rule, even when something as tangible and terrifying as a missing thermonuclear bomb was concerned.

Still, although Bayesian updating was not used at Palomares, the success of the search strengthened Craven's faith in scientific searches and the potential of Bayes' rule. He and his team had learned how to compute subjective presearch hypotheses and weight their importance. They realized that the future of Bayesian search methods depended crucially on computer power and the portability of computerized information. This was not an insignificant realization. Richardson had been the only member of his graduate class in pure mathematics to take a computer course, and computer computation was still thought of as cowardly. Within months, though, Wagner, Associates acquired a punched-tape terminal, its first direct access to electronic computation. The next time they were called, the Bayesians would have better tools.

The navy got a dramatic opportunity to use Bayes' rule two years later, in the spring of 1968, when two attack submarines, one Soviet and the other U.S., disappeared with their crews within weeks of each other. As head of the Deep Submergence Systems Project, Craven was responsible for the search for both subs. Despite Bayes' limited role in the H-bomb search, both Craven and Richardson remained convinced the method was scientifically valid.

The first submarine to disappear was a diesel-fueled and missile-armed Soviet K-129, the source of Tom Clancy's fictionalized bestseller *The Hunt for Red October*. The U.S. Navy was alerted to its loss by a massive Soviet search in the Pacific off the Kamchatka peninsula along a major route frequented by its submarines. About the same time, U.S. underwater sensors recorded a "good-sized bang." The noise was far less than the sound of a sub imploding upon itself, but it occurred at a curious place, far from the Soviet search operation and on the International Date Line, at 40 degrees north and precisely 180 degrees longitude. Because the date line is a human artifact, the noise suggested a man-made event. Craven, one of a handful of U.S. personnel who knew of the "extremely classified" affair, hired Wagner, Associates for a full-scale probability analysis without ever telling them what they were looking for. Forty years later, Richardson still did not know he had worked on the search for the Soviet submarine.

Craven could think of only three plausible scenarios for the K-129's disappearance: "First, that the sound had nothing to do with the lost submarine. Second, that the sound was made by the submarine but that it did not sink and like Jules Verne's *Nautilus* was still gliding beneath the sea." Third, that the sub's watertight compartments were open when the crisis occurred and the vessel flooded so fast it did not collapse. Craven reasoned that if the sound recorded on the International Date Line came from the submarine, "then it *was indeed not where it was supposed to be*, which was why the Soviets could not find it."

Johnson, distracted during the tumultuous last months of his presidency, authorized a search for the Russian sub on the hypothesis that it might be a rogue, even though the prospect of finding it was slim. Eventually Craven concluded that the sub—armed with ballistic missiles and crewed by about 100 people—was indeed "a rogue, off on its own, in grave disobedience of its orders . . . [and possibly planning to attack Hawaii]. Since the Soviets didn't know how far off course their sub had been, *the Soviets would have had no idea that their ship was a rogue unless we told them.*"[12] U.S. authorities informed the Soviet leader, Leonid Brezhnev, of the bang's location, and in the face of evidence that his military might be out of control he could see détente as an attractive option. Later, Americans photographed the K-129 but were unable to raise it.

In May 1968, a few weeks after the Soviet sub sank, the U.S.S. Scorpion, a nuclear-powered attack sub, disappeared with its crew of 99 in the Atlantic Ocean. The Scorpion was cruising west, toward home, somewhere along a 3,000-mile submarine route between Spain and the East Coast of the United States. It was reportedly armed with two nuclear torpedoes. According to a study made in 1989, Scorpion's reactor and torpedos would be among at least eight nuclear reactors and 50 nuclear warheads to have been lost at sea; of these, 43 were on sunken Soviet submarines and eight originated with U.S. military activities. With the Scorpion's final resting place unknown, the military launched a full-scale search.

Craven and Andrews, by now the world's leading search experts, rapidly reassembled their H-bomb search crew. At first, the hunt stretched across the Atlantic Ocean. After some bureaucratic sleuthing, though, Craven learned that an ultrasecret listening post for "an unnamed agency" had recorded mysterious "blips" in extremely deep water about 400 miles southwest of the Azores. The location of the blips corresponded with the sub's expected itinerary and drastically narrowed the search area from a 3,000-mile-long rectangle to three or four square miles. Thanks to Craven, the investigation took a spectacular leap forward.

Craven organized a full-blown Bayesian hunt for Scorpion from the very beginning. When the H-bomb was lost off the Spanish coast, Craven had turned almost accidentally to Bayes in the hope of deflecting Congressional displeasure in case of failure. This time, the navy moved tentatively but with growing faith to exploit the method.

"Craven had confidence in the scientific approach from the very beginning but, to put it mildly, that wasn't everybody's thing," Richardson said. "Tough" is the word most often used to describe Craven, and for the next five months, from June to October 1968, he staunchly defended Bayes against skeptics. Although the H-bomb search in Palomares had failed to combine Bayesian priors and SEPs, Craven was enthusiastic when Richardson proposed doing so this time. A powerful computer in the United States would compute the probabilities of the various presearch hypotheses. Then this prior was to be combined and updated on shipboard with daily search results.

Shortly after the sub disappeared, Richardson was flown to the Azores to observe the surface search for Scorpion and to visit the USNS Mizar, a research ship conducting underwater operations. Personnel from the Naval Research

Laboratory, the Navy Oceanographic Office, and various equipment makers were aboard the *Mizar*, working 12-hour shifts around the clock. Over the next five months, they cruised through the area for weeks at a time, dragging across the ocean bottom a sledlike platform covered with a wide-angle camera, sonars, and magnetometers. The chief scientist on board the *Mizar*, Chester L. "Buck" Buchanan, had originally designed the equipment to find the *Thresher* and had improved it greatly since. He vowed not to shave until the *Scorpion* was found.

Scorpion searchers faced even more uncertainties than the H-bomb hunters had along the Mediterranean coast: a remote location 400 miles from land-based navigation systems, an ocean floor two miles down, and no eyewitness accounts pinpointing the *Scorpion's* location. Navigational systems also introduced large errors and uncertainties. Two land-based radio networks, Loran and the new global Omega, were too imprecise to be useful, satellite fixes were available only irregularly, and transponders anchored to the ocean bottom were often indistinguishable from one another.

When Richardson arrived on board the *Mizar*, he found the ship following orders from Washington to search off Point Oscar—over and over again. Craven's early analysis of acoustic data suggested that the *Scorpion* might have settled near Oscar. Using Bayes, however, Richardson tried to show graphically that they had oversearched Point Oscar and that very little probability remained of finding *Scorpion* there. Despite his brilliant demonstration, the search around Oscar continued. Washington would have to issue orders to change operations, and this would require persuasion based upon calculation of a detailed probability map, that is, a Bayesian prior.

"In all the operations that I've ever operated in you have strong personalities with their own ideas, and you have to argue—unless somebody [like Craven] in Washington shoves it down their throat," Richardson said. "Otherwise, you have to convince people. And they have to come to their own conclusions that it's the right way to do it." Besieged by Craven, authorities in Washington later ordered that the prior probability map be treated as an important factor in the search.

On July 18, 1968, a month after the *Scorpion's* disappearance, Craven gave "a brain dump" for Richardson and a new Wagner employee, Lawrence D. ("Larry") Stone. Craven reported everything he had learned from his experts, and Captain Andrews presented a submariner's view of what a sub might do under various circumstances. Working in Washington, Craven and Andrews outlined nine scenarios that might explain how *Scorpion* sank. Then they assigned a weight to each according to how believable it was. This was the same approach Craven had used in the H-bomb search. Each scenario simulated *Scorpion's* movements and multiple uncertainties as to its course, speed, and position at the time of the blip.

One high-priority scenario was based on a mysterious piece of bent metal found by the *Mizar* during a quick survey of the region before the start of systematic searching. The metal was so shiny that it could not have lain very long on the sea bottom, and it was far away from the overly investigated Point Oscar.

Richardson and Stone carried their copious notes to Wagner, Associates' headquarters to quantify Craven and Andrews's assumptions and compute a prior "probability map" of the sub's location on the ocean floor. First, they established a search grid around the blip that Craven had identified as the probable location of the *Scorpion's* explosion. Each cell in the grid measured one mile north–south and 0.84 miles east–west, for a total of 140 square miles.

At Richardson's suggestion, the stateside search team made a key decision to use Monte Carlo methods to model the sub's movements before and after the accident. Physicists on the Manhattan Project had pioneered Monte Carlo techniques for tracking the probable paths of neutrons in a chain reaction explosion. Richardson substituted "little hypothetical submarines" for the neutrons. Academic Bayesians would not adopt Monte Carlo methods for another 20 years.

Starting with Craven's probable location of the explosion (the blip), a mainframe computed the probabilities that, in its death throes, the submarine changed course and traveled, for example, another mile in any of several random directions. Using Thomas Bayes' simplification, Richardson began by considering each of those directions as equally probable. Then, making a point at each new possible location, the computer repeated the process to produce new points, reiterating the procedure 10,000 times to make 10,000 points on the seafloor where the sub might have settled.

The use of Monte Carlo simulation to generate numbers based on Craven's presearch scenarios and weighting represented a big advance in search work. According to Richardson, "The nice thing with Monte Carlo is that you play a game of let's pretend, like this: first of all there are ten scenarios with different probabilities, so let's first pick a probability. The dice in this case is a random number generator in the computer. You roll the dice and pick a scenario to work with. Then you roll the dice for a certain speed, and you roll the dice again to see what direction it took. The last thing is that it collided with the bottom at an unknown time so you roll dice for the unknown time. So now you have speed, direction, starting point, time. Given them all, I know precisely where it [could have] hit the bottom. You have the computer put a point there. Rolling dice, I come up with different factors for each scenario. If I had enough patience, I could do it with pencil and paper. We calculated ten thousand points. So you have ten thousand points on the bottom of the ocean that represent equally likely positions of the sub. Then you draw a grid, count the points in each cell of the grid, saying that 10% of the points fall in this cell, 1% in that cell, and those percentages are what you use for probabilities for the prior for the individual distributions."

The 10,000 points were calculated on a mainframe computer in a small Princeton company that encoded classified data and punched it into paper tape. Such computers were available only on the mainland in the 1960s. Still in the future were so-called portable modems, 45-pound backbreakers for dialing into phone lines.

As cumbersome as it seems today, the time-shared mainframe made Bayes' repetitive calculations feasible. It calculated the coordinates of Scorpion's 10,000 possible locations and then counted the number of points that fell in each cell of the search grid. Lacking any sort of display screens, the computer printed out the numbers on ordinary teletype paper tape. Then the data were relayed over insecure public telephone lines to Richardson and Stone in Paoli. It was the only practical way of incorporating all the presearch data accumulated by Craven and Andrews in Washington into a detailed probability map.

Richardson later felt guilty about calculating only a "skimpy" 10,000 points, but at the time it seemed like a big number. Today's computers refine detail even in low-probability areas. When the 10,000-point map was finished, it described the initial probabilities in 172 cells covering 140 square miles. Two cells, E5 and B7, stood out like rock stars under spotlights. With presearch simulation "hits" of 1,250 and 1,096, respectively, they were by far the most likely resting places for Scorpion and its crew. The next 18 most likely cells had far lower probabilities, between 100 and 1,000; and most cells (which had scores below 100) seemed almost irrelevant. The map was based on hours of conversations with Craven and Andrews, their scenarios, and their weighting. Unlike the H-bomb analyses of two years before, this map represented a real scientific advance, primarily because of the Monte Carlo calculations of the Scorpion's possible movements.

By late July, the map was ready, and Washington ordered that it be treated as an important factor in the search. It was now time for Bayesian updating, with data on the effectiveness of the fleet's search effort in each cell.

Aboard the *Mizar*, mathematicians from Wagner, Associates accumulated and recorded the effectiveness of each day's search. Stone and, later, two young students—Steven G. Simpson, a mathematics Ph.D. candidate from MIT, and James A. Rosenberg, a Drexel University undergraduate co-op student—worked on the area identified by Craven's scenarios about 2,000 yards from Buchanan's shiny piece of metal. Calculating by hand, they estimated the capabilities of the fleet's cameras, sonars, and magnetometers and combined them into a single number expressing the effectiveness of the search conducted in each cell of the sea-bottom grid. These numbers would eventually become the second component in Bayes' formula. Each morning the students had the unenviable task of tactfully advising a succession of naval commodores on the effectiveness of their searches: "Well, sir, I think it would be better if you did this rather than that."

Psychologically, a search can be difficult. Until the target is found, every day represents a failure. As Stone put it, "Bayes says that the longer you search without finding the target, the worse your prospects are, because the remaining time to detect the target gets longer, not shorter."[13] On the other hand, those with confidence in Bayes' rule could trace their progress. "Areas that you search go down in probability," Richardson explained, "and areas that you haven't searched go up. So your updated probabilities get higher where you haven't been looking. . . . And generally, it's always most optimal to keep looking in the highest probability area. The next day you have a high probability area somewhere else, probably not where you searched, and they pop up somewhere else the third day, and you just keep doing it and doing it and doing it. And unless you've made some drastic mistake, you'll eventually find what you're looking for."

As it was during the H-bomb hunt, the biggest problem turned out to be overestimates of the sensors' capabilities. Most of them had never been tested or evaluated systematically for how well they could detect a piece of metal to the left or right of their detectors. Thinking over the problem, Richardson realized, "So you had two uncertainties and, if your objective is to come up with the mathematical expression for how to allocate resources optimally, that's an interesting point."

As naval commanders came and went during the *Mizar*'s five cruises to the search site, Bayes' rule became the group memory of the search and its coordinating principle. For the first time, the rule was used from beginning to end of a long search. Unfortunately, not everyone saw the Monte Carlo prior map as a powerful tool for directing Mizar's search. It took almost a month to get the map to the scene of operations for its Bayesian updating. Communications between land and sea were so poor that Stone finally hand-carried the map to the Azores on August 12. Not until the research ship's fourth and fifth cruises in October—five months after Scorpion disappeared—did the detailed prior distributions based on Craven's and Andrews' scenarios become available.

The Mizar's fifth and last cruise was originally planned to test the sensors, refine the underwater tracking system, and study the contours of the ocean bottom. By this time Craven had organized acoustical studies to more precisely calibrate the location of the blip recorded by the supersecret sensors. Small depth charges were exploded in the ocean at precisely known positions, and their sounds were used to refine the information recorded by naval listening posts during the Scorpion's last moments. Every day Craven's acoustical analyses edged the most likely spot for the Scorpion closer to Buchanan's shining piece of metal.

In late October the increasingly impatient and by now heavily bearded Buchanan finally got approval to investigate the shiny metal. As Mizar's sled made its 74th run over the ocean floor, its magnetometer spiked high at several anomalies in cell F6. Returning to the area on October 28, Mizar struggled to pinpoint the spot again. Finally its cameras revealed, lying on the sea bottom, partially buried in sand, the submarine Scorpion. A poorly functioning sonar detector had previously passed right over the sub without finding it. Word that Buchanan was shaving his beard spread rapidly to the United States.

Richardson was back in the States when he got the phone call. "They gave me the location in code," he said, "and I plotted it up, and at first I thought it was going to plot right smack in the middle of that high probability square, and I was really excited." Instead, it was 260 yards away, close to the mysterious piece of shiny metal found at the beginning of the search. The scrap was later determined to be a Scorpion fragment. Still, Richardson joked ruefully, 260 yards off in a 140-square mile area of open sea was "close enough for government work."

Years later, Captain Andrews argued that Bayes was only one day and a half mile behind Buchanan. If Buchanan had not returned the Mizar to the shiny metal that day, Craven's presearch probabilities, updated with his later acoustical studies, would have found Scorpion first.

On November 1, five months after the search began, Rosenberg, the co-op student from Drexel, hand-carried the photos of *Scorpion* back to the United States. Excluding time spent studying a misleadingly magnetic, hull-shaped rock, the search sled had located the submarine after scanning 1,026 miles of ocean bottom at a speed of one knot for the equivalent of 43 days, two days earlier than Bayesian predictions.

President Johnson was told "the highest probability was that the sinking was caused by an accident on board the submarine."[14] This time he may have listened to probabilities.

Analyses of the sounds made by the *Scorpion* suggested that it had been traveling east, rather than west, when it sank. Twenty years later Craven learned that the sub could have been destroyed by a "hot-running torpedo." Other subs in the fleet had replaced their defective torpedo batteries, but the navy wanted *Scorpion* to complete its mission first. If *Scorpion* had fired a defective torpedo, it would have missed its target and probably turned back and struck the sub that had launched it.

Anxious to document the methods used in the search for *Scorpion*, the Office of Naval Research commissioned Stone to write *Theory of Optimal Search*. Published in 1975, it is an unabashedly Bayesian book incorporating applied mathematics, statistics, operations research, optimization theory, and computer programs. Cheaper and more powerful computers were transforming Bayesian searches from mathematical and analytical problems to algorithms for software programs. Stone's book became a classic, important for the military, the Coast Guard, fishermen, police, oil explorers, and others.

While Stone was writing his book, the United States agreed to help Egypt clear the Suez Canal of unexploded ammunition from the Yom Kippur war with Israel in 1973. The explosives made dredging dangerous. Using the SEPs developed in Palomares, it was possible to measure the search effectiveness to get the probability that, if a bomb had been there, it would have been spotted. But how could anyone estimate the number of bombs remaining in the canal when no one knew how many were there to begin with? Wagner, Associates chose three priors with different probability distributions to express high, middle, and low numbers. Next, using the handy system of conjugate priors described by Raiffa and Schlaifer in 1961, they declared that each prior would have a posterior with the same class of probability distributions. This produced three tractable distributions (Poisson, binomial, and negative binomial) complete with those statistical desiderata, mean values and standard deviations. Computing became "a piece of cake," Richardson reported, but it proved impossible to explain the system to hardened ordinance-disposal specialists with missing fingers. In the end, no one talked about Bayes at Suez.

Up to this point, postwar Bayes had searched only for stationary objects like bombs in a canal, or H-bombs and submarines on the ocean floor. Technically, these were simple problems. But shortly after the Suez Canal was cleared and *Theory of Optimal Search* was published, intensive efforts were made to adapt Bayesian methods to moving targets: civilian boats adrift in predictable currents and winds.

The technology was a perfect match for U.S. Coast Guard rescue co-ordinators like Joseph Discenza, whose job in the late 1960s was to answer the telephone when someone called saying, "My husband went out fishing with my son, and they're not back."[15] After checking area ports of call for the boat, he used a Coast Guard Search and Rescue Manual to estimate by hand the target's location and its probable drift.

"Like a dog with a bone in his teeth," Discenza started to computerize the Coast Guard's manual.[16] He studied search theory and earned a master's degree at the Naval Postgraduate School in Monterey, California, and a Ph.D. at New York University. Along the way Discenza discovered that ever since the Second World War the Coast Guard had been using the Bayesian search theory developed by Koopman to find U-boats in the open ocean. Discenza filled in the corporate memory gap between the 1940s and the 1970s. "The Coast Guard was very Bayesian. Even when they're doing it manually, they're doing Bayes," Stone said. But until Discenza, they were like early casualty actuaries, using Bayes' rule without realizing it.

Joining forces with Discenza, Wagner's company designed a computerized search system based on Bayesian principles for the Coast Guard. A natural outgrowth of the H-bomb and *Scorpion* searches, it combined clues about a vessel's original location and subsequent movements into a series of self-consistent scenarios and then weighted them as to their likelihood.

The Coast Guard ruled that estimating probabilities and weights should be a group decision. Each individual involved should weight the scenarios privately before they were averaged or combined by consensus. Above all, no scenario should be discarded. "To leave out subjective information is to throw away valuable information because there is no unique or 'scientific' way to quantify it," Stone urged.[17]

What if a ship in distress radioed its position but a small plane reported seeing it an hour later a hundred miles away? One or the other had made an error in position, but neither report should be ignored; both should be assigned relative reliabilities. As Stone commented, "Discarding one of the pieces of information is in effect making the subjective judgment that its weight is zero and the other weight is one."

Bayesian updating and, at Richardson's insistence, Monte Carlo techniques were incorporated into the Coast Guard system in 1972, almost two decades before university theorists popularized the method or the term "filters." The Monte Carlo methods estimated an enormous number of possible latitudes, longitudes, velocities, times, and weights for each lost ship to pinpoint 10,000 possible target locations.

Stone also used a Bayesian procedure, an early version of a Kalman filter, to separate and concentrate the data or signals according to specified criteria and to weigh each possible path of a target's motion according to its believability. The technique did not become popular among academics until the 1990s, but it saved military and space contractors immense amounts of time in the 1960s because their computers had little memory or power. Before Rudolf E. Kalman and Richard Bucy invented the procedure in 1961, each original observation had to be completely recalculated every time a new one appeared; with the filter, new observations could be added without having to do everything all over again. Kalman vehemently denied that Bayes' theorem had anything to do with his invention, but Masanao Aoki proved mathematically in 1967 that it can be derived directly from Bayes' rule. Today, it is known as a Kalman or a Kalman-Bucy filter.

Once Monte Carlo methods and filters were adopted, even untenable and highly improbable paths produced valuable information and helped searchers determine which of the remaining paths were more likely. As more information arrived from sensors, Coast Guard aircraft, weather reports, tide tables, and charts of prevailing currents and winds, the data were converted into likelihood functions and then combined with priors about the target's movements to predict its probable location. As data accumulated with each iteration, the filter concentrated a relatively small number of highly probable paths.

The Coast Guard's system was up and running in 1974 when a tuna boat sank off Long Beach, California. Two days later, purely by chance, a freighter found 12 of its survivors in a lifeboat. Using their new technology, the Coast Guard calculated backward from the chance rescue to the tuna boat's probable capsize point and then forward again using probability maps of ocean currents and Bayesian updating. Armed with a Bayesian probability map, the Coast Guard rescued three more men the next day. Another successful search took place two years later after a ship capsized and sank while crossing the Pacific. Five sailors took to sea in two life rafts. Twenty-two days later, also by chance, two survivors were found in one of the rafts. Six days after that, the same Coast Guard program found a third survivor, who had been adrift for 28 days.

Bayes had found stationary objects lodged on the seafloor and had tracked boats drifting with predictable ocean currents and the wind. But what about locating and following evasive prey, a Soviet submarine, say, or a moving target operated by human beings? Could Bayes accommodate human behavior?

"It's the Cold War, and there are submarines out there that are a threat to the U.S.," recalled Richardson. "They're moving targets, so why not do something in antisubmarine warfare. . . . It started in the 1970s and continued for two decades, and I personally did a lot of work searching for subs in the Atlantic Ocean and the Mediterranean."

When the future vice admiral John "Nick" Nicholson took command of the U.S. submarine fleet in the Mediterranean in 1975, he secured a $100,000 grant from the ONR to bring Richardson to Naples for a year. It was one of ONR's biggest contracts, and navy accountants considered it a waste of money. But the Mediterranean was full of Soviet and NATO ships and submarines eyeing one another; the Soviets alone had 50 vessels, including ten submarines. By the early 1970s the "Med" was so crowded that the U.S. and Soviet governments signed a pact to reduce collisions. When the U.S. Navy began routine tracking of Soviet subs in the Mediterranean in 1976, Nicholson thought Richardson and Bayes' rule could help.

Starting from scratch, Richardson cranked intelligence information into an antiquated computer in Naples: previous submarine tracks; particular types of Soviet sub that were apt to take a particular route and perform certain maneuvers; and reports from sonobuoys, passive acoustic listening devices that were dropped by aircraft into fixed underwater surveillance systems. Unlike Koopman's purely objective analysis of radio transmissions and submarine tracks during the Second World War, Richardson was making subjective assessments of the behavior of Soviet officers. He was also using real-time feedback of actual search results, something that submarine hunters in the Second World War would have regarded as science fiction. To all this intelligence data Richardson added the islands, sea mounts, and tight passages in the region's constricted geography. These natural obstacles became surprisingly helpful features.

By definition, tracking involves uncertainties and estimations that are far from ideal. Parameters change as new data appear, and "to make matters worse, the data can be remarkably uninformative and obtained from a number of different sources," as Stone wrote.[18] An optical scanner might spot a distant periscope emerging a foot above the horizon for 10 seconds but fail to identify it as a submarine. Operators watching radar signals on their computer screens could not always distinguish a sub from a surface ship. Arrays of acoustic hydrophones extended over the seafloor for hundreds of miles to detect low-frequency acoustic signals emitted by submarines, but their data were often highly ambiguous. Different targets, for example, radiated acoustic signals at the same or nearly the same frequency. Even the ocean distorted sounds. Sound waves bend with every change in water temperature, and the roar of breaking waves affects signal-to-noise ratios. Bayes' common currency—probabilities—fused information gathered from these various sources. Amid such vague and ephemeral reports, Bayes' rule was in its element.

One summer day in 1976 a Soviet nuclear-powered submarine slipped through the Strait of Gibraltar and entered the Mediterranean. It was a 5,600-ton Echo II class vessel, armed with cruise missiles that could be fired from the surface. The U.S. fleet tracked it as far as Italy before losing it. No one could tell when it would pass through the Sicily Straits into the eastern Mediterranean.

Besides the submarines under his command, Nicholson had been assigned four antisubmarine destroyers that pulled experimental sleds packed with trailing-wire sonar detectors. Arranging his forces across the Sicily Straits so the destroyers would have a chance to detect the Soviet sub passing through, Nicholson waited tensely. "The wait went on longer and longer than all of our operations [intelligence] people were expecting," Nicholson related years later. "Tony kept working his program and he kept saying, 'I still think there's an x percent possibility that it hasn't gone through yet.'"[19]

Nicholson's superiors were pressuring him to move his submarines and surface ships over to the eastern Mediterranean to look for the sub there. But he was an old hand at pressure; he had been executive officer and navigator of the second nuclear submarine to go under the Arctic ice cap to the North Pole and had commanded the first nuclear sub to go from the Pacific to the North Pole in the winter.

Richardson, still poring over the old computer, urged Nicholson to ignore his commander. "I think we should take at least one to two more days," he said. He estimated the probability that the sub had not yet slipped through the straits at about 55%. Nicholson didn't know how much confidence to place in Richardson's system. It was new, it included subjective assessments of human behavior, and it was being used for real-time decision making. But Richardson was filling a void other intelligence and operations experts could not. Making a bold decision that could have destroyed his career, Nicholson decided to wait. "And, lo and behold, we made contact and were able to track the guy through the Strait." The Sixth Fleet was jubilant, and, as Richardson described the reaction of the brass, "most everyone became a believer" in Bayesian search methods.

Thanks to the trailing-wire sonar detectors, every time the Soviet submarine came to the surface in the eastern Mediterranean one of Nicholson's destroyers was cruising nearby. Their skippers were under orders not to come too close to the sub, but, as Nicholson says, "These destroyer guys don't listen very well."

One rather clear Sunday morning the Soviet submarine surfaced with its sail (a metallic structure covering periscopes and masts) about four feet out of the water. Waiting nearby was one of Nicholson's destroyers, the 3,400-ton *Voge*. To everyone's surprise, the Soviet sub turned toward the *Voge* and charged at full speed.

As Nicholson recounts the story, "Everybody on the ship was taking pictures of this thing, the submarine moving along at 20 knots with its sail out of the water, when the skipper of the surface ship slowed for some reason. The submarine skipper apparently didn't keep his eye right on it, and the first thing you know, the submarine rammed right into the *Voge*. We believe the Soviet captain was trying to cut the trailing-wire sonar that had given him such fits."

The sub was badly damaged, and its captain was relieved of his command that same night. The *Voge* was towed to France for repairs. The incident proved the value of Richardson's tracking and of trailing-wire sonar systems on surface ships. Later, Bayesian methods tracked Soviet submarines in the Atlantic and Pacific, although after the breakup of the USSR in 1991 Russian out-of-area submarine deployments were greatly reduced.

"The antisub warfare work was pretty much the highlight of things that were really Bayesian," Richardson reflected. ". . . It was like being back in Spain again. I was ten or fifteen years older, sitting up all night, running my computer and briefing the admiral in the morning. . . . That's kind of the ultimate happiness, when you can make things move around in the world based on your ideas."

Meanwhile, the military that had been so slow to embrace Bayesian search theory was exploring its use for identifying asteroids speeding toward Earth and for locating Soviet satellites as they orbited in space. In 1979 NATO held a symposium in Portugal to encourage the solution of "real problems" with Bayesian methods.

Most of the attendees were from the military, but Richardson and Stone gave talks about their submarine hunts, while others spoke about search and rescue and oil deposit exploration. Among the civilian attendees was Ray Hilborn, a newly minted zoology Ph.D. who was interested in saving the fish populations in the world's oceans. He had gotten his first exposure to simple Bayesian applications at the East–West think tank run by Raiffa in Vienna six years earlier.

Hilborn was struck by the fact that people at the NATO conference dealt with practical problems that required making decisions. His own job involved setting legal limits on fishing for particular species, and, listening to the speeches, he said to himself, "God, this really is the way to ask the questions I want to ask. Everyone who is actually involved in the real world does things in a Bayesian way. The limit of [frequentist] approaches just isn't obvious until you actually have to make some decisions. You have to be able to ask, 'What are the alternative states of nature, and how much do I believe they're true?' [Frequentists] can't ask that question. Bayesians, on the other hand, can compare hypotheses."[20] It would take him almost 10 years to find a fisheries problem for Bayes, but Hilborn was a patient man.

part V
victory

eureka!

As the computer revolution flooded the modern world with data, Bayes' rule faced one of its biggest crises in 250 years. Was an eighteenth-century theory—discovered when statistical facts were scarce and computation was slow and laborious—doomed to oblivion? It had already survived five near-fatal blows: Bayes had shelved it; Price published it but was ignored; Laplace discovered his own version but later favored his frequency theory; frequentists virtually banned it; and the military kept it secret.

By 1980 anyone studying the environment, economics, health, education, or social science was tap-tapping data into a terminal connected to a mainframe computer. "Input" became a verb. Medical records, for example, included dozens of measurements of every patient, ranging from age, gender, and race to blood pressure, weight, heart attacks, and smoking history. Wine statistics included chemical measurements and quality scores for every vintner, varietal, and vintage.

But who knew which of the 20-odd attributes of a patient or a wine were important? Researchers needed to analyze more than one unknown at once, calculate the relationships among multiple variables, and determine the effect that a change in one had on others. Yet real-life facts did not fall into tidy bell-shaped curves, and each time the variables were refined, more unknowns cropped up. Computers were generating a multivariate revolution and spawning a plague of unknowns called the curse of dimensionality. Statisticians had to wonder whether a method ideal for tossing a few gold coins could adapt to the new world.

Bayesians were still a small and beleaguered band of a hundred or more in the early 1980s. Computations took forever, so most researchers were still limited to "toy" problems and trivialities. Models were not complex enough. The title of a meeting held in 1982, "Practical Bayesian Statistics," was a laughable oxymoron. One of Lindley's students, A. Philip Dawid of University College London, organized the session but admitted that "Bayesian computation of any complexity was still essentially impossible. . . . Whatever its philosophical credentials, a common and valid criticism of Bayesianism in those days was its sheer impracticability."[1]

The curse of dimensionality plagued both Bayesians and frequentists. Many in the academic statistical community still debated whether to indulge in computer-intensive analysis at all. Most statisticians of the era were mathematicians, and many confused their beloved old calculators—their manual Brunsvigas and electric Facits—with the new electronic computers. They tried to analyze the new data with methods designed for old calculating tools. One statistician boasted that his calculating procedure consisted of marching into his university's computer center and saying, "Get on with it."[2] Thanks to pioneers like Robert Schlaifer and Howard Raiffa, Bayesians held sway in business schools and theoretical economics, while statistics departments were dominated by frequentists, who focused on data sets with few unknowns rather than on those packed with unknowns.

As a result, many statistics departments watched from the sidelines as physical and biological scientists analyzed data about plate tectonics, pulsars, evolutionary biology, pollution, the environment, economics, health, education, and social science. Soon engineers, econometricians, computer scientists, and information technologists acquired the cachet that humdrum statisticians seemed to lack. Critics sniffed that statistics departments were isolated, defensive, and on the decline. Leading statistical journals were said to be so mathematical that few could read them and so impractical that few would want to. The younger generation seemed to think that computers and their algorithms could replace mathematics entirely.

In what could have been a computational breakthrough, Lindley and his student Adrian F. M. Smith showed Bayesians how to develop models by breaking complex scientific processes into stages called hierarchies. The system would later become a Bayesian workhorse, but at the time it fell flat on its face. The models were too specialized and stylized for many scientific applications. It would be another 20 years before Bayesian textbooks taught hierarchical models. Mainstream statisticians and scientists simply did not believe that Bayes could ever be practical. Indicative of their attitude is the fact that while Thomas Bayes' clerical ancestors were listed in Britain's Dictionary of National Biography he himself was not.

Yet amazingly, amid these academic doubts, a U.S. Air Force contractor used Bayes to analyze the risk of a *Challenger* space shuttle accident. The air force had sponsored Albert Madansky's Bayesian study at the RAND Corporation during the Cold War, but the National Aeronautics and Space Administration (NASA) still distrusted subjective representations of uncertainty. Consequently, it was the air force that sponsored a review in 1983 of NASA's estimates of the probability of a shuttle failure. The contractor, Teledyne Energy Systems, employed a Bayesian analysis using the prior experience of 32 confirmed failures during 1,902 rocket motor launches. Using "subjective probabilities and operating experience," Teledyne estimated the probability of a rocket booster failure at 1 in 35; NASA's estimate at the time was 1 in 100,000. Teledyne, however, insisted that "the prudent approach is to rely on conservative failure estimates based on prior experience and probabilistic analysis."[3] On January 28, 1986, during the shuttle's twenty-fifth launch, the *Challenger* exploded, killing all seven crew members aboard.

The disparity between the military's sometime acceptance of Bayes and the academic statistical community's refusal to embrace it is still puzzling. Did the military's top-secret experience with Bayes during the Second World War and the Cold War give it confidence in the method? Was the military less afraid of using computers? Or did it simply have easier access to powerful ones? Given that many sources dealing with the Second World War and the Cold War are still classified, we may never know the answers to these questions.

Several civilian researchers tackling hitherto intractable problems concerning public health, sociology, epidemiology, and image restoration did experiment during the 1980s with computers for Bayes. A major controversy about the effect of diesel engine emissions on air quality and cancer inspired the first attempt. By the 1980s cancer specialists had solid data about the effects of cigarette smoke on people, laboratory animals, and cells but little accurate information about diesel fumes. William H. DuMouchel from MIT's mathematics department and Jeffrey E. Harris from its economics department and Massachusetts General Hospital teamed up in 1983 to ask, "Could you borrow and extrapolate and take advantage of information from non-human species for humans?"[4] Such meta-analyses, combining the results of similar trials, were too complex for frequentists to address, but DuMouchel was a disciple of Smith and his hierarchical work with Lindley. Harris was not a statistician and did not care what method he used as long as it answered the question. Adopting hierarchical Bayes, they borrowed information from laboratory tests on mice, hamster embryo cells, and chemical substances. They even incorporated experts' opinions about the biological relevance of non-humans to humans and of cigarette to diesel smoke. Bayes let them account formally for their uncertainties about combining information across species.

Microcomputers were not widely available. Many of the researchers studying the new acquired immune deficiency syndrome (AIDS) epidemic, for example, were making statistical calculations by hand, and mathematical shortcuts were still being published for them. Harris programmed the diesel project in APL, a language used for matrix multiplications, and sent it via teletype to MIT's computer center. He drew illustrations on poster boards, added captions by pressing on wax letters, and arranged for an MIT photographer to take their pictures.

Thanks to mice and hamster studies, DuMouchel and Harris were able to conclude that even if light-duty diesel vehicles captured a 25% market share over 20 years, the risk of lung cancer would be negligible for the typical urban resident compared to the typical pack-a-day cigarette smoker. The smoker's risk was 420,000 times worse. Today, Bayesian meta-analyses are statistically old hat, but DuMouchel and Harris made Bayesians salivate for more big-data methods—and for the computing power to deal with them.

While lung cancer researchers explored Bayes, Adrian Raftery was working at Trinity College in Dublin on a well-known set of statistics about fatal coal-dust explosions in nineteenth-century British mines. Previous researchers had used frequency techniques to show that coal mining accident rates had changed over time. They assumed, however, that the change had been gradual. Raftery wanted to check whether it had been gradual or abrupt. First, he developed some heavy frequentist mathematics for analyzing the data. Then, out of curiosity, he experimented with Bayes' rule, comparing a variety of theoretical models to see which had the highest probability of determining when the accidents rates actually changed. "I found it very easy. I just solved it very, very quickly," Raftery recalled. And in doing so he discovered a remarkable, hitherto unknown event in British history. Raftery's Bayesian analysis revealed that accident rates plummeted suddenly in the late 1880s or early 1890s. A historian friend suggested why. In 1889, British miners had formed the militant Miners' Federation (which later became the National Union of Mine Workers). Safety was their number one issue. Almost overnight, coal mines got safer.

"It was a Eureka moment," Raftery said. "It was quite a thrill. And without Bayesian statistics, it would have been much harder to do a test of this hypothesis."[5] Frequency-based statistics worked well when one hypothesis was a special case of the other and both assumed gradual behavior. But when hypotheses were competing and neither was a special case of the other, frequentism was not as helpful, especially with data involving abrupt changes—like the formation of a militant union.

Raftery wound up publishing two papers in 1986 about modeling abrupt rate changes. His first, frequentist paper was long, dense, and virtually unread. His second, Bayesian paper was shorter, simpler, and had a much greater impact. Raftery's third 1986 paper ran just 1-1/4 pages and had an immediate effect on sociologists. The article appeared just as many sociologists were about to give up on frequentism's controversial p-values. A typical sociologist might work with data sets about thousands of individuals, each with hundreds of variables such as age, race, religion, climate, and family structure. Unfortunately, when researchers tried to determine the relevance of those variables using frequentist methods developed by Karl Pearson and R. A. Fisher for 50 to 200 cases, the results were often bizarre. Obscure effects became important, or went in opposite directions, or were disproved by later studies. By selecting a single model for large samples, frequentists ignored uncertainties about the model. Yet few social scientists could repeat their surveys or rerun experiments under precisely the same conditions. By the early 1980s many sociologists had concluded that, for testing hypotheses, their intuition was more accurate than frequentism.

Bayes, on the other hand, seemed to produce results that corresponded more closely to sociologists' intuition. Raftery told his colleagues, "The point is that we should be *comparing* the models, not just looking for possibly minor discrepancies between one of them and the data."[6] Researchers really want to know which of their models is more likely to be true, given the data. With Bayes, researchers could study sudden shifts from one stable form to another in biological growth phases, trade deficits and economic behavior, the abandonment and resettlement of archaeological sites, and clinical conditions such as rejection and recovery in organ transplantation and brain waves in Parkinson's disease. Bayesian hypothesis testing swept sociology and demography, and Raftery's short paper is still among the most cited in sociology.

Meanwhile, image processing and analysis had become critically important for the military, industrial automation, and medical diagnosis. Blurry, distorted, imperfect images were coming from military aircraft, infrared

sensors, ultrasound machines, positron emission tomography, magnetic resonance imaging (MRI) machines, electron micrographs, and astronomical
telescopes. All these images needed signal processing, noise removal, and
deblurring to make them recognizable. All were inverse problems ripe for
Bayesian analysis.

The first known attempt to use Bayes to process and restore images
involved nuclear weapons testing at Los Alamos National Laboratory. Bobby
R. Hunt suggested Bayes to the laboratory and used it in 1973 and 1974. The
work was classified, but during this period he and Harry C. Andrews wrote
a book, *Digital Image Restoration*, about the basic methodology; the laboratory
declassified the book and approved its publication in 1976. The U.S. Congress
retained Hunt in 1977 and 1978 to analyze images of the shooting of President
Kennedy. In his testimony, Hunt did not refer to Bayes. "Too technical for a
Congressional hearing," he said later.

At almost the same time that Hunt was working on image analysis for
the military, Julian Besag at the University of Durham in England was using
diseased tomato plants to study the spread of epidemics. Bayes helped him
discern local regularities and neighborly interactions among plants growing
in pixel-like lattice systems. Looking at one pixel, Besag realized he could
estimate the probability that its neighbor might share the same color, a useful
tool for image enhancement. But Besag was not a card-carrying Bayesian,
and his work went largely unnoticed at the time.

A group of researchers with Ulf Grenander at Brown University was
trying to design mathematical models for medical imaging by exploring
the effect one pixel could have on a few of its neighbors. The calculations
involved easily a million unknowns. Grenander thought that once Bayes was
embedded in a realistic problem, philosophical objections to it would fade.

Stuart Geman was attending Grenander's seminar in pattern theory, and
he and his brother Donald Geman tried restoring a blurry photograph of a
roadside sign. The Gemans were interested in noise reduction and in finding ways to capture and exploit regularities to sharpen the lines and edges
of unfocused images. Stuart had majored in physics as an undergraduate
and knew about Monte Carlo sampling techniques. So the Geman brothers
invented a variant of Monte Carlo that was particularly suited to imaging
problems with lots of pixels and lattices.

Sitting at a table in Paris, Donald Geman thought about naming their
system. A popular Mother's Day gift at the time was a Whitman's Sampler
assortment of chocolate bonbons; a diagram inside the box top identified

the filling hidden inside each candy. To Geman, the diagram was a matrix of unknown but enticing variables. "Let's call it Gibbs sampler," he said, after Josiah Willard Gibbs, a nineteenth-century American physicist who applied statistical methods to physical systems.[7]

The dots were starting to connect. But the Gemans, like Besag, operated in a small niche field, spatial statistics. And instead of nibbling at their problem a pixel at a time, the Gemans tried gobbling it whole. Working at pixel levels on a 64 x 64-cell fragment of a photo, they produced too many unknowns for computers of the day to digest. They wrote up their sampler in a formidably difficult paper and published it in 1984 in *IEEE Transactions on Pattern Analysis and Machine Intelligence*. Specialists in image processing, neural networks, and expert systems quickly adopted the method, which, with computers gaining more power every year, also sparked the interest of some statisticians. The brothers spent the next year racing around the globe giving invited talks.

Donald Geman used the Gibbs sampler to improve satellite images; Stuart used it for medical scans. Several years later, statisticians outside the small spatial imaging community began to realize that more general versions could be useful. The Gibbs sampler's flexibility and reliability would make it the most popular Monte Carlo algorithm. Still later the West learned that a Russian dissident mathematician, Valentin Fedorovich Turchin, had discovered the Gibbs sampler in 1971, but his work had been published in Russian-language journals, did not involve computers, and was overlooked.

By 1985 the old argument between Bayesians and frequentists was losing its polarizing zing, and Glenn Shafer of Rutgers University thought it had "calcified into a sterile, well-rehearsed argument." Persi Diaconis made a similar but nonetheless startling observation, one that no one familiar with the battles between Bayesians, Karl Pearson, Ronald Fisher, and Jerzy Neyman could have imagined. "It's nice that our field is so noncompetitive," Diaconis said. "If you take many other fields, like biology, people just slice each other up."[8]

Still, the conviction remained that without more powerful and accessible computers and without user-friendly and economical software, computing realistic problems with Bayes was impossible.

Lindley had been programming his own computers since 1965 and regarded Bayes as ideal for computing: "One just feeds in the axioms and the data and allows the computer to follow the laws of arithmetic." He called it "turning the Bayesian crank." But his student Smith saw something the older man did not: the key to making Bayes useful in the workplace would be computational ease, not more polished theory. Later Lindley wrote, "I consider it a major mistake of my professional life, not to have appreciated the need for computing rather than mathematical analysis."[9]

Ignoring the defensive posture of many statistics departments, Smith launched an offensive in a radically new direction. Smith's friends think of him as a lively, practical man with street smarts, people skills, and a can-do personality, the kind of person more comfortable in running shorts than academic garb. He was certainly willing to do the dirty work needed to make Bayes practical. He learned Italian to help translate de Finetti's two-volume *Theory of Probability* and then got it published. For the first time de Finetti's subjectivist approach was widely available to Anglo-American statisticians. Smith also developed filters, practical computer devices that would ease Bayesian computations enormously.

Next, Smith and three others—Lindley, José M. Bernardo, and Morris DeGroot—organized an international conference series for Bayesians in Valencia, Spain. It has been held regularly since 1979. Smith expected "the usual criticism from non-Bayesians in reaction to whatever I say." Sure enough, frequentists accused Bayesians of sectarian habits, meetings in remote locations, and mock cabarets featuring skits and songs with Bayesian themes. Other disciplines have done the same, of course. The conferences played a vital role in helping to build camaraderie in a small field under attack.

In 1984 Smith issued a manifesto—and italicized it for emphasis: "*Efficient numerical integration procedures are the key to more widespread use of Bayesian methods.*"[10] With computerized data collection and storage, hand analyses were becoming impossible. When microcomputers appeared, attached to fast networks with graphics and vast storage capabilities, data analysts could finally hope to improvise as easily as they had with pencil and paper. With characteristic practicality, Smith set his University of Nottingham students to work developing efficient, user-friendly software for Bayesian problems in spatial statistics and epidemiology.

Intrigued with Smith's projects, Alan Gelfand at the University of Connecticut asked Smith if he could spend his sabbatical year at Nottingham. When Gelfand arrived, Smith suggested they start something new. Gelfand recalled, "He gave me this Tanner and Wong paper, saying, 'This is kind of an interesting paper. There must be something more to it.'" Wing Hung Wong and Martin A. Tanner, at the Universities of Chicago and Wisconsin, respectively, were interested in spatial image analysis for identifying genetic linkages and for scanning the brain using positron emission tomography (PET). Wong had been adapting for Bayesians the EM algorithm, an iterative system secretly developed by the National Security Agency during the Second World War or the early Cold War. Arthur Dempster and his student Nan Laird at Harvard discovered EM independently a generation later and published it for civilian use in 1977. Like the Gibbs sampler, the EM algorithm worked iteratively to turn a small data sample into estimates likely to be true for an entire population.

Gelfand was studying the Wong paper when David Clayton from the University of Leicester dropped by and said, "Oh, the paper by Geman and Geman has something to do with it, I think." Clayton had written a technical report which, although never published, concerned Gibbs sampling. The minute Gelfand saw the Gemans' paper, the pieces came together: Bayes, Gibbs sampling, Markov chains, and iterations. A Markov chain can be composed of scores of links, and for each one a range of possible variables must be sampled and calculated one after another. Anyone studying a rare and subtle effect must compute each chain over and over again to get a big enough number to reveal the rarity. The numbers involved become enormous, and the length and tediousness of the calculations turned off most researchers.

But Gelfand and Smith saw that replacing difficult integration with sampling would be a wonderful computational tool for Bayesians. "It went back to the most basic things you learn in an introductory statistics course," Gelfand explained. "If you want to learn about a distribution or a population, you take samples from it. But don't sample directly." Imaging and spatial statisticians had been looking at local models as a whole, but Gelfand and Smith realized they should build long chains, series of observations generated one or two at a time, one after another. As Gelfand explained, "The trick was to look at simple distributions one at a time but never look at the whole. The value of each one depended only on the preceding value. Break the problem into tiny pieces that are easy to solve and then do millions of iterations. So you replace one high-dimensional draw with lots of low-dimensional draws that are easy. The technology was already in place. That's how you break the curse of high-dimensionality."[11]

Smith and Gelfand wrote their article as fast as they could. The elements of their system were already known, but their grand synthesis was new. Once others thought about it, they'd realize the importance of the method too.

When Smith spoke at a workshop in Quebec in June 1989, he showed that Markov chain Monte Carlo could be applied to almost any statistical problem. It was a revelation. Bayesians went into "shock induced by the sheer breadth of the method."[12] By replacing integration with Markov chains, they could finally, after 250 years, calculate realistic priors and likelihood functions and do the difficult calculations needed to get posterior probabilities.

To outsiders, one of the amazing aspects of Bayes' history is that physicists and statisticians had known about Markov chains for decades. To illustrate this puzzling lapse, some flashbacks are required. Monte Carlo began in 1906, when Andrei Andreyevich Markov, a Russian mathematician, invented Markov chains of variables. The calculations took so long, though, that Markov himself applied his chains only to the vowels and consonants in a Pushkin poem.

Thirty years later, with the beginning of nuclear physics in the 1930s, the Italian physicist Enrico Fermi was studying neutrons in collision reactions. Fermi fought insomnia by computing Markov chains in his head to describe the paths of neutrons in collision reactions. To the surprise of his colleagues the next morning, Fermi could predict their experimental results. With a small mechanical adding machine, Fermi built Markov chains to solve other problems too. Physicists called the chains statistical sampling.

Fermi did not publish his methods, and, according to Jack Good, government censorship kept Markov chains under tight wraps during the Second World War. After the war Fermi helped John von Neumann and Stanislaw Ulam develop the technique for hydrogen-bomb developers using the new ENIAC computer at the University of Pennsylvania. To calculate the critical mass of neutrons needed to make a thermonuclear explosion, Maria Goeppert Mayer, a future Nobel Prize–winning physicist, simulated the process with Markov chains, following one neutron at a time and making decisions at various places as to whether the neutron was most likely to get absorbed, escape, die, or fission. The calculation was too complicated for existing IBM equipment, and many thought the job was also beyond computers. But Mayer reported that it did "not strain the capacity of the Eniac."[13] In 1949 she spoke at a symposium organized by the National Bureau of Standards, Oak Ridge National Laboratory, and RAND for physicists to brief mathematicians and statisticians on hitherto classified applications.

That same year Nicholas Metropolis, who had named the algorithm Monte Carlo for Ulam's gambling uncle, described the method in general terms for statisticians in the prestigious *Journal of the American Statistical Association*. But he did not detail the algorithm's modern form until 1953, when his article appeared in the *Journal of Chemical Physics*, which is generally found only in physics and chemistry libraries. Moreover, he and his coauthors—two husband-and-wife teams, Arianna and Marshall Rosenbluth and Augusta and Edward Teller—were concerned strictly with particles moving around a square. They did not generalize the method for other applications. Thus it was physicists and chemists who pioneered Monte Carlo methods. Working on early computers with between 400 and 80,000 bytes of memory, they dealt with memory losses, unreadable tapes, failed vacuum tubes, and programming in assembly language. Once it took literally months to track down a small programming error. In the 1950s RAND developed a lecture series on Monte Carlo techniques and used them in a specially built simulation laboratory to test case after case of problems too complex for mathematical formulas.

During this difficult period, statisticians were advised several more times to use Monte Carlo either with or without computers. In 1954 two statisticians associated with the British Atomic Energy Research Establishment recommended that readers of the *Journal of the Royal Statistical Society* consult "Poor Man's Monte Carlo" for pen-and-paper calculations; John M. Hammersley and Keith W. Morton said Monte Carlo was as easy as "simple knitting." Lindley described Markov chains in 1965 in a text for college students.

In a particularly poignant case, W. Keith Hastings, a mathematician at the University of Toronto, was approached by chemists who were studying 100 particles interacting with each other while subject to outside forces. Because of the 600 variables in the case, Hastings said he immediately realized the importance of Markov chains for mainstream statistics and devoted all his time to them: "I was excited because the underlying idea goes way back to Metropolis. As soon as I realized it, it was off to the races. I just had to work on it. I had no choice." In 1970 he published a paper generalizing Metropolis's algorithm in the statistical journal *Biometrika*. Again, Bayesians ignored the paper. Today, computers routinely use the Hastings–Metropolis algorithm to work on problems involving more than 500,000 hypotheses and thousands of parallel inference problems.

Hastings was 20 years ahead of his time. Had he published his paper when powerful computers were widely available, his career would have been very different. As he recalled, "A lot of statisticians were not oriented toward computing. They take these theoretical courses, crank out theoretical papers, and some of them want an exact answer."[14] The Hastings–Metropolis algorithm provides estimates, not precise numbers. Hastings dropped out of research and settled at the University of Victoria in British Columbia in 1971. He learned about the importance of his work after his retirement in 1992.

Why did it take statisticians so long to understand the implications of an old method? And why were Gelfand and Smith first? "The best thing I can say for us is that we just sort of stumbled on it. We got lucky," Gelfand says today. "It was just sort of sitting there, waiting for people to put the pieces together."

Timing was important too. Gelfand and Smith published their synthesis just as cheap, high-speed desktop computers finally became powerful enough to house large software packages that could explore relationships between different variables. Bayes was beginning to look like a theory in want of a computer. The computations that had irritated Laplace in the 1780s and that frequentists avoided with their variable-scarce data sets seemed to be the problem—not the theory itself.

Yet Smith and Gelfand still thought of Monte Carlo as a last resort to be used in desperation for complicated cases. They wrote diffidently, careful to use the B-word only five times in 12 pages. "There was always some concern about using the B-word, a natural defensiveness on the part of Bayesians in terms of rocking the boat," Gelfand said. "We were always an oppressed minority, trying to get some recognition. And even if we thought we were doing things the right way, we were only a small component of the statistical community and we didn't have much outreach into the scientific community."[15]

The Gelfand–Smith paper was an "epiphany in the world of statistics," as Bayesians Christian P. Robert and George Casella reported. And just in case anyone missed their point, they added: "Definition: epiphany n. A spiritual event . . . a sudden flash of recognition." Years later, they still described its impact in terms of "sparks," "flash," "shock," "impact," and "explosion."[16]

Shedding their diffidence, Gelfand and Smith wrote a second paper six months later with Susan E. Hills and Amy Racine-Poon. This time they punctuated their mathematics exuberantly with words like "surprising," "universality," "versatility," and "trivially implemented." They concluded grandly, "The potential of the methodology is enormous, rendering straightforward the analysis of a number of problems hitherto regarded as intractable from a Bayesian perspective."[17] Luke Tierney at Carnegie Mellon tied the technique to Metropolis's method, and the entire process—which physicists had called Monte Carlo—was baptized anew as Markov chain Monte Carlo, or MCMC for short. The combination of Bayes and MCMC has been called "arguably the most powerful mechanism ever created for processing data and knowledge."[18]

When Gelfand and Smith gave an MCMC workshop at Ohio State University early in 1991, they were astonished when almost 80 scientists appeared.

They weren't statisticians, but they had been using Monte Carlo in archaeology, genetics, economics, and other subjects for years.

The next five years raced by in a frenzy of excitement. Problems that had been nightmares cracked open as easily as eggs for an omelet. A dozen years earlier the conference title "Practical Bayesian Statistics" had been a joke. But after 1990 Bayesian statisticians could study data sets in genomics or climatology and make models far bigger than physicists could ever have imagined when they first developed Monte Carlo methods. For the first time, Bayesians did not have to oversimplify "toy" assumptions.

Over the next decade, the most heavily cited paper in the mathematical sciences was a study of practical Bayesian applications in genetics, sports, ecology, sociology, and psychology. The number of publications using MCMC increased exponentially.

Almost instantaneously MCMC and Gibbs sampling changed statisticians' entire method of attacking problems. In the words of Thomas Kuhn, it was a paradigm shift.[19] MCMC solved real problems, used computer algorithms instead of theorems, and led statisticians and scientists into a world where "exact" meant "simulated" and repetitive computer operations replaced mathematical equations. It was a quantum leap in statistics.

When Smith and Gelfand published their paper, frequentists could do a vast amount more than Bayesians. But within years Bayesians could do more than frequentists. In the excitement that followed, Stanford statistician Jun S. Liu, working with biologist Charles E. Lawrence, showed genome analysts that Bayes and MCMC could reveal motifs in protein and DNA. The international project to sequence the human genome, launched in 1990, was generating enormous amounts of data. Liu showed how a few seconds on a workstation programmed for Bayes and iterative MCMC sampling could detect subtle, closely related patterns in protein and nucleic acid sequences. Then he and Lawrence could infer critical missing data pointing to common ancestors, structures, and functions. Soon genomics and computational biology were so overrun with researchers that Gelfand decided to look elsewhere for another research project.

Between 1995 and 2000 Bayesians developed particle filters like the Kalman filter and real-time applications in finance, image analysis, signal processing, and artificial intelligence. The number of attendees at Bayesian conferences in Valencia quadrupled in 20 years. In 1993, more than two centuries after his death, Thomas Bayes finally joined his clerical relatives in the *Dictionary of National Biography*.

Amid the Bayesian community's frenzy over MCMC and Gibbs sampling, a generic software program moved Bayesian ideas out into the scientific and computer world.

In an example of serendipity, two groups 80 miles apart worked independently during the late 1980s on different aspects of the same problem. While Smith and Gelfand were developing the theory for MCMC in Nottingham, Smith's student, David Spiegelhalter, was working in Cambridge at the Medical Research Council's biostatistics unit. He had a rather different point of view about using Bayes for computer simulations. Statisticians had never considered producing software for others to be part of their jobs. But Spiegelhalter, influenced by computer science and artificial intelligence, decided it was part of his. In 1989 he started developing a generic software program for anyone who wanted to use graphical models for simulations. Once again, Clayton was an important influence. Spiegelhalter unveiled his free, off-the-shelf BUGS program (short for Bayesian Statistics Using Gibbs Sampling) in 1991.

BUGS caused the biggest single jump in Bayesian popularity. It is still the most popular software for Bayesian analyses, and it spread Bayesian methods around the world.

"It wasn't a very big project," Spiegelhalter admits. "It was a staggeringly basic and powerful idea, relating Gibbs sampling to a graph to write generic programs."[20] Its simple code remains almost exactly the same today as it was in 1991.

Ecologists, sociologists, and geologists quickly adopted BUGS and its variants, WinBUGS for Microsoft users, LinBUGS for Linux, and OpenBUGS. But computer science, machine learning, and artificial intelligence also joyfully swallowed up BUGS. Since then it has been applied to disease mapping, pharmacometrics, ecology, health economics, genetics, archaeology, psychometrics, coastal engineering, educational performance, behavioral studies, econometrics, automated music transcription, sports modeling, fisheries stock assessment, and actuarial science. A Bayesian visiting a marine laboratory was surprised to discover all its scientists using BUGS with nary a statistician in sight.

Medical research and diagnostic testing were among the earliest beneficiaries of Bayes' new popularity. Just as the MCMC frenzy appeared to be moderating, Peter Green of Bristol University showed Bayesians how to compare the elaborate hypotheses that scientists call models. Before 1996 anyone making a prediction about the risk of a stroke had to focus on one model at a time. Green showed how to jump between models without spending an infinite time on each. Previous studies had identified 10 possible factors involved in strokes. Green identified the top four: systolic blood pressure, exercise, diabetes, and daily aspirin.

Medical testing, in particular, benefited from Bayesian analysis. Many medical tests involve imaging, and Larry Bretthorst, a student of the Bayesian physicist Ed Jaynes, improved nuclear magnetic resonance, or NMR, signal detection by several orders of magnitude in 1990. Bretthorst had studied imaging problems to improve the detection of radar signals for the Army Missile Command.

In 1991 a public frightened about the AIDS epidemic demanded universal screening for human immunodeficiency virus (HIV). Biostatisticians quickly used Bayes to demonstrate that screening the entire population for a rare disease would be counterproductive. Wesley O. Johnson and Joseph L. Gastwirth showed that a sensitive test like that for HIV virus would tell many patients they were infected with HIV when in fact they were not. The media publicized several suicides of people who had received a positive HIV test result but not realized they did not necessarily have the virus. Scaring healthy people and retesting them with more sophisticated procedures would have been extremely costly.

In much the same way, but more controversially, a Bayesian approach showed that an expensive MRI test for breast cancer might be appropriate for a woman whose family had many breast cancer patients but inappropriate for giving every woman between 40 and 50 years of age. A woman who has a mammogram every year for 10 years can be almost 100% sure of getting one false positive test result, and the resulting biopsy can cost $1,000 to $2,000. In the case of prostate cancer, the screening test for high blood levels of prostate-specific antigen (PSA) is highly accurate when it comes to identifying men with the cancer. Yet the disease is so rare that almost everyone who gets a positive test result is found not to have the cancer at all. (See appendix B for how to calculate a Bayesian problem involving breast cancer.)

On the other hand, Bayes also showed that people with negative test results for breast and prostate cancer cannot feel carefree. The PSA test is so insensitive that good news provides almost no assurance that a man does not actually have prostate cancer. The same is true to a lesser extent of mammography: its sensitivity is about 85 to 90%, meaning that a woman who finds a lump only a few months after getting a negative mammogram should still see a doctor immediately. A strict Bayesian gives patients their probabilities for cancer instead of a categorical yes or no.

Because genetics involves extremely rare diseases, imperfect tests, and complicated problems where tiny errors in the data or calculations can affect decisions, Bayesian probabilities are expected to become increasingly important for diagnostic test assessment.

Spiegelhalter spent more than 10 years trying to sell the medical community on BUGS as the mathematical way to learn from experience. He argued that "advances in health-care typically happen through incremental gains in knowledge rather than paradigm-shifting breakthroughs, and so this domain appears particularly amenable to a Bayesian perspective." He contended that "standard statistical methods are designed for summarizing the evidence from single studies or pooling evidence from similar studies, and have difficulties dealing with the pervading complexity of multiple sources of evidence."[21] While frequentists can ask only certain questions, a Bayesian can frame any question.

With the introduction of high-performance workstations in the 1980s it became possible to use Bayesian networks to handle medicine's many interdependent variables, such as the fact that a patient with a high temperature will usually also have an elevated white blood count. Bayesian networks are graphs of nodes with links revealing cause-and-effect relationships. The "nets" search for particular patterns, assign probabilities to parts of the pattern, and update those probabilities using Bayes' theorem. A number of people helped develop Bayesian networks, which were formalized in 1988 in a book by Judea Pearl, a computer scientist at UCLA. By treating cause and effect as a quantifiable Bayesian belief, Pearl helped revive the field of artificial intelligence.

Ron Howard, who had become interested in Bayes while at Harvard, was working on Bayesian networks in Stanford's economic engineering department. A medical student, David E. Heckerman, became interested too and for his Ph.D. dissertation wrote a program to help pathologists diagnose lymph node diseases. Computerized diagnostics had been tried but abandoned decades earlier. Heckerman's Ph.D. in bioinformatics concerned medicine, but his software won a prestigious national award in 1990 from the Association for Computing Machinery, the professional organization for computing. Two years later, Heckerman went to Microsoft to work on Bayesian networks.

The Federal Drug Administration (FDA) allows the manufacturers of medical devices to use Bayes in their final applications for FDA approval. Devices include almost any medical item that is not a drug or biological product, items such as latex gloves, intraocular lenses, breast implants, thermometers, home AIDS kits, and artificial hips and hearts. Because they are usually applied locally and improved a step at a time, new models of a device should come equipped with objective prior information.

Pharmaceuticals are different. Unlike devices, pharmaceuticals are generally one-step, systemic discoveries so that potentially an industry could subjectively bias Bayes' prior hunch. Thus the FDA has long resisted pressure from pharmaceutical companies that want to use Bayes when applying for approval to sell a drug in the United States.

According to Spiegelhalter, however, the same battle seems to have subsided in England. Drug companies use WinBUGS extensively when submitting their pharmaceuticals for reimbursement by the English National Health Service. The process is, in Spiegelhalter's words, "very Bayesian without using the B-word" because it uses judgmental priors about a drug's cost effectiveness. International guidelines also allow Bayesian drug applications, but these guidelines are widely considered too vague to be effective.

Outside of diagnostic and medical device testing, Bayes' mathematical procedures have had little impact on basic clinical research or practice. Working doctors have always practiced an intuitive, nonmathematical form of Bayes for diagnosing patients. The biggest unknown in medicine, after all, is the question, What is causing the patient's symptoms? But traditional textbooks were organized by disease. They said that someone with, for instance, measles probably has red spots. But the doctor with a speckled patient wanted to know the inverse: the probability that the patient with red spots has measles. Simple Bayesian problems—for example, What is the probability that an exercise echocardiogram will predict heart disease?—started appearing on physicians' licensing examinations in 1992.

One of the few times physicians make even rough Bayesian calculations occurs when a patient has symptoms that could involve a life-threatening heart attack, a deep venous thrombosis, or a pulmonary embolism. To assess the danger, a physician assigns points for each of the patient's risk factors and adds the points. In the heart attack algorithm, the score determines the probability that within the next two weeks the patient will die, have a heart attack, or need coronary arteries opened. The points for thrombosis and embolism tell whether a patient has a low, medium, or high risk of developing a clot and which test can best produce a diagnosis. It is expected that software will be available soon to automatically tell doctors and patients the effect of a particular test result on a diagnosis.

Outside of medicine, endangered populations of ocean fish, whales, and other mammals were among the first to benefit from Bayes' new computational heft. Despite the U.S. Marine Mammal Protection Act of 1972, only a few visible and highly publicized species of whales, dolphins, and other marine mammals had been protected. Some exploited populations, including several whale species in the Antarctic, collapsed while being "managed." Having strong, abundant information about a species, frequentists and Bayesians could reach similar decisions, but when evidence was weak—as it often is in the case of marine mammals—only Bayes incorporated uncertainties about the data at hand and showed clearly when more information was needed.

Most whale populations rebounded during the 1980s, but in 1993 two government biologists, Barbara L. Taylor and Timothy Gerrodette, wrote, "At least part of the blame for the spectacular [past] overexploitation of the great whales can be placed on scientists being unable to agree . . . [on a] clear way to treat uncertainty. . . . In certain circumstances, a population might go extinct before a significant decline could be detected."[22] During the administration of Bill Clinton, the Wildlife Protection Act was amended to accept Bayesian analyses alerting conservationists early to the need for more data.

Scientists advising the International Whaling Commission were particularly worried about the uncertainty of their measurements. Each year the commission establishes the number of endangered bowhead whales Eskimos can hunt in Arctic seas. To ensure the long-term survival of the bowheads, scientists compute 2 numbers each year: the number of bowheads and their rate of increase. The whales, perhaps the longest-lived mammals on Earth, can grow to more than 60 feet in length, weigh more than 60 tons, and eat 2 tons of food a day. They spend only about 5% of their time on the ocean surface because they can submerge for 30 minutes at a time, and they use their enormous heads to ram through ice when they need to surface for air. In spring, teams of scientists stood on tall perches to spot bowheads rounding Point Barrow, Alaska, on their annual migration into the western Arctic. The count was fraught with uncertainties.

Scientists representing an entire spectrum of opinion, from Greenpeace to whaling nations, worried that a lack of trustworthy data on bowhead populations was opening the species to too much risk. During a weeklong meeting to discuss the problem in 1991, their chair asked, "What can we do?"[23] There was complete silence. The scientists were the world's leading bowhead experts, but none of them could answer the question.

When Judith Zeh, the committee chair, got back to the department of statistics at the University of Washington in Seattle, she talked with Raftery, who had recently moved there from Dublin. Not surprisingly, after his experience analyzing coal mining accidents, Raftery thought Bayes might help. Using it, the committee could assign uncertainties to all their data and augment visual sightings with recordings of whales vocalizing near underwater hydrophones.

Providentially, the spring of 1993 was a rewarding year for bowhead counting, and sightings plus vocalizing showed that the whales were almost assuredly increasing at a healthy rate. Their recovery indicated that protecting other great whale populations from commercial whaling might help them recover too.

The entire process—involving rival Bayesian and frequentist methods and whaling factions that often profoundly disagreed—could have been wildly contentious. But times were changing. Pragmatism ruled. Making full-scale Bayesian analyses to combine visual and acoustical data was expensive, and thus, because they confirmed previous frequentist studies, they were discontinued. Raftery moved on to using Bayes for 48-hour weather forecasting.

Other wildlife researchers picked up the Bayesian banner. When Paul R. Wade decided in 1988 to use Bayes for his Ph.D. thesis, he said, "I was off in this small area of marine mammal biology but I felt as if I were in the center of a revolution in science." Ten years later, at the National Oceanic and Atmospheric Administration, he was comparing frequentist and Bayesian analyses of a small, isolated population of 200 or 300 beluga whales in the Arctic and Sub-Arctic waters of Cook Inlet, Alaska. The legal take by native hunters was roughly 87 whales a year. Frequentist methods would have required seven years of data collection to assess whether this catch was sustainable. With Bayes, five years of data showed that the beluga population was almost certainly declining substantially, and the experiment could stop. "With a small population, even a two-year delay can be important," Wade said.[24] In May 1999 a hunting moratorium went into effect for the Cook Inlet belugas.

Meanwhile, a committee of the National Research Council in the National Academy of Sciences strongly recommended the aggressive use of Bayesian methods to improve estimates of marine fish stocks too. Committee members emphasized in 1998 that, because the oceans are vast and opaque, wildlife managers need realistic measurements of the uncertainties in their observations and models. Otherwise, policymakers cannot gauge potential risks to wildlife. Today many fisheries journals demand Bayesian analyses.

Lindley had predicted that the twenty-first century would be a Bayesian era because the superior logic of Bayes' rule would swamp frequency-based methods. David Blackwell at Berkeley disagreed, saying, "If the Bayesian approach does grow in the statistical world, it will not be because of the influence of other statisticians but because of the influence of actuaries, engineers, business people, and others who actually like the Bayesian approach and use it."[25] It appeared that Blackwell was right: pragmatism could drive a paradigm shift. Philosophies of science had not changed. The difference was that Bayes finally worked.

Diaconis had been wondering for years, "When is our time?" In 1997, he decided, "Our time is now."[26]

Smith became the first Bayesian president of the Royal Statistical Society in 1995. Three years later he stunned his friends by quitting statistics to become an administrator of the University of London. A proponent of evidence-based medicine, he wanted to help develop evidence-based public policy too. Dismayed colleagues chastised him for abandoning Bayes' rule. But Smith told Lindley that all the problems of statistics had been solved. We have the paradigm, he said, and with MCMC we know how to implement it. He told Diaconis that there was nothing else to do with statistical problems but to plug them into a computer and turn the Bayesian crank.

In 2008, when Smith became scientific adviser to the United Kingdom's Minister of Innovation, Universities, and Skills, a Royal Society spokesman volunteered that three statisticians have become prime ministers of Great Britain.[27]

17.

rosetta stones

Two and a half centuries after Bayes and Laplace discovered a way to apply mathematical reasoning to highly uncertain situations, their method has taken wing, soaring through science and the Internet, burrowing into our daily lives, dissolving language barriers, and perhaps even explaining our brains. Gone are the days when a few driven individuals searched orphanages and coded messages for data and organized armies of women and students to make tedious calculations. Today's Bayesians revel in vast archives of Internet data, off-the-shelf software, tools like MCMC, and computing power so cheap it is basically free.

The battle between Bayesian and frequentist forces has cooled. Bayesianism as an all-encompassing framework has been replaced by utilitarian applications and computation. Computer scientists who joined the Bayesian community cared about results, not theory or philosophy. And even theorists who once insisted on adhering strictly to fundamental principles now accept John Tukey's view from the 1950s: "Far better an approximate answer to the right question, . . . than an exact answer to the wrong question." Researchers adopt the approach that best fits their needs.

In this ecumenical atmosphere, two longtime opponents—Bayes' rule and Fisher's likelihood approach—ended their cold war and, in a grand synthesis, supported a revolution in modeling. Many of the newer practical applications of statistical methods are the results of this truce.

As a collection of computational and statistical machinery, Bayes is still driven by Bayes' rule. The word "Bayes" still entails the idea, shared by de Finetti, Ramsey, Savage, and Lindley, that probability is a measure of belief and that it can, as Lindley put it, "escape from repetition to uniqueness." That said, most modern Bayesians accept that the frequentism of Fisher, Neyman, and Egon Pearson is still effective for most statistical problems: for simple and standard analyses, for checking how well a hypothesis fits data, and as the foundation of many modern technologies in areas such as machine learning.

Prominent frequentists have also moderated their positions. Bradley Efron, a National Medal of Science recipient who wrote a classic defense of frequentism in 1986, recently told a blogger, "I've always been a Bayesian." Efron, who helped develop empirical Bayesian procedures while remaining a committed frequentist, told me that Bayes is "one of the great branches of statistical inference. . . . Bayesians have gotten more tolerant these days, and frequentists are seeing the need to use Bayesian kinds of reasoning, so maybe we are headed for some kind of convergence."

Bayes' rule is influential in ways its pioneers could never have envisioned. "Neither Bayes nor Laplace," Robert E. Kass of Carnegie Mellon observed, "recognized a fundamental consequence of their approach, that the accumulation of data makes open-minded observers come to agreement and converge on the truth. Harold Jeffreys, the modern founder of Bayesian inference for scientific investigation, did not appreciate its importance for decision-making. And the loyalists of the 1960s and 1970s failed to realize that Bayes would ultimately be accepted, not because of its superior logic, but because probability models are so marvelously adept at mimicking the variation in real-world data."

Bayes has also broadened to the point where it overlaps computer science, machine learning, and artificial intelligence. It is empowered by techniques developed both by Bayesian enthusiasts during their decades in exile and by agnostics from the recent computer revolution. It allows its users to assess uncertainties when hundreds or thousands of theoretical models are considered; combine imperfect evidence from multiple sources and make compromises between models and data; deal with computationally intensive data analysis and machine learning; and, as if by magic, find patterns or systematic structures deeply hidden within a welter of observations. It has spread far beyond the confines of mathematics and statistics into high finance, astronomy, physics, genetics, imaging and robotics, the military and antiterrorism, Internet communication and commerce, speech recognition, and machine translation. It has even become a guide to new theories about learning and a metaphor for the workings of the human brain.

One of the surprises is that Bayes, as a buzzword, has become chic.

Stanford University biologist Stephen H. Schneider wanted a customized cancer treatment, called his logic Bayesian, got his therapy, went into remission, and wrote a book about the experience. Stephen D. Unwin invented a personal "faith-belief factor" of 28% to boost the 67% "Bayesian probability" that God exists to 95%, and his book hit the bestseller list. A fashionable expression, "We're all Bayesians now," plays on comments made years ago by Milton Friedman and President Richard Nixon that "We're all Keynesians now." And the CIA agent in a Robert Ludlum thriller tells the hero, "Lucky? Obviously you haven't heard anything I've said. It was a matter of applying Bayes' Theorem to estimate the conditional probabilities. Giving due weight to the prior probabilities and . . ."[1]

It must be conceded that not everyone shares this enthusiasm. Some important fields of endeavor remain opposed. Perhaps the biggest irony is that partisan politics have kept the American census anti-Bayesian, despite Laplace's vision that enlightened governments would adopt it.

Anglo-American courtrooms are also still largely closed to Bayes. Among the few exceptions was a case in 1994 where Bayes was used to demonstrate that New Jersey state troopers singled out African American drivers for traffic stops. During a rape trial in the 1990s, British lawyers tried teaching judges and juries how to assess evidence using Bayesian probability; judges concluded that the method "plunges the jury into inappropriate and unnecessary realms of theory and complexity."[2] Forensic laboratory science in Great Britain and Europe is a different story. Unlike the FBI Laboratory in the United States, the Forensic Sciences Services in Britain have followed Lindley's advice and now employ Bayesian methods extensively to assess physical evidence. Continental European laboratories have developed a quantitative measure for the value of various types of evidence, much as Turing and Shannon used Bayes to develop bans and bits as units of measurement for cryptography and computers. Bayes—tactfully referred to in forensic circles as the "logical" or "likelihood ratio" approach—has been applied successfully in cases where numbers were available, particularly in DNA profiling. Because DNA databanks involve probabilities about almost unimaginably tiny numbers— one in 20 million, say, or one in a billion—they may eventually open more courtroom doors to Bayesian methods.

Bayes made headlines in 2000 by augmenting DNA evidence with statistical data to conclude that Thomas Jefferson had almost certainly fathered six children by his slave Sally Hemings. DNA evidence from Jefferson's and Hemings's families had already offered strong evidence that the third president and the author of the Declaration of Independence fathered Hemings's youngest son. But Fraser D. Neiman, the director of archaeology at Jefferson's Monticello plantation, studied whether Hemings's other conceptions fell during or close to one of Jefferson's sporadic visits to Monticello. Then he used Bayes to combine the prior historical testimony and DNA evidence with probable hypotheses based on Jefferson's calendar. Assuming a 50–50 probability that the prior evidence was true, Fraiser concluded it was nearly certain—99% probable—that Jefferson had fathered Hemings's six children.

In economics and finance Bayes appears at multiple levels, ranging from theoretical mathematics and philosophy to nitty-gritty moneymaking. The method figured prominently in three Nobel Prizes awarded for theoretical economics, in 1990, 1994, and 2004. The first Nobel involved the Italian Bayesian de Finetti, who anticipated the Nobel Prize–winning work of Harry Markowitz by more than a decade. Mathematical game theorists John C. Harsanyi and John Nash (the latter the subject of a book and movie, *A Beautiful Mind*) shared a Bayesian Nobel in 1994. Harsanyi often used Bayes to study competitive situations where people have incomplete or uncertain information about each other or about the rules. Harsanyi also showed that Nash's equilibrium for games with incomplete or imperfect information was a form of Bayes' rule.

In 2002 Bayes won perhaps not an entire Nobel Prize but certainly part of one. Psychologists Amos Tversky, who died before the prize was awarded, and Daniel Kahneman showed that people do not make decisions according to rational Bayesian procedures. People answer survey questions depending on their phrasing, and physicians choose surgery or radiation for cancer patients depending on whether the treatments are described in terms of mortality or survival rates. Although Tversky was widely regarded as a philosophical Bayesian, he reported his results using frequentist methods. When James O. Berger of Duke asked him why, Tversky said it was simply a matter of expedience. During the 1970s it was more difficult to publish Bayesian research. "He just took the easy way out," Berger said.

Alan Greenspan, former chairman of the Federal Reserve, said he used Bayesian ideas to assess risk in monetary policy. "In essence, the risk-management approach to monetary policymaking is an application of Bayesian decision-making," Greenspan told the American Economic Association in 2004.[3] The audience of academic and government economists gasped; few experts in finance analyze empirical data with Bayes.

Economists were still catching their breaths when Martin Feldstein, professor of economics at Harvard, stood up at the same meeting and delivered a crash course in Bayesian theory. Feldstein had been President Ronald Reagan's chief economic advisor and was president of the National Bureau of Economic Research, a leading research organization. He learned Bayesian theory at the Howard Raiffa–Robert Schlaifer seminars at Harvard Business School in the 1960s. Feldstein explained that Bayes lets the Federal Reserve weigh a low-probability risk of disaster more heavily than a higher-probability risk that would cause little damage. And he likened Bayes to a man who has to decide whether to carry an umbrella even when the probability of rain is low. If he carries an umbrella but it does not rain, he is inconvenienced. But if he does not carry an umbrella and it pours, he will be drenched. "A good Bayesian," Feldstein concluded, "finds himself carrying an umbrella on many days when it does not rain."[4]

Four years later rain flooded the financial markets and banking. Greenspan, who by then had retired from the Federal Reserve, told Congress he had not foreseen the collapse of the real-estate lending bubble in 2008. He did not blame the theory he used but his economic data, which "generally covered only the past two decades, a period of euphoria . . . [instead of] historic periods of stress."[5]

But did Greenspan actually employ Bayesian statistics to quantify empirical economic data? Or were Bayesian concepts about uncertainty only a handy metaphor? Former Reserve Board governor Alan S. Blinder of Princeton thought the latter, and when he said so during a talk, Greenspan was in the audience and did not object.

In pragmatic contrast to abstract Bayes at the Nobel ceremonies and philosophical Bayes at the Federal Reserve, the rule stands behind one of the most successful hedge funds in the United States. In 1993 Renaissance Technologies hired away from IBM a Bayesian group of voice recognition researchers led by Peter F. Brown and Robert L. Mercer. They became comanagers of RenTech's portfolio and technical trading. For several years, their Medallion Fund, limited to former and current employees, averaged annual returns of about 35%. The fund bought and sold shares so rapidly one day in 1997 that it accounted for more than 10% of all NASDAQ trades.

To search for the nonrandom patterns and movements that will help predict markets, RenTech gathers as much information as possible. It begins with prior knowledge about the history of prices and how they fluctuate and correlate with each other. Then the company continuously updates that prior base. As Mercer explained, "RenTec gets a trillion bytes of data a day, from newspapers, AP wire, all the trades, quotes, weather reports, energy reports, government reports, all with the goal of trying to figure out what's going to be the price of something or other at every point in the future. . . . We want to know in three seconds, three days, three weeks, three months. . . . The information we have today is a garbled version of what the price is going to be next week. People don't really grasp how noisy the market is. It's very hard to find information, but it is there, and in some cases it's been there for a long, long time. It's very close to science's needle-in-a-haystack problem."

Astronomers, physicists, and geneticists also use Bayes to discern elusive phenomena almost drowning in unknowns. A scientist may face hundreds of thousands of variables without knowing which produce the best predictions. Bayes lets them estimate the most probable values of their unknowns.

When Supernova 1987A exploded, astronomers detected precisely 18 neutrinos. The particles originated from deep inside the star and were the only clues about its interior, so the astronomers wanted to extract as much information as possible from this minuscule amount of data. Tom Loredo, a graduate student at the University of Chicago, was told to see what he could learn. Because the supernova was a one-of-a-kind opportunity, frequency-based methods did not apply. Loredo began reading papers by Lindley, Jim Berger, and other leading Bayesians and discovered that Bayes would let him compare various hypotheses about his observations and choose the most probable. His Ph.D. thesis from 1990 wound up introducing modern Bayesian methods to astronomy.

Since then, Bayes has found a comfortable niche in high-energy astrophysics, x-ray astronomy, gamma ray astronomy, cosmic ray astronomy, neutrino astrophysics, and image analysis. In physics, Bayes is hunting for elusive neutrinos, Higgs-Boson particles, and top quarks. All these problems deal with needles in haystacks, and Loredo now uses Bayes at Cornell University in a new field, astrostatistics.

In much the same way, biologists who study genetic variation are limited to tiny snippets of information almost lost among huge amounts of meaningless and highly variable data in the chromosomes. Computational biologists searching for genetic patterns, leitmotifs, markers, and disease-causing misspellings must extract the weak but important signals from the deafening background noise that masquerades as information.

Susan Holmes, a professor in Stanford's statistics department, works in computational and molecular biology on amino acids. Some are extremely rare, and if she used frequentist methods she would have to quantify them with a zero. Adopting the cryptographic technique used by Turing and Good at Bletchley Park, she tries to crack the genetic code by assigning missing species a small possibility.

Given that the DNA in every biological cell contains complete instructions for making every kind of protein in the body, what differentiates a kidney cell from a brain cell? The answer depends on whether a particular gene is turned on or off and whether the genes work together or not. Holmes assembles huge microarrays of genetic data filled with noise and other distractions that may hide a few important signals from the turned-on genes. Each microarray consists of many genes arrayed in a regular pattern on a small glass slide or membrane; with it, she can analyze the expression of thousands of genes at once.

"It's very tenuous," she says. "[Imagine that] you have a city at night like Toronto or Paris with a very dense population and lots of buildings, and at 2 a.m. you look at which lights are lit up in all the buildings. Then at 3 and 4 a.m., you look again. So you develop a pattern of which rooms are lit up, and from that you infer who in the city knows who. That's how sparse the signal is and how far you have to jump, to see which genes are working together. You don't even have phone connections. But the image of something lighting up is a little bit like the image of microarrays. Microarrays have so much noise, it seems crazy. You just have rustles, whispers of signals, and then lots of noise. You spend a lot of time looking at a lot of data." Because prior information is needed to assemble the networks, many microarrays are analyzed using Bayesian methods.

Daphne Koller, a leader in artificial intelligence and computational biology at Stanford, also works on microarrays. She wanted to see not only which genes have turned on or off, but also what controls and regulates them. By looking at the activity levels of genes in yeast, she figured out how they are regulated. Then she switched to mouse and human cells to determine the differences in genetic regulation between healthy people and patients with cancer or Type II diabetes, particularly the metabolic (insulin resistance) syndrome.

On the vexed issue of priors, Koller considers herself a relaxed middle-of-the-roader. In contrast, Bayesian purists like Michael I. Jordan of Berkeley and Philip Dawid of Cambridge object to the term "Bayesian networks"; they regard Judea Pearl's nomenclature as a misnomer because Bayesian networks do not always have priors and Bayes without priors is not Bayes. But Koller insists that her networks fully qualify as Bayesian because she carefully constructs priors for their variables.

Koller's fascination with uncertainty has led her from genetics to imaging and robotics. Images typically have variable and ambiguous features that are embedded in background clutter. The human visual system sends ten million signals per second to the brain, where a billion neurons strip off random fluctuations and irrelevant, ambiguous information to reveal shape, color, texture, shading, surface reflections, roughness, and other features. As a result, human beings can look at a blurry, distorted, noisy pattern and instantly recognize a tomato plant, a car, or a sheep. Yet a state-of-the-art computer trained to recognize cars and sheep may picture only nonsensical rectangles. The difference is that the human brain has prior knowledge to integrate with the new images.

"It's mind-boggling," says Koller. The problem is not computer hardware; it is writing the software. "A computer can easily be trained to distinguish a desert from a forest, but where the road is and where it's about to fall off a cliff, that's much harder."

To explore such imaging problems, Sebastian Thrun of Stanford built a driverless car named Stanley. The Defense Advanced Research Projects Agency (DARPA) staged a contest with a $2 million prize for the best driverless car; the military wants to employ robots instead of manned vehicles in combat. In a watershed for robotics, Stanley won the competition in 2005 by crossing 132 miles of Nevada desert in seven hours.

While Stanley cruised along at 35 mph, its camera took images of the route and its computer estimated the probability of various obstacles. As the robot navigated sharp turns and cliffs and generally stayed on course, its computer could estimate with 90% probability that a wall stood nearby and with a 10% probability that a deep ditch was adjacent. In the unlikely event that Stanley fell into the ditch, it would probably have been destroyed. Therefore, like the Bayesian economist who carries an umbrella on sunny days, Stanley slowed down to avoid even unlikely catastrophes. Thrun's artificial intelligence team trained Stanley's sensors, machine-learning algorithms, and custom-written software in desert and mountain passes.

Thrun credited Stanley's victory to Kalman filters. "Every bolt of that car was Bayesian," Diaconis said proudly. After the race, Stanley retired in glory to its own room in the Smithsonian National Museum of American History in Washington.

The next year a Bayesian team from Carnegie Mellon University and General Motors won another $2 million from DARPA by maneuvering a robot through city traffic while safely avoiding other cars and obeying traffic regulations. Urban planners hope driverless cars can solve traffic congestion. Another Carnegie Mellon team relied on Bayes' rule and Kalman filters to win international robotic soccer championships involving fast-moving multirobot systems.

The U.S. military is heavily involved in imaging issues. Its Automatic Target Recognition (ATR) technology is a heavy user of Bayesian methods for robotic and electronic warfare, combat vehicles, cruise missiles, advanced avionics, smart weapons, and intelligence, surveillance, and reconnaissance. ATR systems employ radar, satellites, and other sensors to distinguish between, for example, a civilian truck and a missile launcher. Some ATR computer programs start with Bayes' controversial 50–50 odds, even though these can have a strong impact on rare events and better information may be available. Echoing generations of critics, at least one anonymous ATR analyst regards Bayes as "an affront, a cheap easy trick. It depends on an initial hunch. And yet it turns out to be an effective approximation that seems to solve many of the world's problems. So Bayes' rule is wrong . . . except for the fact that it works." Other approaches have been computationally more expensive and did not produce better answers.

Besides imaging problems, the military involves Bayes in tracking, weapons testing, and antiterrorism. Reagan's Ballistic Missile Defense applied a Bayesian approach to tracking incoming enemy ballistic missiles. Once it was sufficiently probable that a real missile had been detected, Bayes allowed sensors to communicate only their very latest data instead of recalculating an entire problem from scratch each time. The National Research Council of the National Academy of Sciences strongly urged the U.S. Army to use Bayesian methods for testing weapons systems, specifically the Stryker family of light, armored vehicles. Many military systems cannot be tested in the large sample sizes required by frequentist methods. A Bayesian approach allows analysts to combine test data with information from similar systems and components and from earlier developmental tests. Terrorist threats are generally estimated with Bayesian techniques. Even before the attacks of September 11, 2001, Digital Sandbox of Tyson's Corner, Virginia, used Bayesian networks to identify the Pentagon as a possible target. Bayes combined expert and subjective opinions about possible events that had never occurred.

The United States is not the only country trying to predict terrorism. As

Britain considered building a national data bank to detect potential terrorists, Bayes raised the same alarm it had against mass HIV screening. Terrorists are so rare that the definition of a terrorist will have to be extremely accurate or else many, many people will be identified as dangerous when they are, in fact, not at all.

On the Internet Bayes has worked its way into the very fiber of modern life. It helps to filter out spam; sell songs, books, and films; search for web sites; translate foreign languages; and recognize spoken words. David Heckerman, who used Bayesian networks to diagnose lymph node diseases for his Ph.D. thesis, has the modern practitioner's wide-open attitude about Bayes: "The whole thing about being a Bayesian is that all probability represents uncertainty and, anytime you see uncertainty, you represent it with probability. And that's a lot bigger than Bayes' theorem."

In 1992 Heckerman moved from Stanford to Microsoft, where he founded and manages the Machine Learning and Applied Statistics Group of Microsoft Research. The problems there are very different. Because Stanford had plenty of experts but little data, he says he built Bayesian nets with priors based on expert opinion: "But Microsoft had lots of data and only a few experts, so we got into combining expert knowledge with data." One of Microsoft's first Bayesian applications helped parents with sick children type in their children's symptoms and learn the best course of action. In 1996 Bill Gates, cofounder of Microsoft, made Bayes headline news by announcing that Microsoft's competitive advantage lay in its expertise in Bayesian networks.

That same year Heckerman, Robert Rounthwaite, Joshua Goodman, Eric Horvitz, and others began investigating Bayesian antispam techniques. Remember vVi-@-gra, l0w mOrtg@ge rates, PARTNERSHIP INVESTMENT, and !!!! PharammcyByMAIL? Advertisements that are unwanted and often pornographic and fraudulent are sent to millions without their permission. Spam soon accounted for more than half of all mail on the Internet, and some e-mail users spent half an hour a day weeding it out.

Bayesian methods attack spam by using words and phrases in the message to determine the probability that the message is unwanted. An e-mail's spam score can soar near certainty, 0.9999, when it contains phrases like "our price" and "most trusted"; coded words like "genierc virgaa"; and uppercase letters and punctuation like !!! or $$$. High-scoring messages are automatically banished to junk mail files. Users refine their own filters by reading low-scoring messages and either keeping them or sending them to trash and junk files. This use of Bayesian optimal classifiers is similar to the technique used by Frederick Mosteller and David Wallace to determine who wrote certain *Federalist* papers.

Bayesian theory is firmly embedded in Microsoft's Windows operating system. In addition, a variety of Bayesian techniques are involved in Microsoft's handwriting recognition; recommender systems; the question-answering box in the upper right corner of a PC's monitor screen; a data-mining software package for tracking business sales; a program that infers the applications that users will want and preloads them before they are requested; and software to make traffic jam predictions for drivers to check before their commute.

Bayes was blamed—unfairly, say Heckerman and Horvitz—for Microsoft's memorably annoying paperclip, Clippy. The cartoon character was originally programmed using Bayesian belief networks to make inferences about what a user knew and did not know about letter writing. After the writer reached a certain threshold of ignorance and frustration, Clippy popped up cheerily with the grammatically improper observation, "It looks like you're writing a letter. Would you like help?" Before Clippy was introduced to the world, however, non-Bayesians had substituted a cruder algorithm that made Clippy pop up irritatingly often. The program was so unpopular it was retired.

Bayes and Laplace would probably be appalled to learn that their work is heavily involved in selling products. Much online commerce relies on recommender filters, also called collaborative filters, built on the assumption that people who agreed about one product will probably agree on another. As the e-commerce refrain goes, "If you liked this book/song/movie, you'll like that one too." The updating used in machine learning does not necessarily follow Bayes' theorem formally but "shares its perspective." A $1-million contest sponsored by Netflix.com illustrates the prominent role of Bayesian concepts in modern e-commerce and learning theory. In 2006 the online film-rental company launched a search for the best recommender system to improve its own algorithm. More than 50,000 contestants from 186 countries vied over the four years of the competition. The AT&T Labs team organized around Yehuda Koren, Christopher T. Volinsky, and Robert M. Bell won the prize in September 2009.

Interestingly, although no contestants questioned Bayes as a legitimate method, almost none wrote a formal Bayesian model. The winning group relied on empirical Bayes but estimated the initial priors according to their frequencies. The film-rental company's data set was too big and too filled with unknowns for anyone to—almost instantaneously—create a model, assign priors, update posteriors repeatedly, and recommend films to clients. Instead, the winning algorithm had a Bayesian "perspective" and was laced with Bayesian "flavors." However, by far the most important lesson learned from the Netflix competition originated as a Bayesian idea: sharing.

Volinsky had used Bayesian model averaging for sharing and averaging complementary models while working in 1997 on his Ph.D. thesis about predicting the probability that a patient will have a stroke. But the Volinsky and Bell team did not employ the method directly for Netflix. Nevertheless, Volinsky emphasized how "due to my Bayesian Model Averaging training, it was quite intuitive for me that combining models was going to be the best way to improve predictive performance. Bayesian Model Averaging studies show that when two models that are not highly correlated are combined in a smart way, the combination often does better than either individual model." The contest publicized Bayes' reputation as a fertile approach to learning far beyond mere Bayesian technology.

Web users employ several forms of Bayes to search through the billions of documents and locate what they want. Before that can happen, though, each document must be profiled or categorized, organized, and sorted, and its probable interconnectedness with other documents must be calculated. At that point, we can type into a search engine the unrelated keywords we want to appear in a document, for example, "parrots," "madrigals," and "Afghan language." Bayes' rule can winnow through billions of web pages and find two relevant ones in 0.31 seconds. "They're inferential problems," says Peter Hoff at the University of Washington. "Given that you find one document interesting, can you find other documents that will interest you too?"

When Google starts projects involving large amounts of data, its giant search engines often try naïve Bayesian methods first. Naïve Bayes assumes simplistically that every variable is independent of the others; thus, a patient's fever and elevated white blood cell counts are treated as if they had nothing to do with each other. According to Google's research director, Peter Norvig, "There must have been dozens of times when a project started with naïve Bayes, just because it was easy to do and we expected to replace it with something more sophisticated later, but in the end the vast amount of data meant that a more complex technique was not needed."

Google also uses Bayesian techniques to classify spam and pornography and to find related words, phrases, and documents. A very large Bayesian network finds synonyms of words and phrases. Instead of downloading dictionaries for a spelling checker, Google conducted a full-text search of the entire Internet looking for all the different ways words can be spelled. The result was a flexible system that could recognize that "shaorn" should have been "Sharon" and correct the typo.

While Bayes has helped revolutionize modern life on the web, it is also helping to finesse the Tower of Babel that has separated linguistic communities for millennia. During the Second World War, Warren Weaver of the Rockefeller Foundation was impressed with how "a multiplicity of languages impedes cultural interchange between the peoples of the earth and is a serious deterrent to international understanding."[6] Struck by the power of mechanized cryptography and by Claude Shannon's new information theory, Weaver suggested that computerized statistical methods could treat translation as a cryptography problem. In the absence of computer power and a wealth of machine-readable text, Weaver's idea lay fallow for decades.

Ever since, the holy grail of translators has been a universal machine that can transform written and spoken words from one language into any other. As part of this endeavor, linguists like Noam Chomsky developed structural rules for English sentences, subjects, verbs, adjectives, and grammar but failed to produce an algorithm that could explain why one string of words makes an English sentence while another string does not.

During the 1970s IBM had two competing teams working on a similar problem, speech recognition. One group, filled with linguists, studied the rules of grammar. The other group, led by Mercer and Brown, who later went to RenTech, was filled with mathematically inclined communications specialists, computer scientists, and engineers. They took a different tack, replaced logical grammar with Bayes' rule, and were ignored for a decade.

Mercer's ambition was to make computers do intelligent things, and voice recognition seemed to be the way to make this happen. For both Mercer and Brown speech recognition was a problem about taking a signal that had passed through a noisy channel like a telephone and then determining the most probable sentence that the speaker had in mind. Ignoring grammar rules, they decided to figure out the statistical probability that words and phrases in one language would wind up as particular words or phrases in another. They did not have to know any language at all. They were simply

computing the probability of a single word given all the words that pre-
ceded it in a sentence. For example, by looking at pairs of English words
they realized that the word after "the" was highly unlikely to be "the" or
"a," somewhat more likely to be "cantaloupe," and still more likely to be
"tree."

"It all hinged on Bayes' theorem," Mercer recalled. "We were given
an acoustic output, and we'd like to find the most probable word sequence,
given the acoustic sequence that we heard." Their prior knowledge consisted
of the most probable order of the words in an English sentence, which they
could get by studying enormous amounts of English text.

The big problem throughout the 1970s was finding enough data. They
needed bodies of text focused on a fairly small topic, but nothing as adult
as the New York Times. At first they worked their way through old, out-of-
copyright children's books; 1,000 words from a U.S. Patent Office experiment
with laser technology; and 60 million words of Braille-readable text from
the American Printing House for the Blind.

At an international acoustic, speech, and signal meeting the group wore
identical T-shirts printed with the words "Fundamental Equation of Speech
Recognition" followed by Bayes' theorem. They developed "a bit of swagger,
I'm ashamed to say," Mercer recalled. "We were an obnoxious bunch back
in those days."

In a major breakthrough in the late 1980s they gained access to French
and English translations of the Canadian parliament's daily debates, about 100
million words in computer-readable form. From them, IBM extracted about
three million pairs of sentences, almost 99% of which were actual translations
of one another. It was a Rosetta Stone in English and French. "You had a day's
worth of English and the corresponding day's worth of French, so things were
lined up to that extent, but we didn't know that this sentence or word went
with this or that sentence or word. For example, while the English shouted,
'Hear! Hear!,' the French said, 'Bravo!' So we began working on getting a better alignment of the sentences. We were using the same methods as in speech
recognition: Bayes' theorem and hidden Markov models." The latter are particularly useful for recognizing patterns that involve likely time sequences,
for example, predicting one word in a sentence based on the previous word.

In a landmark paper in 1990 the group applied Bayes' theorem to full
sentences. There was a small probability that the sentence, "President Lincoln was a good lawyer," means "Le matin je me brosse les dents" but a
relatively large probability that it means "Le president Lincoln était un bon

avocat." After that paper, several of the leading machine translation systems incorporated Bayes' rule.

In 1993, lured by lucre and the challenge, Mercer and Brown moved from IBM and machine translation to RenTech, where they became vice presidents and co–portfolio managers for technical trading. So many members of IBM's speech recognition group joined them that critics complain they set back the field of machine translation five years.

After the 9/11 disaster and the start of the war in Iraq, the military and the intelligence communities poured money into machine translation. DARPA, the U.S. Air Force, and the intelligence services want to ease the burden on human translators working with such little-studied languages as Uzbek, Pashto, Dari, and Nepali.

Machine translation got still another boost when Google trawled the Internet for more Rosetta Stone texts: news stories and documents published in both English and another language. United Nations documents alone contributed 200 billion words. By this time the web was churning out enormous amounts of text, free for the asking. Combing English words on the web, Google counted all the times that, for example, a two-word sequence in English meant "of the." To determine which words in the English sentence correspond to which words in the other language, Google uses Bayes to align the sentences in the most probable fit.

The blue ribbons Google won in 2005 in a machine language contest sponsored by the National Institute of Standards and Technology showed that progress was coming, not from better algorithms, but from more training data. Computers don't "understand" anything, but they do recognize patterns. By 2009 Google was providing online translations in dozens of languages, including English, Albanian, Arabic, Bulgarian, Catalan, Chinese, Croatian, Czech, Danish, Dutch, Estonian, Filipino, Finnish, and French.

The Tower of Babel is crumbling.

Even as Bayes' rule was improving human communications, it was returning full circle to the fundamental question that had occupied Bayes, Price, and Laplace: How do we learn? Using Bayes' rule, more than half a million students in the United States learn the answer to that question each year: we combine old knowledge with new. Approximately 2,600 secondary schools teach algebra and geometry with Bayesian computer programs developed at Carnegie Mellon University since the late 1980s. The software also teaches French, English as a second language, chemistry, physics, and statistics.

The programs, called Cognitive Tutors, are based on John R. Anderson's idea that Bayes is a surrogate for the way we naturally learn, that is to say, gradually. The ability to accumulate evidence is an optimal survival strategy, but our brains cannot assign a high priority to everything. Therefore, most students must see and work many times with a mathematical concept before they can retrieve and apply it at will. Our ability to do so depends on how frequently and recently we studied the concept.

In addition to viewing Bayes as a continuous learning process, Cognitive Tutors depend on Bayes' theorem for calculating each student's "skillometer," the probability that the individual has mastered a topic and is ready for a new challenge. Ten years after this double-edged Bayesian approach was launched, its students were learning as much or more than conventionally taught pupils—in a third the time.

The flowering of Bayesian networks, neural nets, and artificial intelligence nets has helped neuroscientists study how the brain's neurons process information that arrives directly and indirectly, a little at a time, in tiny, often contradictory packets. As a computational tool and learning theory, Bayes has been involved in mapping the brain and analyzing its circuitry as well as in decoding signals from neurons and harnessing them to make better prostheses and robots.

Hundreds of megabits of sensory information bombard the waking brain every second. From that stream of data, 10 billion nerve cells extract information and correct prior understanding several times every 100 milliseconds. Discerning which sensory stimulus has caused which neuronal response is a difficult problem: the neurons fire unpredictably, scientists cannot monitor all of them at once, and the brain combines cues from multiple sources. The vision regions of our brains, for example, create three-dimensional objects and scenes. To do so, they rely on our prior knowledge about the regularities in our environment—for example, that light generally shines from above and that straight lines and 90-degree angles are apt to be man-made. But our brains refine that knowledge with new data pouring in about depth, contours, symmetry, lines of curvature, texture, shading, reflectance, foreshortening, and motion.

In 1998 the neurostatistician Emery N. Brown of MIT and Massachusetts General Hospital realized that Bayesian methods might deal with these uncertainties. Using Kalman filters, he and the MIT neuroscientist Matthew A. Wilson described a rat's brain as it processed information about the animal's location in its environment. Approximately 30 so-called place neurons in the hippocampus keep a rat informed about its location. As a laboratory rat foraged in a box scattered randomly with chocolate tidbits, electrodes implanted in its brain imaged some of the place neurons as they fired. A Bayesian filter sequentially updated the rat's position in the box. The researchers could see neither the rat nor its box, but by watching the neurons fire they could track the rat's movements. Thanks to Bayes, Brown could reconstruct the path of the chocolate-loving rat with only a fifth or a tenth of the neurons that previous methods had required.

To explore the practicality of using the living brain to power prostheses and robots, Brown's statistical method was replicated with a few dozen motor neurons, Bayesian algorithms, and Bayesian particle filters. The goal is to develop an artificial arm that can smoothly reach, rotate the hand, move fingers independently, and grasp and retrieve objects. Illustrating the possibilities of the approach, a rhesus monkey in Andrew B. Schwartz's laboratory at the University of Pittsburgh gazed longingly at an enticing treat. With his arms immobilized in plastic tubes and his mouth salivating, the motor neurons in his brain fired repeatedly, activating a robotic arm. The monkey's control was so precise it could reach out with the robotic arm, nab the treat, and fetch it back to eat. Frequentist methods can deal with simple backwards-and-forwards motion, but Bayesian neurostatisticians believe their algorithms will be powerful and flexible enough to control a robotic arm's position, rotation, acceleration, velocity, momentum, and grasp.

These attempts to capitalize on all the information available in neurons raise questions: What does the brain itself do? Does it maximize the information it gets from an uncertain world by performing Bayesian-like computations? In discussions of these questions, Bayes has become more than just an aid for data analysis and decision making. It has become a theoretical framework for explaining how the brain works. In fact, as such, the "Bayesian Brain" has become a metaphor for a human brain that mimics probability.

In our struggle to survive in an uncertain and changing world, our sensory and motor systems often produce signals that are incomplete, ambiguous, variable, or corrupted by random fluctuations. If we put one hand under a table and estimate its location, we can be off by up to 10 centimeters. Every time the brain generates a command for action, we produce a slightly different movement. In this confusing world, Bayes has emerged as a useful theoretical framework. It helps explain how the brain may learn. And it demonstrates mathematically how we combine two kinds of information:

our prior beliefs about the world with the error-fraught evidence from our senses.

As Lindley emphasized years ago, if we are certain about the evidence relayed by our senses, we rely on them. But when faced with unreliable sensory data, we fall back on our prior accumulation of beliefs about the world.

When Daniel Wolpert of Cambridge University tested the theory with a virtual tennis game, he showed that players unconsciously combine their prior knowledge of bouncing balls in general with sensory data about a particular ball coming over the net. In short, they unconsciously behave like good Bayesians. In addition, Wolpert said, the nice thing about Bayes was that it did not produce a single number. It made multiple predictions about every possible state given the sensory data. Thus, the tennis ball would most probably bounce in a particular spot—but there was also a reasonable chance it would fall elsewhere.

According to Bayes, the brain stores a wide range of possibilities but assigns them high and low probabilities. Color vision is already known to operate this way. We think we perceive red, but we actually see an entire spectrum of colors, assign the highest probability to red, **and** keep in mind the outside possibilities that the color could be pink or purple.

Wolpert concluded that Bayesian thinking is basic to everything a human does, from speaking to acting. The biological brain has evolved to minimize the world's uncertainties by thinking in a Bayesian way. In short, growing evidence suggests that we have Bayesian brains.

Given Bayes' contentious past and prolific contributions, what will the future look like? The approach has already proved its worth by advancing science and technology from high finance to e-commerce, from sociology to machine learning, and from astronomy to neurophysiology. It is the fundamental expression for how we think and view our world. Its mathematical simplicity and elegance continue to capture the imagination of its users.

But what about the years to come? Brute computer force can organize stunning quantities of information, but it clusters and searches for documents crudely, according to keywords. Only the brain examines documents and images according to their meaning and their content. Which approach will be more useful? Will computers become so powerful that huge amounts of data alone will teach us everything? Will scientists no longer need to theorize or hypothesize before experimenting or gathering their data? Or will Bayesian organizational principles remain fundamental? Current strategies for designing computers that could perform at biological levels exploit such ancient principles as reusable parts, hierarchical structures, variations on themes, and regulatory systems.

The jumping off point for this debate is Bayes and its priors, says Stuart Geman, whose Gibbs sampler helped launch the modern Bayesian revolution: "In this debate, there is no more powerful argument for Bayes than its recognition of the brain's inner structures and prior expectations." The old controversies between Bayesians and frequentists have been reframed in terms of, Do we use probabilities or not? Old or new, the issues are similar, if not identical, Geman says. And in its new guise, Bayesian learning and its priors occupy the heart of the debate.

Can we look forward to a time when computers can compete with our biological brains for understanding? Will they be programmed with Bayes? Or with something else?

Whatever the outcome of the revolution, Diaconis insists that Bayes will play a role. "Bayes is still young. Probability did not have any mathematics in it until 1700. Bayes grew up in data-poor and computationally poor circumstances. It hasn't settled down yet. We can give it time.

"We're just starting."

epilogue

Bayes' rule intervened spectacularly in April 2011 to help solve one of the most mysterious accidents in aviation history by pointing to the wreckage of a plane that had disappeared almost two years earlier.

Late on the night of May 31, 2009, Air France Flight 447 took off from Rio de Janeiro bound for Paris, met an intense high-altitude electrical storm over the South Atlantic in the early morning hours, and disappeared without a trace with 228 aboard.

The jet was beyond radar range, almost 600 miles from the Brazilian coast, when an automatic GPS location message system transmitted via satellite the plane's last known position at 2:10 a.m. June 1. Four and a half minutes later, the messages ceased, and officials of the French civil aviation safety agency realized that the plane must have crashed within about 45 miles of its last known position. Unfortunately, that meant the Airbus could be anywhere within 6,500 square miles of the South Atlantic.

To look for survivors and other signs of AF 447, the Bureau d'Enquêtes et d'Analyses (BEA), the French equivalent of the U.S. Federal Aviation Administration, immediately organized a surface search by Brazilian and French air forces and nearby ships. Almost a week passed, however, before they started to see floating debris and the bodies of 33 people about 45 miles north of the plane's last reported position.

An undersea search would be needed, and the BEA appointed Olivier Ferrante to coordinate it. Ferrante, a Frenchman with a master's degree in aviation engineering, is also a licensed pilot and certified diver. He describes himself as a member of a closely knit, international community of civil aviation experts who often meet and work together. He had already coordinated an underwater recovery operation off Sharm el-Sheikh in Egypt after a Flash Airlines Boeing 737 crashed there. Ferrante had also spent two years working on a joint data-sharing and risk-modeling project with the FAA in Renton, Washington.

Ferrante would spend the next two years coordinating the longest, most difficult, most high-tech, and most expensive undersea search ever launched. Not even the government minister in charge of the BEA thought that AF 447 would ever be found. Aviation and oceanographic specialists from government agencies, laboratories, and industries around the world would help. The cost would total an estimated $45 million split among Air France (25%), Airbus (25%), and BEA (50%).

For the first month, searchers raced against time to locate the plane's two emergency distress beacons. These cigar-sized pingers are designed to withstand a high-impact crash and begin emitting a high-frequency signal the moment they touch water but their batteries were certified to last only 30 days. The pingers were important because they were attached to the plane's two "black boxes" containing flight data and cockpit voice recorders. If BEA was to learn what caused the tragedy, the black boxes had to be found.

Despite their name, the black boxes used throughout the aviation industry are actually red and white. But they are the size of shoeboxes, and they were lost somewhere in an undersea area the size of Switzerland. The region is also as mountainous as Switzerland; it is located near the mid-Atlantic ridge between the tectonic plates of the Americas and Africa and is so remote that scientists had not yet charted its seabed.

The U.S. Navy sent the only two hydrophones in the world that can operate 20,000 feet below the ocean's surface. Sweeping along the plane's projected trajectory, their acoustic sensors actually passed over the plane's debris while the pingers were still supposed to be working. According to on-site inspections, the hydrophones were working properly but they detected no signals—other than whale vocalizing at roughly the same frequencies. So the hydrophones moved on, and the search continued.

Once all hope of hearing the pingers died, the only way to detect the plane's wreckage was sonar: detectors emitting sound waves which—once they hit something solid underwater—would bounce back for analysis and imaging.

The sonar experts were looking for an anomaly about 325 yards long, the length of the debris field created when a B-52 crashed at comparable depths off Guam in 2008. Still, the debris from an airplane crash is very small compared to underwater mountain ranges. And the sonar picked up objects that turned out to be fossilized rocks and seashells and oil drums thrown overboard by passing ships. "Geologists were very excited," Ferrante recalled drily, "but we weren't."

During two more fruitless cruises the next summer, experts from international oceanographic institutes, including the one at Woods Hole, Massachusetts, dragged side-scanning sonar equipment north of the plane's trajectory. Ferrante and Larry Stone, a Bayesian consultant, were stunned at the decision to search there. The oceanographers had chosen the area after analyzing how debris from the plane might have drifted. Although Bayesian software used by U.S. Coast Guard can often trace floating objects backward along known currents to an accident, the six-day delay in finding debris from the accident made reverse-drift calculations difficult. Complicating the problem were the turbulent wind and water currents close to the equator during the seasonal start of counter-currents in June. When the French Navy dropped nine drift buoys near the accident site in early June 2010, the buoys quickly floated every which way.

After the failure of the oceanographers' searches, the BEA decided to try one more cruise. If the plane was not found then, BEA planned to wait for another round of equipment improvements before trying again.

Lessons had already been learned. As soon as an emergency occurs on a plane transporting the public on long hauls over water, its position should be reported immediately and more often than the then-standard 10-minute intervals; doing that would reduce the size of any search area. The first aircraft to arrive over the zone of an accident should drop drift-measurement buoys. Pinger batteries should last 90 days instead of 30 and should emit signals at lower frequencies that would reach the ocean surface and be detectable by most of the world's naval vessels.

Meanwhile, Ferrante had been conferring with Johan Strumpfer, a South African who had successfully investigated a similar crash involving a Boeing 747 plane from Taipei in 1987. Strumpfer told Ferrante about Bayesian search theory and a Virginia military contractor that used it.

In July 2010, one year after the crash, BEA hired Metron, Inc., of Reston, Virginia, to launch an exhaustive Bayesian review of the entire search effort. Metron's Larry Stone had first worked on Bayesian search theory with Tony Richardson while looking for the lost nuclear submarine U.S.S. *Scorpion* and had used it for other searches since.

Metron assembled an AF 447 analysis team with Stone, Strumpfer, Tom Kratzke, and Colleen Keller to follow the basic Bayesian methodology that Richardson and John Craven had developed for the *Scorpion* hunt: organizing scenarios, evaluating unsuccessful searches, and estimating all the available data and their associated uncertainties.

The group started off by incorporating everything that was known prior to the search about airplane flight dynamics and winds and currents in the area. Included were data from Russia about nine commercial aircraft accidents involving loss of control; all these planes had crashed within 23—not 45—miles of the point where the emergency began. Metron assigned 70% probability to the credibility of the flight data. The positions and recovery times of the bodies found drifting between June 6 and June 9, 2009, were also incorporated into the prior probability but they were assigned only a 30% probability because of the turbulent equatorial waters. All this information was organized into consistent scenarios and their uncertainties quantified and weighted.

To update this pre-search information, all available data from the air, surface, and underwater searches were assembled. Sensor experts, the sea state, visibility, underwater geography, and water column conditions were used to estimate the performance of the sensors. The oceanographers had already searched the north so thoroughly with sonar that those areas could be ruled out. Finally, Bayes' rule was used to update the prior pre-search information with the search data.

But Metron's team, Strumpfer in particular, had strong doubts about whether the two pingers had functioned properly after the crash. So they produced two analyses. The first assumed that the pingers had worked after the crash. An alternative analysis assumed that the pingers had been damaged on impact and malfunctioned. Introducing that uncertainty produced a highest-priority area that sonar had never explored: the very center of the circle around the plane's last known position. The U.S. Navy's acoustic sensors had searched there during the first weeks after the accident. But when they heard nothing, it was assumed that the pingers must be somewhere else, and the oceanographers' cruises had gone to the north.

In April 2011, after two years of fruitless searches, the hunt began anew. But this time, Bayes was providing what Ferrante called an external, neutral, and rational eye. It had considered all the data available both before and after the crash. It had produced probability maps and a methodical plan for using available equipment efficiently every day of the search until the plane was found. And although it made assumptions, they were clearly labeled. Following Bayes' recommendation, the search started in the middle of the highest probability area and worked out from the plane's last known position.

Just one week later, the long-sought breakthrough occurred. On April 2, pictures taken with side-scanning sonar revealed a concentration of backscattered data over a 2,000 by 600 foot area. The next day, more pictures revealed the plane's engine, wing, fuselage panel, and landing gear. AF 447 lay almost 2.5 miles below the surface of the ocean, 7.5 miles north-northeast of the last position it had transmitted two years before.

Bayes had helped end a two-year-long search with one week of underwater searching. And remarkably, given the fact that even the word Bayes had been too controversial to mention for decades of the twentieth century, the French government publicly credited Bayes' rule for pointing to the place where AF 447 was found within one week.

Almost immediately, Ferrante returned to the site with a salvage ship. Both the flight data recorder and the cockpit voice recorder were retrieved in working order. One pinger was located but it had a small nick at one end, and even with new batteries, it did not work; the back of the plane where the black boxes were stowed had absorbed the worst of the impact. The second pinger was never found. If either had functioned as expected, the search for AF 447 would have ended in one month.

Recordings in the black boxes and studies of the recovered plane parts revealed that the plane's speed sensors had probably iced up during the storm. When the plane stalled, its pilots were confronted with conflicting speed data.

As for the hunt for the plane itself, it had been, in Ferrante's words, "a huge collective effort." It ended on June 16, 2011, two years and two weeks after AF 447 crashed, when a ship bearing the remains of 104 victims recovered from the ocean depths docked in Bayonne, France, where families waited for their loved ones.

appendix a

dr fisher's casebook: the doctor sees the light

by michael j. campbell

As one gets older one's thoughts turn naturally to religion and I have been pondering the religious metaphors in statistics. Clearly the frequentists are metaphorical Catholics; dividing results into "significant" and "nonsignificant" instead of dividing sins into "mortal" (i.e. significant) and venial. Randomisation is the grace that saves the world. In confession the priest is interested in the frequency with which one committed a sin (I can imagine passing the priest a bar-chart of how many times I swore, or was uncharitable, rather than giving him a verbal list—so much more informative!) After confession frequentists/Catholics are forgiven and so, having rejected a null hypothesis at $p < 0.05$, once it is published they are free to use 0.05 as the limit again. The frequentist prayer is "Our Fisher, who art in Heaven". Their saints are Pearson and Neyman. Instead of Heaven and Hell they have the Null and Alternative hypotheses, and in their Creed instead of "Do you reject Satan?" they have "Do you reject the null hypothesis?".

On the other hand Bayesians are born-again fundamentalists. One must be a "believer" and Bayesians can often pinpoint the day when Bayes came into their lives, when they dropped these childish frequentist ways (or even "came out"). Clearly the Reverend Thomas Bayes is their spiritual guide and leader, and he even imitated the Christian God by not publishing in his own lifetime (mind you, I have heard non-Bayesians wish that some of his followers had done likewise). Bayesians divide the world into people who believe and those who do not and will ask complete strangers at statistics conferences "are you a Bayesian?" as if it were an important defining characteristic. On finding a non-Bayesian, they will express amazement at the things the non-Bayesian does, point out the certainties of their own beliefs and attempt to convert the non-believer.

Then there are the sects. The agnostics are those who think that non-parametric statistics are the answer to everything. Similarly the bootstrappers cannot see why you need to bring God into it at all. There is the "bell-curve" cult, who think everything can be explained by reference to the Normal distribution. The simulators think God is purely a human invention.

Where do I put myself? Well, in typically woolly English fashion, I regard myself as Anglican. I believe in statistics as a way of finding the truth and am happy to adopt whatever means will help me to get there. I can see dangers in extremism in either direction and so try to steer a "middle way". I still use p-values and confidence intervals but temper them with prior beliefs. I like the idea of "empirical Bayes" where one uses earlier studies to inform one's priors. I can see the advantages of Bayesian methods for modelling complex systems and attaching uncertainty to parameters and think that in many ways it reflects scientific inference better. However, I prefer simply to call myself a believer, and not to attach labels to these beliefs.

Talking of religion, I am reminded of a strip of cartoons about Bayesians that appeared some time ago. They showed a series of monks. One was looking lost, one was dressed as a soldier, one was holding a guide book and one had his tongue stuck out. They were, respectively, a vague prior, a uniform prior, an informative prior and, of course, an improper prior . . .

appendix b
applying bayes' rule

Mammograms and Breast Cancer

In 2009 a U.S. government task force on breast cancer screening advised most women in their forties not to have annual mammograms. The public reaction was immediate—and in large part—enraged. Here's a simple version of the Bayesian calculation that lay at the very heart of the controversy.

A 40-year-old woman without any symptoms or family history of breast cancer has a mammogram as part of a routine checkup. A week later she receives a letter saying her test result was abnormal. She needs additional testing. What is the probability she actually has breast cancer?

Quite low.

Many beginning statistics students—and many physicians—find this surprising because mammograms, as a screening test, are reasonably accurate. They identify roughly 80% of 40-year-old women who have breast cancer at the time of their exam, and they provide positive test results to only about 10% of women without the disease.

However, breast cancer is relatively rare. And Bayes' rule takes background disease rates into account as prior knowledge. As a result, Bayes highlights the fact that not everyone who gets a positive test for a disease actually has that disease. It also underscores the fact that the probability of breast cancer is higher in a woman who finds a lump in her breast than in a woman who has a mammogram as part of a routine checkup.

To apply Bayes' rule to problems, here is the general equation:

$$P(A|B) = \frac{P(A)\ P(B|A)}{P(B)}$$

where A is a hypothesis and B is data.

To illustrate:

$$\begin{pmatrix} \text{Probability of} \\ \text{cancer given} \\ \text{a positive} \\ \text{mammogram} \end{pmatrix} = \begin{pmatrix} \text{Probability of} \\ \text{a positive mam-} \\ \text{mogram among} \\ \text{cancer patients} \end{pmatrix} \times \frac{\text{(Probability of}}{\text{(Probability of a}} \frac{\text{having breast cancer)}}{\text{positive mammogram)}}$$

According to this formula, we need three pieces of information, which will all go on the right-hand side of the equation:

1. The probability of having breast cancer: This is our prior knowledge about the background disease rate of breast cancer among women in their forties at the time they get a mammogram. According to *Cancer* and *Jama*, this is approximately 4/10 of 1%. Thus out of every 10,000 women in their forties who have mammograms, we can estimate that approximately 40 actually have the disease. The number: 40/10,000.

2. The probability of a breast cancer patient getting a positive mammogram: According to the National Cancer Institute and evidence from mammography, approximately 32 of those 40 women with breast cancer will get a positive test result from the mammogram. The number: 32/40.

3. The probability of getting a positive mammogram: The total number of women who get positive results (whether or not cancer is present) include women with cancer and women who are falsely informed that they have the disease. Mammograms give a positive ("abnormal") result to some women who do not have the disease; they are called false positives. For mammography, this rate is quite high, approximately 10%, according to the *New England Journal of Medicine*. Thus out of 10,000 women in their forties, 996 will get a letter telling them they have an abnormal test result. To rule out breast cancer, these women will need more mammography, ultrasound, or tissue sampling, perhaps even a biopsy. To this number

must be added the 32 breast cancer patients per 10,000 who will get a positive mammogram. The total number: 1028/10,000 or a little more than 10% of the women screened.

Inserting these numbers into the formula, we get the following:

$$\frac{P(A)\,P(B|A)}{P(B)} = \frac{\left(\frac{40}{10,000}\right) \times \left(\frac{32}{40}\right)}{\left(\frac{1028}{10,000}\right)}$$

Doing the arithmetic produces 0.03, or 3%. Thus the probability that a woman who tests positive has breast cancer is only 3%. She has 97 chances out of 100 to be disease free.

None of this is static. Each time more research data become available, Bayes' rule should be recalculated.

As far as Bayes is concerned, universal screening for a disease that affects only 4/10 of 1% of the population may subject many healthy women to needless worry and to additional treatment which in turn can cause its own medical problems. In addition, the money spent on universal screening could potentially be used for other worthwhile projects. Thus Bayes highlights the importance of improving breast cancer screening techniques and reducing the number of false positives. Another fact also points to the need for better mammography: negative test results miss 1 in 5 cancers.

The Case of Sally Clark

Starting in 1996, four British mothers were arrested for murder because two or more of their infants had died in their cribs. Three of the mothers were convicted and sentenced to life imprisonment. Yet by 2006, all four mothers were declared innocent of crimes that had never occurred.

Sally Clark, a lawyer who lost her first son at 11 weeks and her second at 8 weeks, was convicted in 1999. A prominent pediatrician, Sir Roy Meadow, testified for the prosecution about Sudden Infant Death Syndrome, known as SIDS in the United States and cot death in Britain. Citing a government study, Meadow said the incidence of one SIDS death was one in 8,500 in a family like Clark's—stable, affluent, nonsmoking, with a mother more than 26 years old.

Then, despite the fact that some families are predisposed to SIDS, Meadow assumed erroneously that each sibling's death occurred independently of the other. Multiplying 8,500 by 8,500, he calculated that the chance of two children dying in a family like Sally Clark's was so rare—one in 73 million—that they must have been murdered.

A Bayesian analysis would have shown that the children most probably died of SIDS.

According to Bayes' rule, a highly unlikely event can happen but it must be compared with other highly unlikely events. Thus, the question before the court should have been: Did the Clark babies more likely die of natural causes or murder?

Here's how the Bayesian analysis was done.

First, we look at natural causes of sudden infant death. The chance of one random infant dying from SIDS was about 1 in 1,300 during this period in Britain. Meadow's argument was flawed and produced a much slimmer chance of natural death. The estimated odds of a second SIDS death in the same family was much larger, perhaps 1 in 100, because family members can share a common environmental or genetic propensity for SIDS.

Second, we turn to the hypothesis that the babies were murdered. Only about 30 children in England, Scotland, and Wales out of 650,000 births annually were known to have been murdered by their mothers. The number of double murders must be much lower, estimated at 10 times less likely.

We take as the hypothesis H to be updated as: Sally Clark's two children died of SIDS. The data D will be that both children died suddenly and unexpectedly. The starting probabilities that go into Bayes' Rule are $P(H)$, and its opposite $P(\bar{H})$. As discussed above,

$P(H)$ is taken to have the value

$$P(H) = \frac{1}{1,300} \times \frac{1}{100} = 0.0000077$$

The alternative $P(\bar{H})$ is then given by

$$P(\bar{H}) = 1 - P(H) = 0.9999923$$

Now we look at the data. It appears in the probabilities $P(D|H)$ and $P(D|\bar{H})$. The first of these equals one, since it is the probability that the children died suddenly and expectedly given that they died of SIDS. $P(D|\bar{H})$ is the probability that a random pair of siblings die suddenly and expectedly but not from SIDS. The prosecution will equate that with murder. The estimate made in the paragraph above was

$$P(D|\bar{H}) = \frac{30}{650,000} \times \frac{1}{10} = 0.0000046.$$

The goal is to estimate $P(H|D)$, the probability that the cause of death was SIDS, given their unexplained deaths. Bayes' rule provides the formula

$$P(H|D) = \frac{P(H)P(D|H)}{P(H)P(D|H) + P(\bar{H})P(D|\bar{H})}$$

Inserting the numbers one finds

$$P(H|D) = \frac{1 \times 0.0000077}{1 \times 0.0000077 + 0.999992 \times 0.0000046} \approx 0.6$$

Thus, it is more likely that the infants died of SIDS.

Clark spent three years in prison before medical evidence freed her. A pathologist had withheld evidence that her second son had a bacterial blood infection known to cause sudden infant death. With the case against Clark in shreds, the other mothers were eventually freed too. But broken by the deaths of two sons and her conviction and imprisonment, Sally Clark died a few years later of acute alcohol intoxication.

I am indebted to Ray Hill, Professor of Mathematics at Salford University, Manchester, England, for his help with this example. It is based on Helen Joyce's *Plus Magazine* article, http://plus.maths.org/content/beyond-reasonable -doubt, and Ray Hill (2004) Multiple sudden infant deaths—coincidence or beyond coincidence? in *Paediatric and Perinatal Epidemiology* (18) 320–26 and (2005) Reflections on the cot death cases in *Significance* (2:1).

Subjective Priors

The heart of the Bayesian-frequentist controversy is the fact that, when only a few data are involved, the outcome of a Bayesian computation depends on prior opinion. In such a case, Bayes' theorem can lead to a subjective, rather than an objective, assessment of a situation. Modern practitioners want large amounts of data because they overwhelm prior opinion.

Here's an illustration of the controversy that bedeviled Bayes for most of the twentieth century. In it, two people look at the same problem and form different subjective notions of its prior probability. Then, even though they see the same experimental data, they reach different conclusions.

Tim and Susan are accosted in a bar by a stranger who pulls three coins out of his pocket; he says that two coins are legitimate (that is, each has a head and a tail) but that the third is false (it has two heads). The stranger offers to pick at random any one of the three coins, flip it three times, let Tim and Susan observe the results, and then let them guess whether the tossed coin was honest or false.

Both Tim and Susan accept that the stranger has two honest coins and one false coin.

The question that will affect the prior is: Will the stranger randomly select the coin to be tossed? The ultimate question, of course, is whether the tossed coin was the false one.

To use mathematical notation, we designate P for probability; F for the false two-headed coin; the prior probability that the tossed coin is false is P(F); D is the new data to be collected later, and Tim and Susan's new opinion based on the new data will be P(F|D).

They appear in Bayes' theorem as:

$$P(F|D) = \frac{P(F)P(D|F)}{P(F)P(D|F) + P(\bar{F})P(D|\bar{F})}$$

Tim's first impression is that the stranger is honest and will select one coin randomly from the first three. Thus, Tim's prior probability that the false coin was selected is P(F) = 1/3. Here is Bayes' theorem with Tim's prior opinion.

$$P(F|D) = \frac{\frac{1}{3}P(D|F)}{\frac{1}{3}P(D|F) + \frac{2}{3}P(D|\bar{F})}$$

Susan, however, suspects that the stranger will not randomly choose one of the three coins. So Susan does not agree with Tim's 1/3 prior probability. If Susan is totally convinced that the guy at the bar will deliberately choose the false coin, then her P(F) = 1.

Here is Bayes' formula with Susan's first prior:

$$P(F|D) = \frac{1 \cdot P(D|F)}{1 \cdot P(D|F) + 0 \cdot P(D|\bar{F})}$$

On the other hand, if Susan is totally sure that the guy at the bar will select one of the fair coins, then her prior is P(F) = 0.

Bayes' theorem with Susan's other prior:

$$P(F|D) = \frac{0 \cdot P(D|F)}{0 \cdot P(D|F) + 1 \cdot P(D|\bar{F})}$$

Depending on how much Susan doubts the stranger's honesty, she could have many prior probabilities between 0 and 1.

In any case, Tim and Susan wind up with very different prior opinions.

Now to gather experimental data. The stranger chooses a coin, flips it three times, and each time it comes up heads. If the coin has two heads, the probability of this happening is one. In symbols, P(D|F) = 1. But if it's an honest coin, the probability of this happening is 1/8.

For example, using Tim's prior, it's:

$$P(F|D) = \frac{\frac{1}{3} \cdot 1}{\frac{1}{3} \cdot 1 + \frac{2}{3} \cdot \frac{1}{8}} = \frac{4}{5}$$

Thus, Tim's new belief is that there is an 80% probability that the stranger selected the false coin.

Given Susan's priors, it could be:

$$P(F|D) = \frac{0 \cdot 1}{0 \cdot 1 + 1 \cdot \frac{1}{8}} = 0$$

or

$$P(F|D) = \frac{1 \cdot 1}{1 \cdot 1 + 0 \cdot \frac{1}{8}} = 1.$$

Thus, Susan's new belief that the stranger actually selected the false coin ranges between 0 and 100%. Susan and Tim reach different conclusions about the stranger's coin because their prior opinions were different and because they did not have much observational data for updating the priors.

Thanks to Albert Madansky for this problem. He was inspired by Jon Allen Paulos' review of The Theory That Would Not Die in the New York Times Book Review. Thanks also to Thomas Herzog for comments.

Hemophilia

Does a healthy woman whose brother has hemophilia carry the gene that could give the disease to her son? Neither she nor her husband has hemophilia.

This case shows how acutely sensitive Bayes' rule is to new observational data and how new evidence updates old. It also demonstrates how even one new fact can dramatically alter the probability of a situation.

We start with the long established fact that a healthy sister with a hemophiliac brother has a 50–50 chance of carrying the hemophilia gene.

First some notation: "C" means that the mother is a carrier. "$\mathcal{C}$" means the opposite, that the mother is not a carrier. "H" means that the son is healthy. The prior probability that the mother is a carrier is $P(C)$. We are trying to find the updated probability $P(C|H)$.

Genetics tells us that, if the mother is a carrier, the probability of her having a healthy boy is 1/2, written $P(H|C) = 1/2$. If she is not a carrier, the son is always healthy, i.e., $P(H|\mathcal{C}) = 1$.

Given genetics and the family history, our starting probabilities are $P(C) = P(\mathcal{C}) = 1/2$.

New information appears when the woman gives birth to her first son, a healthy child. We use Bayes' rule to update the probability of her being a carrier with this new information:

$$P(C|H) = \frac{P(C)P(H|C)}{P(C)P(H|C) + P(\mathcal{C})P(H|\mathcal{C})}$$

Replacing the symbols in Bayes' rule with numbers, we get:

$$\frac{\frac{1}{2} \cdot \frac{1}{2}}{\frac{1}{2} \cdot \frac{1}{2} + \frac{1}{2} \cdot 1}$$

Doing the arithmetic produces:

$$P(C|H) = \frac{1}{3}$$

The probability that the woman is a carrier has now dropped substantially, from 1/2 to 1/3.

Her next child is also a healthy boy. We want to update her medical history with this new piece of information. Fortunately, we need not re-do the entire calculation because Bayes' rule lets us combine a prior with the new information each time a new child is born. Thus, the probability from the previous birth becomes the starting point each time. To do so, we recalculate Bayes' rule using 1/3 as our starting point. Putting the numbers into the formula, we get:

$$P(C|2H) = \frac{\frac{1}{3} \cdot \frac{1}{2}}{\frac{1}{3} \cdot \frac{1}{2} + \frac{2}{3} \cdot 1} = \frac{1}{5}$$

When she has a third healthy son, the probability of her being a carrier drops to 1/9.

We still cannot rule out the possibility of her being a carrier, however. And the instant that she gives birth to even one son with hemophilia, the probability that she is a carrier and has given her child the disease rises instantly—and tragically—to certainty.

A version of this problem appears in *Bayesian Data Analysis* (1995ff) by Andrew Gelman, John B. Carlin, Hal S. Stern, and Donald B. Rubin; and in *Likelihood, Bayesian and MCMC Methods in Quantitative Genetics* (2002) by D. Sorensen and Daniel Gianola. Thanks to Daniel Gianola and John Carlin for this problem.

A Search Example Using Bayesian Updating

The following problem is a bit more difficult than the three previous ones, but it's too interesting to leave out. It comes directly from Henry R. ("Tony") Richardson, who developed Bayesian search theory for the U.S. Navy and used it to locate Soviet submarines. Richardson also developed Monte Carlo methods for U.S. Coast Guard searches decades before the academic world.

The following provides a simple example of the use of "Bayesian updating" in an illustrative search problem.

Suppose that a fisherman has not returned home after a day on the water, and his wife reports him missing to the Coast Guard. The search authorities interview his wife and friends in order to learn about his favorite fishing grounds and his plans for the day. As a result of these discussions, three possible search areas are postulated as shown schematically below:

Search Areas

Search Area 1	Search Area 2	Search Area 3

Using the information provided by the interviews and perhaps other relevant information, the authorities assign the following subjective initial target location probabilities to the search areas:

Initial Target Location Probabilities

$P(1) =$ 5%	$P(2) =$ 70%	$P(3) =$ 25%

Search resources are limited, and so a search is planned that will concentrate on the two highest probability areas. The planned effectiveness of the search in each area is expressed by a "search effectiveness probability." This probability can be thought of as the percentage of the assigned area that can be covered thoroughly. The planned search effectiveness probabilities are as follows:

Planned Search Effectiveness Probabilities

E(1) = 0%	E(2) = 80%	E(3) = 30%

The probability of detection, P_D, for this allocation of effort is calculated by the formula

$$P_D = P(1)E(1) + P(2)E(2) + P(3)E(3),$$

or in more compact notation,

$$P_D = \sum_{k=1}^{3} P(k){\cdot}E(k).$$

This is the probability that the target is in a specified area multiplied by the conditional probability that the target would be detected, given that it is in the specified area, summed over all the areas. Thus, using the data above, the probability of detection is

$$P_D = 63.5\%$$

Now suppose that the search is unsuccessful and that analysis shows that the search effectiveness probabilities actually achieved are

Actual Search Effectiveness Probabilities

E(1) = 0%	E(2) = 90%	E(3) = 20%

i.e., the search in Area 2 was more effective than planned, but the search in Area 3 was less effective than planned.

In order to continue the search it is necessary to revise the target location probabilities by taking into account the fact that the target was not found. This is accomplished by using Bayes' theorem to calculate the updated target location probabilities. If $P'(j)$ denotes the updated probability of target location in the jth search area for $j = 1, 2,$ and 3, then

$$P'(j) = \frac{P(j) \cdot (1 - E(j))}{\sum_{k=1}^{3} P(k) \cdot (1 - E(k))} \ .$$

Using the initial target location probabilities and the *actual* search effectiveness probabilities given above, the updated target location probabilities are calculated as follows:

First, compute the single denominator that will be used in each case, i.e.,

$$\sum_{k=1}^{3} P(k) \cdot (1 - E(k)) = (.05) \cdot (1) + (.70) \cdot (.10) + (.25) \cdot (.80) = .32.$$

Then complete the calculations and obtain

$$P'(1) = \frac{(.05) \cdot (1)}{.32} = .16$$

$$P'(2) = \frac{(.70) \cdot (.10)}{.32} = .22$$

and

$$P'(3) = \frac{(.25) \cdot (.80)}{.32} = .62.$$

In summary one has

Initial Target Location Probabilities

P(1) =	P(2) =	P(3) =
5%	70%	25%

Actual Search Effectiveness Probabilities

E(1) =	E(2) =	E(3) =
0%	90%	20%

and

Updated Target Location Probabilities

P′(1) = 16%	P′(2) = 22%	P′(3) = 62%

Notice that the chances of the target being in Area 2 have been reduced substantially, and that the highest updated probability of the target's location is now in Area 3. In short, the search should switch from Area 2 to Area 3 as soon as possible.

notes

1. Causes in the Air

1. Two errors about Bayes' death and portrait have been widely disseminated. First, Bayes died on April 7, 1761, according to cemetery records and other contemporaneous documents gathered by Bayes' biographers, Andrew Dale and David Bellhouse. Bayes was interred on April 15, which is often called the date of his death. The degraded condition of his vault may have contributed to the confusion.

   Second, the often-reproduced portrait of Thomas Bayes is almost assuredly of someone else named "T. Bayes." The sketch first appeared in 1936 in *History of Life Insurance in its Formative Years* by Terence O'Donnell. However, the picture's caption on page 335 says it is of "Rev. T. Bayes, Improver of the Columnar Method developed by Barrett," and Barrett did not develop his method until 1810, a half-century after the death of "our" Rev. Thomas Bayes.

   Bellhouse (2004) first noticed that the portrait's hairstyle is anachronistic. Sharon North, curator of Textiles and Fashion at the Victoria and Albert Museum, London, agrees: "The hairstyle in this portrait looks very 20th century. . . . Clerical dress is always difficult as the robes and bands (collars) change very little over time. However, I would say that the hair of the man . . . is quite wrong for the 1750s. He would have been wearing a wig for a portrait. Clergymen wore a style of the bob wig (which eventually became known as a 'clerical wig'), a short very bushy wig of horsehair powdered white."

2. Dale (1999) 15.

3. All of Bayes' and Price's quotations come from their essay.

2. The Man Who Did Everything

1. For details of Laplace's personal life, I rely on Hahn (2004, 2005). All documents about Laplace's life were thought to have been lost when a descendant's home was destroyed by fire in 1925, but Hahn painstakingly located many original docu-

ments that revealed new facts and corrected previous assumptions about Laplace's life and work.

2. "A dizzying expansion of curiosity" is Daniel Roche's original phrase in his classic, *France in the Enlightenment*.

3. Voltaire 24.

4. Koda and Bolton 21.

5. Stigler (1978) 234–35.

6. Laplace (1774) OC (8) 27; Laplace (1783/1786) OC (11) 37, and Stigler (1986) 359.

7. Laplace (1776) 113. For English translation, see Hahn in Lindberg and Numbers (1986) 268–70.

8. Laplace (1783) OC (10) 301.

9. Laplace in Dale's translation (1994) 120, in section titled "Historical note on the probability calculus."

10. Gillispie (1997) 23.

11. Laplace (1782–85) OC (10) 209–340.

12. Laplace (1778–81) OC (9) 429 and (1783/1786) OC (10) 319.

13. Laplace (1778–81) OC (9) 429.

14. "Easy to see . . . obvious:" Laplace (1778/1781) OC (9) 383–485. The student was Jean-Baptiste Biot.

15. Stigler (1986) 135.

16. Gillispie (1997) 81.

17. Laplace (1783/1786) OC (10) 295–338.

18. Hald (1998) 236 and, for a detailed discussion of Laplace's birth studies, 230–45.

19. Laplace in *Philosophical Essay on Probabilities*, Dale's translation 77.

20. Hahn (2004) 104.

21. Sir William Herschel wrote a firsthand account in his diary. See Dreyer vol. I, lxii, and Hahn in Woolf.

22. Laplace from *Exposition du Système du Monde* in Crosland 90.

23. Glenn Shafer interview.

24. Laplace, *Essai Philosophique*, translated in Hahn (2005) 189 and in Dale (1995) 124.

3. Many Doubts, Few Defenders

1. Clerke 200–203.

2. Bell ix and 172–82.

3. David 30.

4. Gillispie (1997) 67, 276–77.

5. Pearson (1929) 208.

6. Porter (1986) 36.

7. Mill in Gigerenzer et al. (1989) 33.

8. Quoted by Dale (1998) 261.

9. G. Chrystal in Hald (1998) 275.

10. *Le procès Dreyfus* vol. 3, 237–31.

11. Molina in Bailey (1950) 95–96.

12. Rubinow (1914) 13.

13. Rubinow (1917) 35.

14. Rubinow (1914–15) 14.

15. Rubinow (1917) 42.

16. Rubinow (1914–15) 14.

17. Anonymous in Pruitt (1964) 151.

18. Whitney (1918) 287.

19. Pruitt 169.

20. Ibid., 170.

21. Ibid.

22. Pearson in MacKenzie (1981) 204.

23. J. L. Coolidge in Hald (1998) 163.

24. Kruskal 1026.

25. Savage (1976) 445–46.

26. Leonard Darwin in MacKenzie (1981) 19.

27. Fisher in Box (2006) 127.

28. Fisher (1925) 1.

29. Kruskal 1026.

30. Ibid., 1029.

31. Fisher in Kotz and Johnson I 13.

32. Fisher in Gill 122.

33. Fisher (1925) 9–11.

34. Hald (1998) 733.

35. Savage (1976) 446.

36. E. Pearson in Reid 55–56.

37. Perks 286.

38. Fisher in Neyman Supplement 154–57.

39. Tukey, according to Brillinger e-mail.

40. aip.org/history/curie/scandal. Accessed April 18, 2006.

41. De Finetti (1972).

42. Lindley letter to author.

43. Essen-Möller in Kaye (2004).

44. Huzurbazar 19.

45. Lindley (1983) 14.

46. Howie 126.

47. Ibid., 210.

48. Jeffreys (1939) 99.

49. Lindley (1991) 11.

50. Ibid., 391.

51. Jeffreys (1938) 718.

52. Jeffreys (1939) v.

53. Goodman (2005) 284.

54. Howie 165.

55. Box (1978) 441.

56. Lindley (1986a) 43.

57. Jeffreys (1961) 432.
58. Lindley (1983) 8.
59. Lindley (1991) 10.

4. Bayes Goes to War

1. Churchill 598.
2. Peter Twinn in Copeland (2006) 567.
3. Atkinson and Feinberg 36.
4. D. G. Kendall in ibid., 48.
5. Alastair Denniston in Copeland (2006) 57 and (2004) 219.
6. Patrick Mahon in Copeland (2004) 271.
7. Ibid., 279.
8. Max Newman in Gandy and Yates 7.
9. Copeland (2006) 379.
10. Copeland (2004) 258.
11. Copeland (2006) 379.
12. Copeland (2004) 281.
13. Good (1979) 394.
14. Anonymous to author.
15. Britton 214.
16. Hinsley and Stripp 155.
17. Good interview.
18. Michie's draft chapter and Good interview.
19. Copeland (2004) 279.
20. Ibid., 287–88.
21. Ibid., 292.
22. Ibid., 289.
23. Ibid., 260.
24. For the entire letter episode, ibid., 336–37.
25. Shiryaev (1991) 313.
26. Ibid.
27. Kolmogorov (1942).
28. Arnold.
29. Copeland (2006) 383.
30. Copeland (2006) 380–82.
31. Turing (1942).
32. Shannon in Kahn (1967) 744.
33. Waddington 27.
34. Koopman (1946) 771.
35. Koopman (1980) 17.
36. Ibid., 18.
37. Ibid., 60–61.
38. Andresen 82–83.
39. Copeland (2006) 80–81.

40. Michie in Copeland (2006) 380.

41. Ibid., 244.

42. Edward H. Simpson letter to author.

43. Ibid.

44. Good in Britton 221.

45. Hodges (2000) 290.

46. Dennis Lindley letter to author.

47. Hilton 7.

5. Dead and Buried Again

1. Good interview.

2. Olkin, Ingram (1992) A conversation with Churchill Eisenhart. *Statistical Science* (7) 512–30; Sampson et al. 135.

3. John W. Pratt interview.

4. Perks 286.

5. DeGroot (1986a) 40–53.

6. Kotz and Johnson I xxxviii.

7. Anonymous in Reid 273.

6. Arthur Bailey

1. Biographical details are from interviews and correspondence with his son and daughter-in-law, Robert A. and Shirley Bailey.

2. Bailey (1942, 1943) 31–32.

3. Hewitt (1969) 80.

4. Bailey (1950) 7.

5. Ibid., 31–32.

6. Ibid., 7–9.

7. Pruitt 165.

8. Bailey (1950) 8.

9. *PCAS* 37 94–115.

10. The Longley-Cook episode is in Carr 241–43.

11. Charles C. Hewitt Jr. interview.

12. Hans Bühlmann letter to author.

7. From Tool to Theology

1. Stephan et al., 953.

2. Good interview.

3. Reid 216.

4. Fisher (1958) 274.

5. Reid 256.

6. Ibid., 226.

7. Ibid., 274.

8. Good in Kotz and Johnson I 380.

9. Fienberg (2006) 19.

10. Lindley in Smith (1995) 312.

11. Donald Michie in Copeland (2006) 240.

12. Stephen Fienberg interview.

13. George E. P. Box interview.

14. Smith (1995) 308.

15. Sampson (1999) 126–27.

16. Ibid., 128.

17. Kotz and Johnson I 520.

18. Lindley (1989) 14.

19. Savage (1956).

20. Lindley in Erickson 49.

21. Savage in Fienberg (2006) 16–19.

22. Schrödinger 704.

23. Savage in Erickson 297.

24. David Spiegelhalter interview.

25. Robert E. Kass interview.

26. Anonymous.

27. Maurice G. Kendall 185.

28. Kruskal in Brooks, online.

29. Savage in Lindley letter to author.

30. Rivett.

31. Lindley letter to author.

32. Smith (1995) 312.

33. Lindley letter to author.

8. Jerome Cornfield, Lung Cancer, and Heart Attacks

1. Marvin Hoffenberg interview.

2. Ibid.

3. Cornfield (1975) 14.

4. *Memorial Symposium* 55.

5. Gail 9.

6. Ibid.

7. Gail 10.

8. Stories from *Memorial Symposium* 52 and 56.

9. Cornfield (1962) 58.

10. Gail 5.

11. Cornfield (1967) 41.

12. Cornfield (1975) 9–11.

13. *Memorial Symposium* 52.

14. Ellen Cornfield interview.

9. There's Always a First Time

1. Jardini 119.
2. Harken.
3. Albert Madansky interview.
4. Iklé (1958) 3.
5. Ibid., 73.
6. Ibid., 8, 114.
7. Iklé (1958) 74.
8. Madansky interview.
9. Lindley (1985) 104.
10. Madansky interview.
11. Ibid.
12. Ibid.
13. Iklé (1958) 54.
14. Ibid., 53–54.
15. Madansky interview.
16. Iklé (1958) 153.
17. Iklé (2006) 46–47.
18. Ibid.
19. Ibid.

10. 46,656 Varieties

1. Good (1971) 62–63.
2. Glenn Shafer interview.
3. Lindley letter to author.
4. Box interview.
5. Ibid.
6. Efron (1977) and interview.
7. Box interview.
8. Box (2006) 555–56.
9. Bross (1962) 309–10.
10. Savage (1962) 307.
11. Ericson (1981) 299.
12. Box interview.
13. "What I . . . me . . . it was . . . angry . . . heads." Lindley to Smith (1995) 310–11.
14. Bennett 36.
15. Smith (1995) 311.
16. Box interview.
17. Ibid.
18. Homer Warner interview.
19. Leahy (1960) 50.
20. Tribe (1971a) 1376.

11. Business Decisions

1. Fienberg (1990) 206.
2. Schleifer interview.
3. Pratt interview.
4. Pratt, Raiffa, Schlaifer (1965) 1.1.
5. Savage (1956) letter.
6. Schlaifer letter of August 22, 1956.
7. Memorial Service (1994).
8. Arthur Schleifer interview.
9. Raiffa in Fienberg (2008) 137.
10. Ibid., 138.
11. Ibid., 139.
12. Ibid., 141.
13. Fienberg (2006) 10.
14. Raiffa (1968) 283.
15. Raiffa interview.
16. Raiffa (2006) 32.
17. Fienberg (2008) 10.
18. Raiffa interview.
19. Memorial Service.
20. Fienberg (2008) 142.
21. Pratt, Memorial Service.
22. Memorial Service.
23. Arthur Schleifer interview.
24. Ibid.
25. Ibid.
26. Fienberg (2006) 18.
27. Raiffa (2006) 48, 51.
28. Raiffa interview.
29. Schleifer interview.
30. Raiffa, Memorial Service.
31. Raiffa (1968).
32. Lindley in Smith (1995) 312.
33. McGinnis.
34. Raiffa and Pratt (1995).

12. Who Wrote *The Federalist*?

1. Most of the quotations from Mosteller and Wallace in this chapter come from their book, published in 1964 and 1984 under different titles. Exceptions will be noted.
2. David L. Wallace interview.
3. Fienberg et al. 147.
4. Ibid., 192.
5. Petrosino.

6. Kolata 397.

7. DeGroot (1986c) 322.

8. Albers et al. (1990) 256–57.

9. Kolata 398.

10. Robert E. Kass interview.

13. The Cold Warrior

1. Bamford 430–31. The author covered this controversy for the Trenton N.J. *Times*.

2. Stephen Fienberg interview and e-mail.

3. Brillinger in Brillinger (2002a) 1549.

4. Anscombe 296.

5. Wheeler in Brillinger (2002b) 193.

6. Descriptions of Tukey come from Anscombe 289, Bradford Murphy interview, and McCullagh 541.

7. Elizabeth Tukey and Tukey in Brillinger (2002a) 1561–2.

8. John Chambers, Bell Labs.

9. Ibid.

10. Tukey (1962) 5, 7.

11. Kotz and Johnson II 449.

12. Anscombe 294.

13. Bell Labs News (1985) (25) 18 and Brillinger (2002a) 1556.

14. Tukey in Brillinger (2002a) 1561.

15. Wallace interview.

16. Fienberg (2006) 24.

17. Wainer 285, Anscombe 290, and Wallace interview.

18. Box and Edgar Gilbert interviews, McCullagh 544 and 554, and Pratt interview, respectively.

19. Tukey (1967) in Jones (4) 589.

20. Tukey in Lyle V. Jones, (III) 108; (IV) xiv; (III) 188; and (III) 394, respectively.

21. Brillinger (2002a) 1561; L. Jones (1986) (IV) 771–2; Casella 312, respectively.

22. Casella 332 and McCullagh 547, respectively.

23. L. Jones (III) 277. "A natural . . . framework": Casella 312.

24. Wallace, in interviews with the author.

25. Gnanadesikan. Also, Tukey taught Bayesian methods to Princeton students in 1954 and 1955, according to Robert J. Ruben MD who attended the classes (private communication).

26. L. Jones (IV) 589.

27. Brillinger interview.

28. Anscombe 300.

29. Wallace interview.

30. Brillinger e-mail.

31. Wallace interview.

14. Three Mile Island

1. In Lyle Jones (IV) 686 and Box and Tiao (1973) 1.
2. Bather 346–47 and Holmes interview, respectively.
3. Both Diaconis and Lindley recalled this incident in interviews.
4. Efron (1978) 232.
5. Lindley to Smith (1995) 313.
6. Apostalakis interview.

15. The Navy Searches

1. Quotations from John Craven are—unless noted otherwise—from interviews with the author.
2. Craven wrote *The Silent War: The Cold War Battle Beneath the Sea* at the navy's behest in two weeks without notes in order to rebut the popular book *Blind Man's Bluff: The Untold Story of American Submarine Espionage* (1998) by Sontag and Drew. In his own book, Craven said he attended Raiffa's lectures; later he told me that he probably heard about Raiffa's work from others at MIT.
3. Wagner (1988).
4. Ibid., 9.
5. Quotations from Henry R. ("Tony") Richardson are from interviews with the author.
6. Capt. Frank A. Andrews e-mail.
7. Richardson interview.
8. Lewis 99–100, 133, 165, 168. Her Pulitzer Prize–winning book is generally regarded as the best on-site source.
9. Craven 173.
10. Lewis 206 and 208.
11. Wagner 10.
12. Craven 205–7.
13. Stone (1975) 54.
14. Craven 202–3.
15. Stone et al. (1999) ix.
16. Joseph H. Discenza interview.
17. Stone (1999) ix and (1983) 209.
18. Ibid., (1999) ix.
19. VAdm. John "Nick" Nicholson interview.
20. Ray Hilborn interview.

16. Eureka!

1. A. Philip Dawid interview.
2. In Donald Owen (1976) 421.
3. Cooke 20.
4. Jeffrey E. Harris interview.
5. Adrian Raftery interview.

6. Raftery (1986) 145–46.

7. Stuart Geman interview.

8. Shafer (1990) 440; and Diaconis in DeGroot (1986c) 334, respectively.

9. Lindley in Diaconis and Holmes (1996) 5 and in letter to the author.

10. AFM Smith (1984) 245, 255.

11. Alan Gelfand interview.

12. Christian Robert and George Casella.

13. Mayer in Householder 19.

14. W. Keith Hastings interview.

15. S. Gelfand interview.

16. Robert and Casella (2008).

17. Gelfand et al. (1990).

18. Gill 332.

19. Kuhn.

20. David Spiegelhalter interview.

21. Spiegelhalter, Abrams, and Myles.

22. Taylor and Gerrodette (1993).

23. Raftery interview.

24. Paul R. Wade interview.

25. Blackwell in DeGroot (1986a).

26. Diaconis and Holmes (1996) 5.

27. Sir John Russell and William Gladstone in the nineteenth century and Harold Wilson in the twentieth.

17. Rosetta Stones

Almost all the quotations in this chapter come from interviews with the author. Exceptions are noted here.

1. Unwin 190; Schneider; Ludlum 394. The phrase "We're all Bayesians now" is sometimes attributed to John Maynard Keynes, but it may have first appeared in 1976 in John C. Henretta and Richard T. Campbell's article, Status Attainment and Status Maintenance: A Study of Stratification in Old Age in *American Sociological Review* (41) 981–92. To complicate matters, Campbell was paraphrasing an earlier popular expression, "We're all Keynesians now," which has been attributed to Milton Friedman in 1966 and which was "popularized" by President Richard Nixon in 1971. I am indebted to Stephen Senn, Michael Campbell, and Wikipedia for helping sort out the origins of the "Keynesians" quotation.

2. Dawid in Swinburne (2002) 84

3. Greenspan.

4. Ibid.

5. *New York Times* January 4, 2009.

6. Weaver 15.

glossary for nonmathematical readers

algorithm a formula defining a sequence of steps in order to solve a problem

analysis a higher branch of mathematics

a priori *see* prior

axiom an assumption upon which a mathematical theory is based

ban a measure of probability expressed in logarithms to the base 10 so that multiplication can be replaced by addition

Bayes' rule a mathematical device combining prior information with evidence from data (its formula appears on p. 31 with a simplified version on p. 257.)

Bayesian network a graphical model that compactly represents probabilities and their relationships. Each random variable is denoted by a node, and a line between two nodes indicates their interdependency.

centering points types of averages, e.g., the mean, median, and mode

change point the point when change occurs in time-ordered data

credibility a measure of the credence that actuaries place in a particular body of claims experience as they set insurance policy rates

cryptography writing and breaking ciphers, communications that third parties cannot understand

curse of dimensionality the explosive growth of data sets as more variables are added

data bits of information that can be represented numerically

fiducial probability R. A. Fisher's controversial attempt to apply probability to unknown parameters without using Bayes' rule or priors

filter a process that makes data immune to the noise in a system and that extracts information from the data

frequency a branch of probability theory that measures the relative frequency of an event that can be repeated over and over again under much the same conditions

generating function a mathematical shortcut for making approximations

hierarchical Bayes a method that develops mathematical models by breaking complex processes into stages called hierarchies

hypothesis a proposition that is to be tested or modified with new evidence

induction drawing conclusions about natural laws or regularities from an observation or experiment; the opposite of deduction

infer deriving natural laws and regularities from a well-defined statement or observation

inverse probability the branch of probability theory that draws conclusions about antecedents or causes of observed events, e.g., Bayes' rule

likelihood principle an approach to using Bayes' theorem without assuming any prior probabilities

likelihood ratio the comparison between the probabilities of an observation when a hypothesis is true and when it is untrue

Markov chain a process that assumes the probability of an event depends only on the immediately preceding event

MCMC a simulation process combining Markov chains and the Monte Carlo procedure

model a mathematical system used to understand another mathematical, physical, biological, or social system

Monte Carlo method a computer method to simulate probability distributions by taking random samples

multivariate containing many unknowns and variables

naïve Bayes a special, fast kind of Bayesian network

null hypothesis a plausible hypothesis that might explain a particular set of data; a null hypothesis can be compared with other alternatives

odds the ratio of the probabilities that an event will either occur or not occur

operations research or operational research a scientific approach to decision making

parameter in a mathematical expression, a quantity that is normally assumed to be constant; the value of the constant, however, can be changed as conditions are changed

posterior in Bayes' theorem, the probability of a conclusion after evidence has been considered

prior the probability of a hypothesis before new data is observed

probability the mathematics of uncertainty; the numerical measure of uncertainty

p-values an experiment provides statistically significant evidence against a hypothesis if the outcome (or a more extreme outcome) has only a small probability (under the hypothesis) of having occurred by chance alone. The p-value associated with an experimental test is the smallest significance level under which the null hypothesis is reflected.

rotors the geared wheels on Enigma machines

sampling the selection of a finite number of observations in order to learn about a much larger statistical population

sequential analysis the continuous analysis of data as they arrive while taking into account the effect of previous data

statistics a branch of applied mathematics that measures uncertainty and examines its consequences

stopping rule a sampling method in which data are evaluated as they are collected; the sampling stops when significant results are obtained

subjective probability Bayesian probability, a measure of personal belief in a particular hypothesis

transforms mathematical tools that change one kind of function into another that is easier to use

bibliography

Abbreviations

JASA	Journal of the American Statistical Association
JRSS	Journal of the Royal Statistical Society
OC	Oeuvres Complètes de Laplace
PCAS	Proceedings of the Casualty Actuarial Society

Part I. Enlightenment and the Anti-Bayesian Reaction

Chapter 1. Causes in the Air

Bayes, Joshua. *Sermons and Funeral Orations.* English Short Title database of 18th-century microfilms. Reels 7358 no. 08; 7324 no. 06; 7426 no. 03; and 7355 no. 08.

Bayes, Thomas. (1731) "Divine benevolence: Or, an attempt to prove that the principal end of the divine providence and government is the happiness of his creatures." London.

———. (1763) *An Introduction to the Doctrine of Fluxions, and Defence of the Mathematicians against the Objections of the Author of the Analyst.* Printed for J. Noon, London. The Eighteenth Century Research Publications Microfilm A 7173 reel 3774 no. 06.

Bayes, Thomas, and Richard Price. (1763) An essay towards solving a problem in the doctrine of chances. By the late Rev. Mr. Bayes, F.R.S. Communicated by Mr. Price, in a letter to John Canton, A.M.F.R.S. A letter from the late Reverend Mr. Thomas Bayes, F.R.S., to John Canton, M.A. and F.R.S. Author(s): Mr. Bayes and Mr. Price. *Philosophical Transactions* (1683–1775) (53) 370–418. Royal Society. The original Bayes–Price article.

Bebb, ED. (1935) *Nonconformity and Social and Economic Life 1660–1800.* London: Epworth Press.

Bellhouse, David R. (2002) On some recently discovered manuscripts of Thomas Bayes. *Historia Mathematica* (29) 383–94.

———. (2007a) The Reverend Thomas Bayes, FRS: A biography to celebrate the tercentenary of his birth. *Statistical Science* (19:1) 3–43. With Dale (2003) the main source for Bayes' life.

————. (2007b) Lord Stanhope's papers on the Doctrine of Chances. *Historia Mathematica* (34) 173–86.

Bru, Bernard. (1987) Preface in *Thomas Bayes. Essai en vue de résoudre un problème de la doctrine des chances*, trans. and ed., J-P Cléro. Paris.

————. (1988) Estimations laplaciennes. Un exemple: La recherche de la population d'un grand empire, 1785–1812. *J. Soc. Stat. Paris* (129) 6–45.

Cantor, Geoffrey. (1984) Berkeley's The Analyst revisited. *Isis* (75) Dec. 668–83.

Chesterfield PDS. (1901) *Letters to His Son: On the Fine Art of Becoming a Man of the World and a Gentleman*. M. Walter Dunne.

Cone, Carl B. (1952) *Torchbearer of Freedom: The Influence of Richard Price on Eighteenth-Century Thought*. University of Kentucky Press.

Dale, Andrew I. (1988) On Bayes' theorem and the inverse Bernoulli theorem. *Historia Mathematica* (15) 348–60.

————. (1991) Thomas Bayes's work on infinite series. *Historia Mathematica* (18) 312–27.

————. (1999) *A History of Inverse Probability from Thomas Bayes to Karl Pearson*. 2d ed. Springer. One of the foundational works in the history of probability.

————. (2003) *Most Honourable Remembrance: The Life and Work of Thomas Bayes*. Springer. With Bellhouse, the main source for Bayes' life.

Daston, Lorraine. (1988) *Classical Probability in the Enlightenment*. Princeton University Press.

Deming WE, ed. (1940) *Facsimiles of Two Papers by Bayes, With Commentaries by W. E. Deming and E. C. Molina*. Graduate School. Department of Agriculture. Washington, D.C.

Earman, John. (1990) Bayes' Bayesianism. *Studies in History and Philosophy of Science* (21) 351–70.

————. (2002) Bayes, Hume, Price, and miracles. In *Bayes's Theorem*, ed., Richard Swinburne. 91–109.

Gillies, Donald A. (1987) Was Bayes a Bayesian? *Historia Mathematica* (14) 325–46.

Haakonssen, Knud. (1996) *Enlightenment and Religion: Rational Dissent in Eighteenth-Century Britain*. Cambridge University Press.

Hacking, Ian. (1990) *The Taming of Chance*. Cambridge University Press.

————. (1991) Bayes, Thomas. *Biographical Dictionary of Mathematicians* vol. 1, Charles Scribner's Sons.

Hald, Anders. (1990) *A History of Probability and Statistics and Their Applications before 1750*. John Wiley and Sons.

————. (1998) *A History of Mathematical Statistics from 1750 to 1930*. John Wiley and Sons. A classic.

Hembry, Phyllis M. (1990) *The English Spa 1560–1815: A Short History*. Athlone Press, Fairleigh Dickinson University Press.

Holder, Rodney D. (1998) Hume on miracles: Bayesian interpretation, multiple testimony, and the existence of God. *British Journal for the Philosophy of Science* (49) 49–65.

Holland JD. (1968) An eighteenth-century pioneer Richard Price, D.D., F.R.S. (1723–1791). *Notes and Records of the Royal Society of London* (23) 43–64.

Hume, David. (1748) *An Enquiry Concerning Human Understanding*. Widely available. Online see Project Gutenberg.

Jacob, Margaret C. (1976) *The Newtonians and the English Revolution 1689–1720*. Cornell University Press.

Jesseph DM. (1993) *Berkeley's Philosophy of Mathematics*. University of Chicago Press.

Klein, Lawrence E. (1994) *Shaftesbury and the Culture of Politeness: Moral Discourse and Cultural Politics in Early Eighteenth-Century England*. Cambridge University Press.

Miller, Peter N. (1994) *Defining the Common Good: Empire, Religion and Philosophy in Eighteenth Century England*. Cambridge University Press.

Owen, David. (1987) Hume versus Price on miracles and prior probabilities: Testimony and the Bayesian calculation. *Philosophical Quarterly* (37) 187–202.

Price, Richard. *Four Dissertations*. 3d ed. (1772). Dissertation IV: On the nature of historical evidence and miracles. Google online.

Sobel, Jordan Howard. (1987) On the evidence of testimony for miracles: A Bayesian interpretation of David Hume's analysis. *Philosophical Quarterly* (37:147) 166–86.

Stanhope G, Gooch GP. (1914) *The Life of Charles Third Earl Stanhope*. Longmans, Green.

Statistical Science (2004) Issue devoted to Thomas Bayes. (19:1) Many useful articles.

Stigler, Stephen M. (1983). Who discovered Bayes's Theorem? *American Statistician* 37 290–96.

———. (1986). The History of Statistics: The Measurement of Uncertainty before 1900. Belknap Press of Harvard University Press. A classic and the place to start for Thomas Bayes.

———. (1999). Statistics on the Table: The History of Statistical Concepts and Methods. Harvard University Press.

Thomas, DO. (1977) *The Honest Mind: The Thought and Work of Richard Price*. Clarendon Press. Thomas is the authority on Price and edited his correspondence. Thomas, DO, and Peach, WB. (–1994) Eds. *The Correspondence of Richard Price*, vols 1-3. Duke University Press.

Watts, Michael R. (1978) *The Dissenters* vols. 1 and 2. Clarendon Press.

Chapter 2. The Man Who Did Everything

Albrecht, Peter. (1998) What do you think of smallpox inoculations? A crucial question in the eighteenth century, not only for physicians. In *The Skeptical Tradition around 1800*, eds., J. van der Zande, RH Popkin. Dordrecht, Kluwer Academic Publishers. 283–96.

Arago, F. (1875) Laplace: Eulogy before the French Academy. Trans. Baden Powell. Smithsonian Institution. Annual report 1874 in Congressional Papers for 43rd Congress. Washington, D.C., 129–68.

Arbuthnot J. (1711) An argument for Divine Providence, taken from the constant regularity observed in the births of both sexes. *Philos. Trans. Roy. Soc., Lond.* (27) 186–90.

Baker, Keith Michael. (1975) *Condorcet: From Natural Philosophy to Social Mathematics*. University of Chicago Press.

Biot, JB. (1850) Une anecdote relative à M. Laplace. *Journal Des Savants* 65–71.

Buffon. (1774) A Monsieur de la Place. *Journal Officiel* May 24, 1879, p. 4262; and *Comptes Rendus Hebdomadaires des Séances de l'Académie des Sciences* (88) 1879, 1019. I am indebted to Roger Hahn for this letter.

Bugge T, Crosland M. (1969) *Science in France in the Revolutionary Era*. Society for the History of Technology and MIT Press.

Clark W, Golinski J, Schaffer S. (1999) *The Sciences in Enlightened Europe*. University of Chicago Press.

Condorcet, Jean-Antoine-Nicolas de Caritat, B. Bru, P. Crépel (1994) *Condorcet, Arithmétique politique Textes rares ou inédits (1767–1789)*. Presses Universitaires de France.

Crosland, Maurice P. (1967) *The Society of Arcueil; A View of French Science at the Time of Napoleon I*. Harvard University Press.

Dale, Andrew I. (1995) *Pierre-Simon Laplace: Philosophical Essay on Probabilities*. Trans. and notes by Dale. Springer-Verlag.

———. (1999) *A History of Inverse Probability from Thomas Bayes to Karl Pearson*. 2d ed. Springer. One of the foundational works in the history of probability.

Daston, Lorraine. (1979) D'Alembert's critique of probability theory. *Historia Mathematica* (6) 259–79.

Dhombres, Jean. (1989) Books: reshaping science. In *Revolution in Print: The Press in France 1775-1800*, eds., R Darnton, D Roche. University of California Press. 177–202.

Doel, Ronald E. (1990) Theories and origins in planetary physics. *Isis* (90) 563–68.

Dreyer JLF, ed. (1912) *The Scientific Papers of Sir William Herschel*, vol. I. London: Royal Society and Royal Astronomical Society.

Dunnington, G. Waldo. *Carl Friedrich Gauss: Titan of Science*. Exposition Press, 1955.

Duveen DI, Hahn R. (1957) Laplace's succession to Bézout's post of examinateur des élèves de l'artillerie. *Isis* (48) 416–27.

Duveen DI, Hahn R (1958) Deux Lettres de Laplace à Lavoisier. *Revue d'histoire des Sciences et de leurs Applications* (11:4) 337–42.

Fourier, Joseph. (1830) *Historical Eulogy of the M. le Marquis de Laplace*. Trans. RW Haskins. Buffalo; and (1831) Eloge historique de M. le marquis de Laplace, prononcé . . . le 15 juin 1829, MASIF (10) lxxxi-cii.

Fox, Robert. (1974) The rise and fall of Laplacian Physics. *Historical Studies in the Physical Sciences* (4) 89–136.

———. (1987) La professionalisation: un concept pour l'historien de la science française au XIXe siècle. *History and Technology* (4) 413–22.

Gillispie, Charles C. (1972) Probability and politics: Laplace, Condorcet, and Turgot. *Proceedings of the American Philosphical Society* (116) 1–20.

———. (1978) Laplace, Pierre-Simon, Marquis De. *Dictionary of Scientific Biography*, Supplement I, Vol. XV. Charles Scribner's Sons.

———. (1979) Mémoires inédits ou anonymes de Laplace. *Revue d'Histoire des Sciences et de Leurs Applications* (32) 223–80.

———. (2004) *Science and Polity in France: The End of the Old Regime. And Science and Polity in France: The Revolutionary and Napoleonic Years*. Princeton University Press.

Gillispie, CC, with Robert Fox and Ivor Grattan-Guinness. (1997) *Pierre-Simon Laplace 1749–1827: A Life in Exact Science*. Princeton University Press.

Greenberg, John. (1986) Mathematical physics in eighteenth-century France. *Isis* (77) 59–78.

Grimaux, Édouard. (1888) *Lavoisier 1743–1794*. Alcan, Paris.

Grimsley, Ronald. (1963) *Jean d'Alembert 1717–83*. Clarendon Press.

Guerlac, Henry. (1976) Chemistry as a branch of physics: Laplace's collaboration with Lavoisier. *Historical Studies in the Physical Sciences* (7) 193–276.

Hahn, Roger. (1955) Laplace's religious views. *Archives Internationals d'Histoire des Sciences* (8) 38–40.

———. (1967a) *Laplace as a Newtonian Scientist*. William A. Clark Memorial Library 1967.

———. (1967b) Laplace's first formulation of scientific determinism in 1773. *Nadbitka. Actes du XIe Congrès International d'Histoire des Sciences*. (2) 167–171.

———. (1969) Élite scientifique et démocratie politique dans la France révolutionnaire. *Dix-Huitième Siècle* (1) 252–73.

———. (1976) Scientific careers in eighteenth-century France. In *The Emergence of Science in Western Europe*, ed., Maurice Crossland. New York: Science History Publications.

———. (1981) Laplace and the vanishing role of God in the physical universe. In *The Analytic Spirit*, ed., Harry Woolf. Cornell University Press. 85–95.

———. (1987a) Changing patterns for the support of scientists from Louis XIV to Napoleon. *History and Technology* (4) 401–11.

———. (1987b) Laplace and Boscovich. *Proceedings of the Bicentennial Commemoration of R. G. Boscovich*, eds., M. Bossi, P. Tucci. Milan.

———. (1989) The triumph of scientific activity: From Louis XVI to Napoleon. *Proceedings of the Annual Meeting of the Western Society for French History* (16) 204–11.

———. (1990) The Laplacean view of calculation. In *The Quantifying Spirit in the 18th Century*, eds., T Frängsmyr, HL Heilbron, RE Rider. University of California Press. 363–80.

———. (1994) Le role de Laplace à l'École Polytechnique. In *La Formation polytechnicienne, 1794–1994*, eds., B. Belhoste, A. Dahan and A. Picon. Seyssel.

———. (1995) Lavoisier et ses collaborateurs: Une Équipe au Travail. In *Il y a 200 Ans Lavoisier*, ed., C. Demeulenaere-Douyère, Paris: Technique et Documentation Lavoisier. 55–63.

———. (2005) *Pierre Simon Laplace, 1749–1827: A Determined Scientist*. Harvard University Press; (2004) *Le Système du Monde: Pierre Simon Laplace, Un Itinéraire dans la Science*. Trans. Patrick Hersant. Éditions Gallimard. These are the same book, the original in English, the translation in French. These books are my primary sources for Laplace's life.

Hald, Anders. (1998) *A History of Mathematical Statistics from 1750 to 1930*. John Wiley and Sons. A classic.

Hankins, Thomas L. (1970) *Jean d'Alembert: Science and the Enlightenment*. Oxford University Press.

———. (1985) *Science and the Enlightenment*. Cambridge University Press.

Harte, Henry H. (1830) *On the System of the World*. English translation of Laplace's *Exposition du système du monde*. Dublin University Press.

Heilbron HL. (1990) Introductory essay and The measure of enlightenment. In *The Quantifying Spirit in the 18th Century*, eds., T Frängsmyr, HL Heilbron, RE Rider. University of California Press. 1–24, 207–41.

Herivel, John. (1975) *Joseph Fourier: The Man and the Physicist*. Clarendon Press.

Holmes, Frederick Lawrence. (1961) *Antoine Lavoisier—The Next Crucial Year; or, the Sources of His Quantitative Method in Chemistry.* Cornell University Press.

Koda H, Bolton A. (2006) *Dangerous Liaisons: Fashion and Furniture in the Eighteenth Century.* Metropolitan Museum of Art and Yale University Press.

Laplace, Pierre Simon. (1878–1912) *Oeuvres Complètes de Laplace.* 14 vols. National Bibliothèque de la France. Available online. In some cases, two dates are given; the first is the year when Laplace read his paper to the academy, the second when it was published. We are indebted to Charles C. Gillispie for rationalizing Laplace's publication dates.

———. (1773) Recherches: 1° sur l'intégration des équations différentielles aux différences finies, and sur leur usage dans la théorie des hasards; 2° Sur le principe de la gravitation universelle, et sur les inégalités séculaires des planètes qui en dépendent. (February 10, 1773) *OC* (8) 69–197, 198–275.

———. (1774a) Mémoire sur la probabilité des causes par les événements. *OC* (8) 27–69. This is Laplace's discovery of inverse probability, his first version of what is now known as Bayes' rule. For the English translation with modern mathematical notation, see Stigler (1986).

———. (1774b) Mémoire sur les suites récurro-récurrentes et sur leurs usages dans la théorie des hasards. *OC* (8) 5–24.

———. (1776) Sur le principe de la gravitation universelle et sur les inégalités séculaires des planètes qui en dépendent. *OC* (8) 201–78.

———. (1778/81) Mémoire sur les Probabilités. *OC* (9) 383–485.

———. (1782/1785) Mémoire sur les approximations des formules qui sont fonctions de très grands nombres. *OC* (10) 209–294.

———. (1783/1786) Mémoire sur les approximations des formules qui sont fonctions de très grands nombres (suite). *OC* (10) 295–338.

———. (1783/1786) Sur les naissances, les mariages, et les morts à Paris, depuis 1771 jusqu'en 1784, et dans toute l'étendue de la France, pendant les années 1781 et 1782. *OC* (11) 35–49.

———. (1787/1788) Sur l'équation séculaire de la Lune. *OC* (11) 243–71.

———. (1788) Théorie de Jupiter et de Saturne. *OC* (11) 95–207 and (Suite) 211–39.

———. (1800) *Séances des Écoles Normales, recueilles par des Sténographes, et revues par les Professeurs,* nouvelle édition, tome sixième, Paris: Imprimerie du Cercle-Social.

———. (1815/1818) Sur l'*Application du calcul des probabilités à la philosophie naturelle. OC* (13) 98–116.

———. (1818) Troisième Supplément, Application du calcul des probabilités aux opérations géodésiques. *OC* (7) 531–80.

———. (1826/1829) Mémoire sur les deux grandes inégalités de Jupiter et de Saturne. *OC* (13) 313–33.

Lind, Vera. (1998) Skepticism and the discourse about suicide in the eighteenth century. In *The Skeptical Tradition around 1700*, eds., J van der Zande, RH Popkin. Kluwer Academic Publishers. 296–313.

Lindberg DC, Numbers RL, eds. (1986) *God and Nature: Historical Essays on the Encounter between Christianity and Science.* University of California Press.

———. (2003) *When Science and Christianity Meet.* University of Chicago Press.

Luna, Frederick A. de. (1991) The Dean Street style of revolution: J.-P. Brissot, jeune philosophe. *French Historical Studies* (17:1) 159–90.

Maréchal, Pierre Sylvain. (VIII) *Dictionnaire des athées anciens et modernes.* Paris. 231–32.

Marmottan, Paul. (1897). *Lettres de Madame de Laplace à Élisa Napoléon, princesse de Lucques et de Piombino.* Paris.

Mazzotti, Massimo. (1998) The geometers of God: Mathematics and reaction in the Kingdom of Naples. *Isis* (89) 674–701.

Numbers, Ronald L. (1977) *Creation by Natural Law: Laplace's Nebular Hypothesis in American Thought.* University of Washington Press.

Orieux, Jean. (1974) *Talleyrand: The Art of Survival.* Alfred A. Knopf. A classic about Laplace's era.

Outram, Dorinda. (1983) The ordeal of vocation: The Paris Academy of Sciences and the Terror, 1793–95. *History of Science* (21) 251–73.

Parcaut M et al. (1979) History of France. *Encyclopaedia Britannica* (7) 611–81.

Pelseneer, Jean. (1946) La religion de Laplace. *Isis* (36) 158–60.

Poirier, Jean-Pierre. (1998) *Lavoisier: Chemist, Biologist, Economist.* Trans. Rebecca Balinski. University of Pennsylvania Press.

Porter, Roy. (2003) Introduction. In *The Cambridge History of Science* (4) *Eighteenth-Century Science,* ed., R Porter. Cambridge University Press.

Rappaport, Rhoda. (1981) The liberties of the Paris Academy of Sciences 1716–1785. In *The Analytic Spirit,* ed., Harry Woolf. Cornell University Press. 225–53. Richards, Joan L. (2006) Historical mathematics in the French eighteenth century. *Isis* (97) 700–713.

Roche, Daniel. (1998) *France in the Enlightenment.* Trans. Arthur Goldhammer. Harvard University Press. A classic.

Sarton, George. (1941) Laplace's religion. *Isis* (33) 309–12.

Shafer, Glenn. (1990) The unity and diversity of probability. *Statistical Science* (5:4) 435–62.

Shepherd W. (1814) *Paris in Eighteen Hundred and Two and Eighteen Hundred and Fifteen.* 2d ed. M. Carey.

Simon, Lao G. (1931) The influence of French mathematicians at the end of the eighteenth century upon the teaching of mathematics in American colleges. *Isis* (15) 104–23.

Stigler SM. (1975) Napoleonic statistics: The work of Laplace. *Biometrika* (62:2) 503–17.

———. (1978). Laplace's early work: Chronology and citations. *Isis* (69) 234–54.

———. (1982) Thomas Bayes's Bayesian Inference. *JRSS, A.* (145) Part 2, 250–58. Bayes' article with modern mathematical notation and Stigler's commentary. The starting place for anyone interested in Bayes.

———. (1983) Who discovered Bayes's Theorem? *American Statistician* (37) 290–96.

———. (1986a) Laplace's 1774 memoir on inverse probability. *Statistical Science* (1) 359–63. Stigler's English translation with modern mathematical notation. The best place to read this famous paper.

———. (1986b). *The History of Statistics: The Measurement of Uncertainty before 1900.* Belknap Press of Harvard University Press. A classic.

———. (2003) Casanova's lottery. *University of Chicago Record* (37:4) 2–5.

Todhunter I. (1865) *A History of the Mathematical Theory of Probability: From the Time of Pascal to that of Laplace.* Cambridge University Press.

Ulbricht, Otto. (1998) The debate about capital punishment and skepticism in late enlightenment Germany. In *The Skeptical Tradition around 1800*, eds., J Van der Zande, RH Popkin. Kluwer Academic Publishers. 315–28.

Union des Physiciens de Caen. (1999) L'Année Laplace. Section Académique de Caen, http://www.udppc.asso.fr/section/caen/caen.htm

Voltaire. (1961) On the Church of England, on the Presbyterians, on Academies. In *Philosophical Letters.* Bobbs-Merrill.

Williams, L. Pearce. (1956) Science, education and Napoleon I. *Isis* (47) 369–82.

Zabell SL. (1988) Buffon, Price, and Laplace: Scientific attribution in the 18th century. *Archive for the History of Exact Sciences* (39) 173–82.

Chapter 3. Many Doubts, Few Defenders

Alexander, R. Amir. (2006) Tragic mathematics: Romantic narratives and the refounding of mathematics in the early nineteenth century. *Isis* (97) 714–26.

Anonymous. (August 27, 1899) Traps Mercier and Maurel: Capt. Freystaetter convicts both of giving false evidence—Bertillon affords more amusement. *New York Times* 1,2. By a courtroom reporter at Dreyfus's trial.

Anonymous. (1964) Edward C. Molina. *American Statistician* (18:3) 36.

Barnard, George A. (1947) Review: [untitled]. *JASA* (42:240) 658–65.

Bell ET. (1937) *Men of Mathematics.* Touchstone 1986 edition.

Bellamy, Paul B. (1997) *A History of Workmen's Compensation 1898–1915: From Courtroom to Boardroom.* Garland Publishing.

Bennett JH, ed. (1990) *Statistical Inference and Analysis: Selected Correspondence of R. A. Fisher.* Clarendon Press.

Bolt, Bruce A. (1991) Sir Harold Jeffreys and geophysical inverse problems. *Chance* (4:2) 15–17.

Box, Joan Fisher. (1978) *R. A. Fisher: The Life of a Scientist.* John Wiley.

Broemling, Lyle D. (2002) The Bayesian contributions of Edmond Lhoste. *ISBA Bulletin* 3–4.

Bru, Bernard. (1993) Doeblin's life and work from his correspondence. *Contemporary Mathematics* (149) 1–64.

———. (1996) Problème de l'efficacité du tir à l'école d'artillerie de Metz. Aspects théoriques et expérimentaux. *Mathématiques et sciences humaines* (136) 29–42.

———. (1999) Borel, Lévy, Neyman, Pearson et les autres. *MATAPLI* (60) 51–60.

———. (2006) Les leçons de calcul des probabilités de Joseph Bertrand: Les Lois du hasard. Journ@l électronique d'Histoire des Probabilités et de la Statistique/Electronic Journal of History of Probability and Statistics. (2:2) www.jehps.net.

Clerke, Agnes Mary. (1911) Laplace. *Encyclopaedia Britannica* (16) 200–203.

Cochran WM. (1976) Early development of techniques in comparative experimentation. In *On the History of Statistics and Probability*, ed., DB Owen. Marcel Dekker. 1–26.

Cook, Alan. (1990) Sir Harold Jeffreys, 2 April 1891–18 March 1989. *Biographical Memoirs of Fellows of the Royal Society* (36) 302–33.

Crépel, Pierre. (1993) Henri et la droite de Henry. *MATAPLI* (36) 19–22.

Dale, Andrew I. (1999) *A History of Inverse Probability from Thomas Bayes to Karl Pearson*. 2d ed. Springer. One of the foundational works in the history of probability.

Daston, Lorraine J. (1987) The domestication of risk: mathematical probability and insurance 1650–1830. In *The Probabilistic Revolution I*, eds., L Krüger, L Daston, M Heidelberger. MIT Press. 237–60.

———. (1994) How probabilities came to be objective and subjective. *Historia Mathematica* (21) 330–44.

Daston L, Galison P. (2007) *Objectivity*. Zone Books.

David, Florence Nightingale. (1962) *Games, Gods and Gambling*. Charles Griffin. 1998 Dover Edition.

Dawson, Cree S., et al. (2000) Operations research at Bell Laboratories through the 1970s: Part 1. *Operations Research* (48) 205–15.

De Finetti, Bruno. (1972) *Probability, Induction and Statistics: The Art of Guessing*. John Wiley and Sons.

Edwards AWF. (1994) R. A. Fisher on Karl Pearson. *Notes and Records of the Royal Society of London* (48) 97–106.

———. (1997) What did Fisher mean by "Inverse Probability" in 1912–1922? *Statistical Science* (12) 177–84.

Efron, Bradley. (1998) R. A. Fisher in the 21st century. *Statistical Science* (13) 95–122.

Estienne JE. (March 10, 1890) "Étude sur les erreurs d'observation." In *Archives de L'Institut de France*. Académie des Sciences.

———. (1892) "La probabilité de plusieurs causes étant connue, à quelle cause est-il plausible d'attribuer l'arrivée de l'évènement?" *Comptes Rendus des Séances de L'Académie des Sciences* (114:semester 1892) 1223.

———. (1905/6) *Loisirs d'Artilleurs*. Berger-Levrault.

Fagen MD, ed. (1975) *The History of Engineering and Science in the Bell System: The History of the Early Years 1875–1925*. Vol. 1. Bell Telephone Laboratories Inc.

Fienberg, Stephen E. (1992) Brief history of statistics in three and one-half chapters: A review essay. *Statistical Science* (7) 208–25.

Filon LNG. (1936) Karl Pearson 1857–1936. *Obituary Notices of the Royal Society* (2) 73–110.

Fisher, Arne. (1916) Note on an application of Bayes' rule in the classification of hazards in experience rating. *PCAS* (3) 43–48.

Fisher, Ronald Aylmer. (1925) *Statistical Methods for Research Workers*. Oliver and Boyd.

Gigerenzer G, Swijtink Z, Porter T, Daston L, Beatty J, Krüger L. (1989) *The Empire of Chance: How Probability Changed Science And Everyday Life*. Cambridge University Press. Many useful articles.

Hacking, Ian. [1989] Was there a probabilistic revolution 1800–1930? In *The Probabilistic Revolution*, eds., L Krüger, LJ Daston, and M Heidelberger. Vol. 1. MIT Press.

Hald, Anders. (1998) *A History of Mathematical Statistics from 1750 to 1930*. John Wiley and Sons. A classic.

Howie, David. (2002) *Interpreting Probability: Controversies and Developments in the Early Twentieth Century*. Cambridge University Press. The Fisher-Jeffreys debate.

Huzurbazar, Vassant S. (1991) Sir Harold Jeffreys: Recollections of a student. *Chance* (4:2) 18–21.

Jeffreys, Bertha Swirles. (1991) Harold Jeffreys: Some reminiscences. *Chance* (4:2) 22–26.

Jeffreys, Harold. (1939, 1948, 1961). *Theory of Probability*. Clarendon Press.

Kass RE, Raftery AE. (1995) Bayes factors. *JASA* (90:430) 773–95.

Kaye, David H. (2007) Revisiting Dreyfus: A more complete account of a trial by mathematics. *Minnesota Law Review* (91:3) 825–35.

Knopoff, Leon. (1991) Sir Harold Jeffreys: The Earth: Its origin, history, and physical constitution. *Chance* (4:2) 24–26.

Kolmogorov AN, Yushkevich AP. (1992) *Mathematics of the 19th Century*, vol. 1. Birkäuser Verlag.

Krüger L, Daston L, Heidelberger M, eds. (1987) *The Probabilistic Revolution*, vol. 1: *Ideas in History*. MIT Press.

Krüger L, Gigerenzer G, Morgan M, eds. (1987) *The Probabilistic Revolution*, vol. 2: *Ideas in History*. MIT Press.

Kruskal, William. (1980) The Significance of Fisher: A review of *R. A. Fisher: The Life of a Scientist*. *Journal of the American Statistical Association* (75) 1019–30.

Le process Dreyfus devant le conseil de guerre de Rennes (7 aout-9 septembre 1899): compte-rendu sténographique in extenso. (1899) http://gallica2.bnf.fr/ark:/12148/bpt6k242524.zoom.r=procès.f335.langEN.tableDesMatieres.

Lightman, Bernard. (2007) *Victorian Popularizers of Science*. University of Chicago Press.

Lindley DV. (1983) Transcription of a conversation between Sir Harold Jeffreys and Professor D. V. Lindley from a videotape made on behalf of the Royal Statistical Society. In St. John's College, Cambridge, UK, Papers of Sir Harold Jeffreys A25.

———. (1986a) On re-reading Jeffreys. In *Pacific Statistical Congress*, eds., IS Francis, BFJ Manly, FC Lam. Elsevier.

———. (1986b) Bruno de Finetti, 1906–1985. *JRSS Series A (General)* (149) 252.

———. (1989) Obituary: Harold Jeffreys, 1891–1989. *JRSS Series A (Statistics in Society)* (152:3) 417–18.

———. (1991) Sir Harold Jeffreys. *Chance* (4:2) 10–14, 21.

Loveland, Jeff. (2001) Buffon, the certainty of sunrise, and the probabilistic *reductio ad absurdum*. *Archives of the History of Exact Sciences* (55) 465–77.

MacKenzie, Donald A. (1981) *Statistics in Britain 1865–1930: The Social Construction of Scientific Knowledge*. Edinburgh University Press.

———. (1989) Probability and statistics in historical perspective. *Isis* (80) 116–24.

Magnello, M. Eileen. (1996) Karl Pearson's Gresham Lectures: W. F. R. Weldon, speciation and the origins of Pearsonian statistics. *British Journal for the History of Science* (29:1) 43–63.

Marie, Maximilien. (1883–88) *Histoire des sciences mathématiques et physiques*. Vol. 10. Paris: Gauthier-Villars. 69–98.

Mellor DH. (1995) Better than the stars: A radio portrait of Frank Ramsey. *Philosophy* (70) 243–62. The original version was broadcast by BBC Radio 3 February 27, 1978. http://www. Dar.cam.ac.uk/~dhm11/RanseyLect.html. Acc. May 21, 2004.

Millman S. (1984) *The History of Communications and Sciences (1925–1980)*. Vol. 5. AT&T Bell Labs.

Miranti, Paul J. Jr. (2002) Corporate learning and traffic management at the Bell System, 1900–1929: Probability theory and the evolution of organizational capabilities. *Business History Review* (76:4) 733–65.

Molina, Edward C. (1913) Computation formula for the probability of an event happening at least C times in N trials. *American Mathematical Monthly* (20) 190–93.

———. (1922) The theory of probabilities applied to telephone trunking problems. *Bell System Technical Journal* (1) 69–81.

———. (1931) Bayes' theorem. *Annals of Mathematical Statistics* (2) 23–27.

———. (1941) Commentary in *Facsimiles of Two Papers by Bayes*, ed. W. Edwards Deming. Graduate School, Department of Agriculture.

———. (1946) Some fundamental curves for the solution of sampling problems. *Annals of Mathematical Statistics* (17) 325–35.

Moore, Calvin C. (2007) *Mathematics at Berkeley: A History*. AK Peters.

Morehead EJ. (1989) *Our Yesterdays: The History of the Actuarial Profession in North America 1809–1979*. Society of Actuaries.

Morgan, Augustus de. (1839) Laplace. *Penny Cylcopaedia of the Society for the Diffusion of Useful Knowledge*. London: 1833–46. (13) 325–28.

Mowbray AH. (1914–15) How extensive a payroll exposure is necessary to give a dependable pure premium? *PCAS* (1) 24–30.

———. (1915) The determination of pure premiums for minor classifications on which the experience data is insufficient for direct estimate. *PCAS* (2) 124–33.

Neyman J. (1934) Statistical problems in agricultural experiment. With discussion. *Supplement to the JRSS*. (2:2) 107–80. (Discussion pp. 154–80.)

———. (1976) The emergence of mathematical statistics: a historical sketch with particular reference to the United States. In *On the History of Statistics and Probability*, ed., DB Owen. Marcel Dekker. 147–94.

Olkin, Ingram. (1992) A conversation with Churchill Eisenhart. *Statistical Science* (7) 512–30.

Otis, Stanley L. (1914–15) A Letter of Historical Interest. *PCAS* (1) 8–9.

Pearson, Egon S. (1925) Bayes' theorem examined in the light of experimental sampling. *Biometrika* (17) 388–442.

———. (1936 and 1937) Karl Pearson: An appreciation of some aspects of his life and work. *Biometrika* (28) 193–257 and (29) 161–248.

———. (1962) Some thoughts on statistical inference. *Annals of Mathematical Statistics*. (33:2) 394–403.

———. (1968) Studies in the history of probability and statistics. XX: Some early correspondence between W. S. Gosset, R. A. Fisher, and Karl Pearson, with notes and comments. *Biometrika* (55:3) 445–57.

Pearson, Karl. (1892) *The Grammar of Science*. W. Scott.

———. (1901) *The Ethic of Freethought and Other Addresses and Essays*. 2d ed. Charles Black.

———. (1912) *Social Problems: Their Treatment, Past, Present, and Future*. Dulau.

———. (1929) Laplace, being extracts from lectures delivered by Karl Pearson. *Biometrika* (21) 202–16.

Perryman, Francis S. (1937) Experience rating plan credibilities. *PCAS* (24) 60–125.

Porter, Theodore M. (1986) *The Rise of Statistical Thinking, 1820–1900.* Princeton University Press.

———. (1995) *Trust in Numbers: The Pursuit of Objectivity in Science and Public Life.* Princeton University Press.

———. (2003) Statistics and physical theories. In *The Cambridge History of Science.* Vol. 5: *The Modern Physical and Mathematical Sciences,* ed. Mary Jo Nye. Cambridge University Press.

———. (2004) *Karl Pearson: The Scientific Life in a Statistical Age.* Princeton University Press.

Pruitt, Dudley M. (1964) The first fifty years. *PCAS* (51) 148–81.

Reed, Lowell J. (1936) Statistical treatment of biological problems in irradiation. In *Biological Effects of Radiation,* Vol. 1, ed. Benjamin M. Duggar, 227–51. McGraw-Hill.

Reid, Constance. (1982) *Neyman—from Life.* Springer-Verlag.

Rubinow, Isaac M. (1913) *Social Insurance.* Henry Holt.

———. (1914–15) Scientific Methods of Computing Compensation Rates. *PCAS* (1) 10–23.

———. (1915) Liability loss reserves. *PCAS* (1:3) 279–95.

———. (1917) The theory and practice of law differentials. *PCAS* (4) 8–44.

———. (1934) A letter. *PCAS* (21).

Schindler GE Jr., ed. (1982) *A History of Switching Technology (1925–1975).* Vol. 3. Bell Telephone Laboratories.

Searle GR, ed. (1976) *Eugenics and Politics in Britain 1900–1914.* Noordhoff International Publishing.

Seddik-Ameur, Nacira. (2003) Les tests de normalité de Lhoste. *Mathematics and Social Sciences/Mathématiques et Sciences Humaines* (162: summer) 19–43.

Stigler, Stephen M. (1986). *The History of Statistics: The Measurement of Uncertainty before 1900.* Belknap Press of Harvard University Press. A classic.

———. (1999). *Statistics on the Table: The History of Statistical Concepts and Methods.* Harvard University Press.

Taqqu, Murad S. (2001) Bachelier and his times: A conversation with Bernard Bru. *Finance and Stochastics* (5) 3–32.

Whitney, Albert W. (1918) The theory of experience rating. *PCAS* (4) 274–92.

Wilhelmsen L. (1958) Actuarial activity in general insurance in the northern countries of Europe. *ASTIN Bulletin* (1) 22–27.

Wilkinson RI. (1955) An appreciation of E. C. Molina. *First International Teletraffic Congress, Copenhagen, June 20th–June 23rd, 1955.* International Teletraffic Congress 30–31.

Willoughby, William Franklin. (1898) *Workingmen's Insurance.* Thomas Y. Crowell.

Zabell, Sandy L. (1989) R. A. Fisher on the history of inverse probability. *Statistical Science* (4) 247–56.

———. (1989) The Rule of Succession. *Erkenntnis* (31) 283–321.

———. (1992) R. A. Fisher and fiducial argument. *Statistical Science* (7) 369–87.

Ziliak ST, McCloskey DN. (2009) *The Cult of Statistical Significance: How the Standard Error Costs Us Jobs, Justice, and Lives.* University of Michigan Press.

Part II. Second World War Era

Chapter 4. Bayes Goes to War

Andresen, Scott L. (Nov.–Dec. 2001) Donald Michie: Secrets of Colossus revealed. *IEEE Intelligent Systems* 82–83.

Arnold, VI. (2004) A.N. Kolmogorov and natural science. *Russian Math. Surveys* (59:1) 27–46.

Barnard, GA & Plackett, RL. (1985) Statistics in the United Kingdom, 1939–45. In AC Atkinson, SE Fienberg, eds., *A Celebration of Statistics*. Springer-Verlag. 31–55.

Barnard GA. (1986) Rescuing our manufacturing industry—some of the statistical problems. *The Statistician* (35) 3–16.

Bauer FL. (2000) *Decrypted Secrets*. Springer.

Burroughs J, Lieberman D, Reeds J. (2009) The secret life of Andy Gleason. *Notices of the American Mathematical Society* (in draft).

Booss-Bavnbek B, Hoeyrup J. (2003) *Mathematics and War*. Birkhäuser Verlag.

Britton JL. (1992) *Collected Works of A. M. Turing: Pure Mathematics*. North-Holland.

Budiansky, Stephen. (2000) *Battle of Wits: The Complete Story of Codebreaking in World War II*. Free Press.

Carter, Frank L. (1998) *Codebreaking with the Colossus Computer*. Bletchley Park Trust.

———. (2008) *Breaking Naval Enigma*. Bletchley Park Trust.

Champagne L, Carl RG, Hill R. (2003) Multi-agent techniques: Hunting U-boats in the Bay of Biscay. *Proceedings of SimTecT* May 26–29, Adelaide, Australia.

Chentsov, Nikolai N. (1990) The unfathomable influence of Kolmogorov. *Annals of Statistics* (18:03) 987–98.

Churchill, Winston. (1949) *Their Finest Hour*. Houghton Mifflin.

Collins, Graham P. (October 14, 2002) Claude E. Shannon: Founder of Information Theory. *Scientific American* 14ff.

Copeland, B. Jack, ed. (2004) *The Essential Turing*. Clarendon Press. Essential essays.

Copeland BJ et al. (2006) *Colossus: The Secrets of Bletchley Park's Codebreaking Computers*. Oxford University Press. Essential essays.

Eisenhart, Churchill. (1977) The birth of sequential analysis (obituary note on retired RAdm. Garret Lansing Schuyler). *Amstat News* (33:3).

Epstein R, Robert G, Beber G., eds. (2008) *Parsing the Turing Test: Philosophical and Methodical Issues in the Quest for the Thinking Computer*. Springer.

Erskine, Ralph. (October 2006) The Poles reveal their secrets: Alastair Denniston's account of the July 1939 meeting at Pyry. *Cryptologia* (30) 204–305.

Fagen MD. (1978) *The History of Engineering and Science in the Bell System: National Service in War and Peace (1925–1975)*. Vol. 2. Bell Telephone Labs.

Feferman AB, Feferman S. (2004) *Alfred Tarski: Life and Logic*. Cambridge University Press.

Fienberg SE. (1985) Statistical developments in World War II: An international perspective. In *A Celebration of Statistics*, eds., AC Atkinson, SE Fienberg. Springer-Verlag. 25–30.

Gandy R.O, Yates C.E.M., eds. (2001) *Collected Works of A. M. Turing: Mathematical Logic*. North-Holland.

Good, Irving John. (1950) *Probability and the Weighing of Evidence*. London: Charles Griffin.

————. (1958) The interaction algorithm and practical Fourier analysis. *JRSS. Series B.* (20) 361–72.

————. (1958) Significance tests in parallel and in series. *JASA* (53) 799–813.

————. (1965) *The Estimation of Probabilities: An Essay on Modern Bayesian Methods.* Research Monograph 30, MIT Press.

————. (1979) Studies in the history of probability and statistics. XXXVII A. M. Turing's statistical work in World War II. *Biometrika* (66:2) 393–96. Reprinted with Introductory Remarks in *Pure Mathematics*, ed., JR Britton, vol. of *Collected Works of A.M. Turing.* North-Holland, 1992.

————. (1983) *Good Thinking: The Foundations of Probability and Its Applications.* University of Minnesota Press.

————. (1984) A Bayesian approach in the philosophy of inference. *British Journal for the Philosophy of Science* (35) 161–66.

————. (2000) Turing's anticipation of Empirical Bayes in connection with the cryptanalysis of the Naval Enigma. *Journal of Statistical Computation and Simulation* (66) 101–11, and in Gandy and Yates.

Hinsley FH, Stripp A, eds. (1993) *Codebreakers: The Inside Story of Bletchley Park.* Oxford University Press.

Hilton, Peter. (2000) Reminiscences and reflections of a codebreaker. In *Coding Theory and Cryptography: From Enigma and Geheimschreiber to Quantum Theory*, ed., WD Joyner. Springer-Verlag. 1–8.

Hodges, Andrew. (1983, 2000) *Alan Turing: The Enigma.* Walker. A classic.

————. The Alan Turing Webpage. http://www.turing.org.uk/turing/.

————. (2000) Turing, a natural philosopher. Routledge. In *The Great Philosophers*, eds., R. Monk and F. Raphael. Weidenfeld and Nicolson.

————. (2002) Alan Turing—a Cambridge Scientific Mind. In *Cambridge Scientific Minds*, eds., Peter Harmon, Simon Mitton. Cambridge University Press.

Hosgood, Steven. http://tallyho.bc.nu/~steve/banburismus.html.

Kahn, David. (1967) *The Codebreakers: The Story of Secret Writing.* Macmillan. A classic.

Kendall, David G. (1991a) Kolmogorov as I remember him. *Statistical Science* (6:3) 303–12.

————. (1991b) Andrei Nikolaevich Kolmogorov. 25 April 1903–20 October 1987. *Biographical Memoirs of Fellows of the Royal Society.* (37) 300–319.

Kolmogorov, Andrei N. (1942) Determination of the center of scattering and the measure of accuracy by a limited number of observations. *Izvestiia Akademii nauk SSSR. Series Mathematics* (6) 3–32. In Russian.

Kolmogorov AN, Hewitt E. (1948) *Collection of Articles on the Theory of Firing.* Rand Publications. Edited by Kolmogorov and translated by Hewitt.

Koopman, Bernard Osgood. (1946) *OEG Report No. 56, Search and Screening.* Operations Evaluation Group, Office of the Chief of Naval Operations, Navy Department, Washington, D.C.

————. (1980) *Searching and Screening: General Principles with Historical Applications.* Pergamon Press.

Kozaczuk, Wladyslaw. (1984) *Enigma.* Trans. Christopher Kasparek. University Publications of America.

Kuratowski, Kazimierz. (1980) *A Half Century of Polish Mathematics, Remembrances and Reflections.* Pergamon Press.

Lee, JAN (1994) Interviews with I. Jack Good and Donald Michie, 1992. http://ei.cs.vt .edu/~history/Good.html. Downloaded February 14, 2006.

Michie, Donald. (unpublished) Turingery and Turing's sequential Bayes Rule. I am indebted to Jack Copeland and the Michie family for letting me read this draft chapter.

Milllman S, ed. (1984) *The History of Communications Sciences (1925–1980).* Vol. 5. AT&T Bell Labs.

Morison, Samuel Eliot. (2001) *The Battle of the Atlantic: September 1939–May 1943.* University of Illinois Press (1947 edition by Little, Brown).

Morse PM, Kimball GE. (1951) *Methods of Operations Research.* Technology Press of MIT and John Wiley and Sons.

Morse PM. (1982) In Memoriam: Bernard Osgood Koopman, 1900–1981. *Operations Research* (30) viii, 417–27.

Newman MHA. (1953) Alan Mathison Turing, 1912–1954. *Biographical Memoirs of Fellows of the Royal Society* (1) 253–63.

Randell B. (1980) The Colossus. In *A History of Computing in the Twentieth Century: A Collection of Essays,* eds., Metropolis N, Howlett J, Rota G-C. Academic Press.

Rejewski M. (1981) How Polish mathematicians deciphered the Enigma. *Annals of the History of Computing* (3) 223.

Rukhin, Andrew L. (1990) Kolmogorov's contributions to mathematical statistics. *Annals of Statistics* (18:3) 1011–16.

Sales, Tony. www.codesandciphers.org.uk/aescv.htm.

Shannon, Claude E. (July, October 1948) A mathematical theory of communication. *Bell System Technical Journal* (27) 379–423, 623–56.

———. (1949) Communication theory of secrecy systems. netlab.cs.ucla.edu/wiki/files/ Shannon1949.pdf. Acc. March 31, 2007.

Shiryaev, Albert N. (1989) Kolmogorov: Life and Creative Activities. *Annals of Probability* (17:3) 866–944.

———. (1991) Everything about Kolmogorov was unusual. *Statistical Science* (6:3) 313–18.

———. (2003) On the defense work of A. N. Kolmogorov during World War II. In *Mathematics and War,* eds., B. Booss-Bavnbek & J. Hoeyrup. Birkhaeuser.

Sloane NJA, Wyner AD., eds. (1993) *Claude Elwood Shannon: Collected Papers.* IEEE Press.

Syrett, David. (2002) *The Battle of the Atlantic and Signals Intelligence: U-Boat Tracking Papers, 1941–1947.* Navy Records Society.

Turing, Alan M. (1942) *Report by Dr. A. M. Turing, Ph.D. and Report on Cryptographic Machinery Available at Navy Department, Washington.* http://www.turing.org.uk/sources/washington. html. Accessed June 2, 2009.

———. (1986) *A. M. Turing's Ace Report of 1946 and Other Papers,* eds., BE Carpenter, RW Doran. MIT Press.

Waddington CH. (1973) *O.R. in World War 2: Operations Research against the U-Boat.* Scientific Books.

Weierud, Frode. http://cryptocellar.web.cern.ch/cryptocellar/Enigma/index.html. A central archives for the history of cryptanalysis during the Second World War.

Welchman, Gordon. (1983) *The Hut Six Story: Breaking the Enigma Codes*. McGraw-Hill.

Wiener, Norbert. (1956) *I Am a Mathematician*. MIT Press.

Zabell SL. (1995) Alan Turing and the Central Limit Theorem. *American Mathematical Monthly* (102:6) 483–94.

Chapter 5. Dead and Buried Again

Box GEP, Tiao GC. (1973) *Bayesian Inference in Statistical Analysis*. Addison-Wesley.

Cox, Gertrude. (1957) "Statistical frontiers." *JASA* (52) 1–12.

DeGroot, Morris H. (1986a) A conversation with David Blackwell. *Statistical Science* (1:1) 40–53.

Erickson WA, ed. (1981) *The Writings of Leonard Jimmie Savage: A Memorial Selection*. American Statistical Association and Institute of Mathematical Statistics.

Fienberg, Stephen E. (2006) When did Bayesian inference become Bayesian? *Bayesian Analysis* (1) 1–40.

Lindley, Dennis V. (1957) Comments on Cox. In *Breakthroughs in Statistics I*, eds., NL Johnson and S Kotz. xxxviii.

Perks, Wilfred. (1947) Some observations on inverse probability including a new indifference rule. *Journal of the Institute of Actuaries* (73) 285–334.

Reid, Constance. (1982) *Neyman—from Life*. Springer-Verlag.

Sampson AR, Spencer B, Savage IR. (1999) A conversation with I. Richard Savage. *Statistical Science* (14) 126–48.

Part III. The Glorious Revival

Chapter 6. Arthur Bailey

Albers, Donald J. (1983) *Mathematical People*. Birkhäuser.

Bailey, Arthur L. (1929) *A Summary of Advanced Statistical Methods*. United Fruit Co. Research Department. Reprinted 1931 as Circular no. 7.

———. (1942, 1943) Sampling Theory in Casualty Insurance, Parts I through VII. *PCAS* (29) 50–93 and (30) 31–65.

———. (1945) A generalized theory of credibility. *PCAS* (32) 13–20.

———. (1948) Workmen's compensation D-ratio revisions. *PCAS* (35) 26–39.

———. (1950) Credibility procedures: Laplace's generalization of Bayes' rule and the combination of collateral knowledge with observed data. *PCAS* (37) 7–23. Six discussions of this paper and the author's reply are in the same volume at 94–115.

———. (1950) Discussion of Introduction to Credibility Theory by L. H. Longley-Cook. Reprint of 1950 discussion in *PCAS* (1963) (50) 59–61.

Bailey, Robert A, Simon LJ. (1959) An actuarial note on the credibility of experience of a single private passenger car. *PCAS* (46) 159–64.

Bailey RA, Simon LJ. (1960) Two studies in automobile insurance ratemaking. *PCAS* (47) 1–19. This was later reprinted in ASTIN Bulletin (1) 192–217.

Bailey RA. (1961) Experience rating reassessed. *PCAS* (48) 60–82.

Borch, Karl. (1963) Recent developments in economic theory and their application to insurance. *ASTIN Bulletin* (2) 322–41.

Bühlmann, Hans. (1967) Experience rating and credibility. *ASTIN Bulletin* (4) 199–207.

Bühlmann H, Straub E. (1970) Credibility for loss ratios. *Bulletin of the Swiss Association of Actuaries*: (70) 111–33. English trans. by C.E. Brooks.

Carr, William HA. (1967) *Perils: Named and Unnamed. The Story of the Insurance Company of North America.* McGraw-Hill.

Cox, Gertrude. (1957) "Statistical frontiers." *JASA* (52) 1–12.

DeGroot, Morris H. (1986a) A conversation with David Blackwell. *Statistical Science* (1:1) 40–53.

Hachemeister, Charles A. (1974) Credibility for regression models with application to trend. In *Credibility: Theory and Applications*, ed., P. M. Kahn, 129–64.

Hewitt, Charles C., Jr. (1964, 1965, 1969). Discussion. *PCAS* (51) 44–45; (52) 121–27; (56) 78–82.

Hewitt CC Jr. (1970) Credibility for severity. *PCAS* (57) 148–71.

———. (1975) Credibility for the layman. In *Credibility (Theory and Applications)*, Academic Press and in *Proceedings of the Berkeley Actuarial Research Conference on Credibility*, September 19–21.

Hickman, James C., and Heacox, Linda. (1999) Credibility theory: The cornerstone of actuarial science. *North American Actuarial Journal* (3:2) 1–8.

Jewell, William S. (2004) Bayesian statistics. *Encyclopedia of Actuarial Science.* Wiley. 153–66.

Kahn, PM. (1975) *Credibility: Theory and Applications.* Academic Press.

Klugman SA, Panjer HH, Willmot GE. (1998) *Loss Models: From Data to Decisions.* John Wiley and Sons.

Longley-Cook, Laurence H. (1958) The casualty actuarial society and actuarial studies in development of non-life insurance in North America. *ASTIN Bulletin* (1) 28–31.

———. (1962) An introduction to credibility theory. *PCAS* (49) 184–221.

———. (1964) Early actuarial studies in the field of property and liability insurance. *PCAS* (51) 140–47.

———. (1972) Actuarial aspects of industry problems. *PCAS* (49) Part II 104–8.

Lundberg, Ove. (1966) Une note sur des systèmes de tarification basées sur des modèles du type Poisson composé. *ASTIN Bulletin* (4) 49–58.

Mayerson, Allen L. (1964) A Bayesian view of credibility. *PCAS* (51) 85–104.

Miller RB, Hickman JC. (1974) Insurance credibility theory and Bayesian estimation. In *Credibility: Theory and Applications*, ed. PM Kahn, 249–70.

Miller Robert B. (1989) Actuarial applications of Bayesian statistics. In *Bayesian Analysis in Econometrics and Statistics: Essays in Honor of Harold Jeffreys*, ed. Arnold Zellner. Robert E. Krieger.

Morris C, Van Slyke L. (1978) Empirical Bayes methods for pricing insurance classes. *Proceedings of the Business and Economics Statistics Section.* Statweb.byu.edu/faculty/gwf/revnaaj.pdf.

Perks, Wilfred. (1947) Some observations on inverse probability including a new indifference rule. *Journal of the Institute of Actuaries* (73) 285–334.

Taylor GC. (1977) Abstract credibility. *Scandinavian Actuarial Journal* 149–68.

———. (1979) Credibility analysis of a general hierarchical model. *Scandinavian Actuarial Journal* 1–12.

Venter, Gary G. (fall 1987) Credibility. *CAS Forum* 81–147.

Chapter 7. From Tool to Theology

Armitage P. (1994) Dennis Lindley: The first 70 years. In *Aspects of Uncertainty: A Tribute to D. V. Lindley*, eds., PR Freeman and AFM Smith. John Wiley and Sons.

Banks, David L. (1996) A Conversation with I. J. Good. *Statistical Science* (11) 1–19.

Dubins LE, Savage LJ. (1976) *Inequalities for Stochastic Processes (How to Gamble If You Must)*. Dover.

Box, George EP, et al. (2006) *Improving Almost Anything*. Wiley.

Box GEP, Tiao GC. (1973) *Bayesian Inference in Statistical Analysis*. Addison-Wesley.

Cramér, H. (1976). Half of a century of probability theory: Some personal recollections. *Annals of Probability* (4) 509–46.

D'Agostini, Giulio. (2005) The Fermi's Bayes theorem. *Bulletin of the International Society of Bayesian Analysis* (1) 1–4.

Edwards W, Lindman R, Savage LJ. (1963) Bayesian statistical inference for psychological research. *Psychological Review* (70) 193–242.

Erickson WA, ed. (1981) *The Writings of Leonard Jimmie Savage: A Memorial Selection*. American Statistical Association and Institute of Mathematical Statistics.

Ferguson, Thomas S. (1976) Development of the decision model. In DB Owen, ed., *On the History of Statistics and Probability*. Marcel Dekker. 333–46.

Fienberg, Stephen E. (2006) When did Bayesian inference become Bayesian? *Bayesian Analysis* (1) 1–40.

Johnson NL, Kotz S, eds. (1997) *Breakthroughs in Statistics*. Vols. 1–3. Springer. Important reprints of twentieth century articles, primarily post-1940s.

Kendall, Maurice G. (1968) On the future of statistics – a second look. *Journal of the Royal Statistical Society Series A* (131) 182–204.

Lindley, Dennis V. (1953) Statistical inference (with discussion). JRSS, Series B (15) 30–76.

———. (1957) A statistical paradox. *Biometrika* (44: 1/2) 187–92.

———. (1968) Decision making. *The Statistician* (18) 313–26.

———. (1980) L. J. Savage—his work in probability and statistics. *Annals of Statistics* (8) 1–24.

———. (1983) Theory and practice of Bayesian statistics. *The Statistician* (32) 1–11.

———. (1986) Savage revisited: Comment. *Statistical Science* (1) 486–88.

———. (1990) Good's work in probability, statistics and the philosophy of science. *J. Statistical Planning and Inference* (25) 211–23.

———. (2000) The philosophy of statistics. *The Statistician* (49) 293–337.

———. (2004) Bayesian thoughts. *Significance* (1) 73–75.

Mathews J, Walker RL. (1965) *Mathematical Methods of Physics*. W. A. Benjamin.

Old, Bruce S. (1961) The evolution of the Office of Naval Research. *Physics Today* (14) 30–35.

Rigby, Fred D. (1976) Pioneering in federal support of statistics research. In DB Owen, ed., *On the History of Statistics and Probability*. Marcel Dekker. 401-18.

Rivett, Patrick. (1995) Aspects of Uncertainty. [Review] *Journal of the Operational Research Society* (46) 663–70.

Sampson AR, Spencer B, Savage IR. (1999) A conversation with I. Richard Savage. *Statistical Science* (14) 126–48.

Savage LJ. (1954) *The Foundations of Statistics*. Wiley.

————. (1962) *The Foundations of Statistical Inference: A Discussion*. London: Methuen.

————. (1976) On rereading R. A. Fisher. *Annals of Statistics* (4) 441–500.

Schrödinger, Erwin. (1944) The statistical law of nature. *Nature* (153) 704-5.

Shafer, Glenn. (1986) Savage revisited. *Statistical Science* (1) 463–85.

Smith, Adrian. (1995) A conversation with Dennis Lindley. *Statistical Science* (10) 305–19.

Stephan FF. et al. (1965) Stanley S. Willks. *JASA* (60:312) 953.

Chapter 8. Jerome Cornfield, Lung Cancer, and Heart Attacks

Anonymous. (1980) Obituary: Jerome Cornfield 1912–1979. *Biometrics* (36) 357–58.

Armitage, Peter. (1995) Before and after Bradford Hill: Some trends in medical statistics. *JRSS, Series A* (158) 143–53.

Centers for Disease Control. (1999) Achievements in public health, 1900–1999: Decline in deaths from heart disease and stroke, United States, 1900–1999. *MMWR Weekly* (48:30) 649–56.

Cornfield, Jerome. (1951) A method of estimating comparative rates from clinical data: Applications to cancer of the lung, breast, and cervix. in *Breakthroughs in Statistics 3* (1993), eds., S. Kotz and NL Johnson. Springer. Introduction by Mitchell H. Gail.

————. (1962) Joint dependence of risk of coronary heart disease on serum cholesterol and systolic blood pressure: A discriminant function analysis. *Federation Proceedings* (21:4) Part II. July–August. Supplement no. 11.

————. (1967) Bayes Theorem. *Review of the International Statistical Institute* (35) 34–49.

————. (1969) The Bayesian outlook and its application. *Biometrics* (25:4) 617–57.

————. (1975) A statistician's apology. *JASA* (70) 7–14.

Doll, Richard. (1994) Austin Bradford Hill. *Biographical Memoirs of Fellows of the Royal Society*. (40) 129–40.

————. (2000) Smoking and lung cancer. *American Journal of Respiratory and Critical Care Medicine* (162:1) 4–6.

————. (1995) Sir Austin Bradford Hill: A personal view of his contribution to epidemiology. *JRSS: Series A (Statistics in Society)* (158) 155–63.

Duncan JW, Shelton WC. (1978) *Revolution in United States Government Statistics 1926–1976*. U.S. Department of Commerce.

Jerome Cornfield Memorial Issue. (March 1982) *Biometrics Supplement, Current Topics in Biostatistics and Epidemiology* (38).

Duncan JW, Shelton WC. (1978) *Revolution in United States Government Statistics 1926–1976*. U.S. Department of Commerce.

Gail, Mitchell H. (1996) Statistics in Action. *JASA* (91:322) 1–13.

Green, Sylvan B. (1997) A conversation with Fred Ederer. *Statistical Science* (12:2) 125–31.

Greenhouse SW, Halperin M. (1980) Jerome Cornfield, 1912–1979. *American Statistician* (34) 106–7.

Greenhouse, Samuel W. (1982) Jerome Cornfield's contributions to epidemiology. *Biometrics Supplement*. 33–45.

Greenhouse SW, Greenhouse JB. (1998) Cornfield, Jerome. *Encyclopedia of Biostatistics*, vol. 1, ed., P. Armitage, T. Colton. 955–59.

Kass RE, Greenhouse JB. Comment: A Bayesian perspective. *Statistical Science* (4:4) 310–17.

Memorial Symposium in honor of Jerome Cornfield. (1981) Jerome Cornfield: Curriculum, vitae, publications and personal reminiscences. From Fred Ederer–Jerome Cornfield Collection, Acc 1999–022, in the History of Medicine Division, National Library of Medicine.

National Cancer Institute. (1994) *Tobacco and the Clinician* (5) 1–22.

Sadowsky DA, Gilliam AG, Cornfield J. (1953) The statistical association between smoking and carcinoma of the lung. *Journal of the National Cancer Institute* (13:5) 1237–58.

Salsburg, David. (2001) *The Lady Tasting Tea: How Statistics Revolutionized Science in the Twentieth Century.* W. H. Freeman.

Truett J, Cornfield J, Kannel W. (1967) A multivariate analysis of the risk of coronary heart disease in Framingham. *Journal of Chronic Diseases* (20:7) 511–24.

Zelen, Marvin. (1982) The contributions of Jerome Cornfield to the theory of statistics in A Memorial symposium in honor of Jerome Cornfield, March 1982. *Biometrics* (38) 11–15.

Chapter 9. There's Always a First Time

Anonymous. (1991) U.S. nuclear weapons accidents; danger in our midst. *Defense Monitor.* Center for Defense Information, World Security Institute. http://www/Milnet.com. Acc. Jan. 25, 2007.

Caldwell, Dan. (1987) Permissive action links. *Survival* (29) 224–38.

Gott, Richard. (1963) The evolution of the independent British deterrent. *International Affairs (Royal Institute of International Affairs 1944-)* (39) 238–52.

Herken, Gregg. (1985) *Counsels of War.* Knopf.

Hounshell, David. (1997) The Cold War, RAND, and the generation of knowledge, 1946–1962. *Historical Studies in the Physical and Biological Sciences* (27) 237–67.

Iklé, Fred Charles. (1958) *The Social Impact of Bomb Destruction.* University of Oklahoma Press.

———. (2006) *Annihilation from Within: The Ultimate Threat to Nations.* Columbia University Press.

Iklé, Fred Charles, with Aronson GJ, Madansky A. (1958) On the risk of an accidental or unauthorized nuclear detonation. RM-2251 U.S. Air Force Project Rand. RAND Corp.

Jardini, David R. (1996) Out of the Blue Yonder: The RAND Corporation's Diversification into Social Welfare Research, 1946–1968. Dissertation, Carnegie Mellon University.

Kaplan, Fred. (1983) *The Wizards of Armageddon.* Simon and Schuster.

Madansky, Albert. (1964) *Externally Bayesian Groups.* RAND Corp.

———. (1990) Bayesian analysis with incompletely specified prior distributions. In *Bayesian and Likelihood Methods in Statistics and Econometrics: Essays in Honor of George A. Barnard,* ed. S. Geisser. North Holland. 423–36.

Mangravite, Andrew. (spring 2006) Cracking Bert's shell and loving the bomb. *Chemical Heritage* (24:1) 22.

Smith, Bruce LR. (1966) *The RAND Corporation: Case Study of a Nonprofit Advisory Corporation.* Harvard University Press.

U.S. Department of Defense. (April 1981) *Narrative Summaries of Accidents Involving U.S. Nuclear Weapons 1950–1980*. http://www.dod.mil/pubs/foi/reading_room/635.pdf. Acc. Jan. 29, 2007. I am indebted to the Center for Defense Information for this reference.

Wohlstetter SJ et al. (April 1954) *Selection and Use of Strategic Air Bases*. RAND Corporation Publication R266.

Wohlstetter, Albert. (1958) The delicate balance of terror. RAND Corp. Publication 1472.

Wyden, Peter. (June 3, 1961) The chances of accidental war. *Saturday Evening Post.*

Chapter 10. 46,656 Varieties

Anonymous. (1965) Bayes-Turing. *NSA Technical Journal*. I think IJ Good is the author.

———. (1971) Multiple hypothesis testing. *NSA Technical Journal*. 63–72.

———. (1972) The strength of the Bayes score. *NSA Technical Journal*. 87–111.

Bather, John. (1996) A conversation with Herman Chernoff. *Statistical Science* (11) 335–50.

Bennett, JH, ed. (1990) *Statistical Inference and Analysis: Selected Correspondence of R. A. Fisher*. Clarendon Press.

DeGroot, MH. (1986c) A conversation with Charles Stein. *Statistical Science* (1) 454–62.

Edwards W, Lindman H, Savage LJ. (1963) Bayesian statistical inference for psychological research. *Psychological Research* (70:3) 193–242.

Efron, Bradley. (1977) Stein's paradox in statistics. *Scientific American* (236) 119–27.

———. (1978) Controversies in the foundations of statistics. *American Mathematical Monthly* (85) 231–46.

Good IJ. (1971) 46656 Varieties of Bayesians. Letter to the Editor. *American Statistician* (25) 62–63.

Jahn, RG, Dunne BJ, Nelson RD. (1987) Engineering anomalies research. *Journal of Scientific Exploration* (1:1) 21–50.

James W, Stein CM. (1961) Estimation with quadratic loss function. *Proc. of the 4th Berkeley Symp. Math. Statist. and Prob.* (1) 361.

Jefferys, William H. (1990) Bayesian analysis of random event generator data. *Journal of Scientific Exploration* (4:2) 153–69.

Leahy FT. (1960) Bayes marches on. *NSA Technical Journal*. (U) 49–61.

———. (1964) Bayes factors. *NSA Technical Journal*. 1–5.

———. (1965) Bayes factors. *NSA Technical Journal*. 7–10.

Robbins, Herbert. (1956) An empirical Bayes approach to statistics. In *Proc. of the 3rd Berkeley Symp. Math. Statist. and Prob. 1954–1955*. (1) University of California Press. 157–63.

Smith, Adrian. (1995) A conversation with Dennis Lindley. *Statistical Science* 10 305–319.

Stein, Charles. (1956) Inadmissibility of the usual estimator for the mean of a multivariate normal distribution. In *Proc. of the 3rd Berkeley Symp. Math. Statist. and Prob., 1954–1955*, vol. I, ed., J. Neyman. University of California Press. 197–206.

Tribe, Laurence H. (1971a) Trial by mathematics: Precision and ritual in the legal process. *Harvard Law Review* (84:6) 1329–93.

———. (1971b) A further critique of mathematical proof. *Harvard Law Review* (84:8) 1810–20.

Zellner, Arnold. (2006) S. James Press and Bayesian analysis. *Macroeconomic Dynamics*. (10) 667–84.

Part IV. To Prove Its Worth

Chapter 11. Business Decisions

Note: HBSA GC File is in the Faculty Biography Collection, Harvard Business School Archives, Baker Library, Harvard Business School.

Aisner, Jim. (1994) Renowned Harvard Business School Professor Robert O. Schlaifer Dead at 79. Harvard University.

Anonymous. (1959) Interpretation and reinterpretation: The Chicago Meeting, 1959. *American Historical Review* (65) 733–86.

Anonymous (March 24, 1962) Math + Intuition = Decision. *Business Week* 54, 56, 60. HBSA Fac. Biography series GC 772.20, Harvard Business School Archives, Baker Library, Harvard Business School.

Anonymous. (November 22, 1963) *Harbus News*.

Anonymous. (November 3, 1971) Yale statistician Leonard Savage Dies; Authored book on gambling. *New Haven Register*, 23.

Anonymous. (October 1985) Schlaifer and Fuller retire. *HBS Bulletin.* 18–19. HBSA GC File, R.O. Schlaifer.

Anonymous. (1992) Schlaifer awarded Ramsey Medal. *Decision Analysis Society Newsletter* (11:2).

Bilstein, Roger E. (1977) Development of aircraft engines and fuels. *Technology and Culture* (18) 117–18.

Birnberg JG. (1964) Bayesian statistics: A review. *Journal of Accounting Research* (2) 108–16.

Fienberg, Stephen E. (2008) The early statistical years: 1947–1967. A conversation with Howard Raiffa. *Statistical Science* (23:1) 136–49.

Fienberg SE, Zellner A. (1975) *Studies in Bayesian Econometrics and Statistics: In Honor of Leonard J. Savage*. North-Holland.

Gottlieb, Morris J. (1960) Probability and statistics for business decisions. *Journal of Marketing* (25) 116–17.

Harvard University Statistics Department. http//www.stat.Harvard.edu/People/Department_History.html. Kemp, Freda. (2001) Applied statistical decision theory: Understanding robust and exploratory data analysis. *The Statistician* (50) 352–53.

Massie, Joseph L. (1951) Development of aircraft engines; development of aviation fuels. *Journal of Business of the University of Chicago* (24) 141–42.

McGinnis, John A. (November 22, 1963) "Only God can make a tree." *Harbus News.* HBSA Faculty Biography Series GC 772.20, Robert O. Schlaifer.

Memorial Service, Robert O. Schlaifer, Friday, December 2, 1994. HBSA GC 772.20 Faculty Biography.

Nocera, John. (1994) *A Piece of the Action: How the Middle Class Joined the Money Class*. Simon and Schuster.

Pratt JW, Raiffa H, Schlaifer R. (1964) The foundations of decision under uncertainty: An elementary exposition. *JASA* (59) 353–75.

———. (1965) *Introduction to Statistical Decision Theory*. McGraw-Hill.

Pratt, John W. (1985) [Savage Revisited]: Comment. *Statistical Science* (1) 498–99.

Raiffa, Howard. (1968) *Decision Analysis: Introductory Lectures on Choices under Uncertainty*. Addison-Wesley.

————. (2002) Tribute to Robert Wilson on his 65th Birthday. Berkeley Electronic Press. http://www.bepress.com/wilson/art2.

————. (2006) A Memoir: Analytical Roots of a Decision Scientist. Unpublished. I am indebted to Dr. Raiffa for letting me read and quote from his manuscript.

Raiffa H, Schlaifer R. (1961) *Applied Statistical Decision Theory*. MIT Press.

Ramsey Award Winners. (1988) Videotaped talk by Howard Raiffa, Ronald Howard, Peter C. Fishburn, and Ward Edwards at the Joint National Meeting of the Operations Research Society of America. San Diego. I am indebted to INFORMS for letting me watch the video.

Savage, Jimmie. (October 1, 1956) Letter to Committee on Statistics Faculty, Chicago. In Manuscripts and Archives, Yale University Library.

Saxon, Wolfgang. (July 28, 1994) Robert O. Schlaifer, 79, managerial economist. *New York Times*.

Schlaifer, Robert. (1936) Greek theories of slavery from Homer to Aristotle. *Harvard Studies in Classical Philology* (47) 165–204.

Schlaifer R, Heron SD. (1950) *Development of Aircraft Engines. Development of Aviation Fuels*. Graduate School of Business Administration, Harvard University.

Schlaifer, Robert O. (1959) *Probability and Statistics for Business Decisions: An Introduction to Managerial Economics under Uncertainty*. McGraw-Hill.

Chapter 12. Who Wrote *The Federalist?*

Albers DJ, Alexanderson GL, Reid C., eds. (1990) *More Mathematical People*. Harcourt Brace Jovanovich.

Brooks, E. Bruce. (2001) Tales of Statisticians: Frederick Mosteller. www.UMass.edu/wsp/statistics/tales/mosteller.html. Acc. December 21, 2004.

Chang, Kenneth. (July 27, 2006) C. Frederick Mosteller, a pioneer in statistics, dies at 89. *New York Times*.

Cochran WG, Mosteller F, Tukey JW. (1954) *Statistical Problems of the Kinsey Report on Sexual Behavior in the Human Male*. American Statistical Association.

Converse, Jean M. (1987) *Survey Research in the United States: Roots and Emergence 1890–1960*. University of California Press.

Fienberg SE, Hoaglin DC, eds. (2006) *Selected Papers of Frederick Mosteller*. Springer.

Fienberg SE et al., eds. (1990) *A Statistical Model: Frederick Mosteller's Contributions to Statistics, Science and Public Policy*. Springer-Verlag.

Hedley-Whyte J. (2007) Frederick Mosteller (1916–2006): Mentoring, A Memoir. *International Journal of Technology Assessment in Health Care* (23) 152–54.

Ingelfinger, Joseph, et al. (1987) *Biostatistics in Clinical Medicine*. Macmillan.

Jones, James H. (1997) *Alfred C. Kinsey: A Public/Private Life*. W. W. Norton.

Kinsey AC, Pomeroy WB, Martin CE. (1948) *Sexual Behavior in the Human Male*. WB Saunders.

Kolata, Gina Bari. (1979) Frederick Mosteller and applied statistics. *Science* (204) 397–98.

Kruskal W, Mosteller F. (1980) Representative sampling, IV: The history of the concept in statistics, 1895–1939. *International Statistical Review* (48) 169–95.

Mosteller, Frederick, et al. (1949) *The Pre-Election Polls of 1948*. Social Science Research Council.

Mosteller F, Tukey J. (1954) Data analysis, including statistics. In *The Collected Works of John W. Tukey*, vol. 4, ed. Lyle V. Jones. Wadsworth and Brooks.

Mosteller F, Rourke REK, Thomas GB Jr. (1961, 1970) *Probability with Statistical Applications*. Addison-Wesley.

Mosteller F, Wallace DL. (1964) *Inference and Disputed Authorship, The Federalist*. Addison-Wesley; and (1984) *Applied Bayesian and Classical Inference: The Case of the Federalist Papers*. Springer-Verlag. These two books are identical but meant for different audiences.

Mosteller F, Wallace DL. (1989) Deciding authorship. In *Statistics: A Guide to the Unknown*, eds., Judith M. Tanur et al. Wadsworth and Brooks/Cole. 115–25.

Petrosino, Anthony. (2004) Charles Frederick [Fred] Mosteller. JamesLindLibrary.org.

Squire, Peverill. (1988) Why the 1936 Literary Digest poll failed. *Public Opinion Quarterly* (52) 125–33.

Zeckhauser RJ, Keeney RL, Sebenius JK. (1996) *Wise Choices: Decisions, Games, and Negotiations*. Harvard Business School Press.

Chapter 13. The Cold Warrior

Anscombe, FR. (2003) Quiet contributor: the civic career and times of John W. Tukey. *Statistical Science* (18:3) 287–310.

Bamford, James. (1983) *The Puzzle Palace: A Report on America's Most Secret Agency*. Penguin.

Bean, Louis H. (1950) The pre-election polls of 1948 (review). *JASA* (45) 461–64.

Bell Labs. Memories of John W. Tukey. http://cm.bell-labs.com/cm/ms/departments/sia/tukey/ Acc. Feb. 27, 2007

Bell Telephone Labs. (1975–85) *The History of Engineering and Science in the Bell System*. Vols. 1–7.

Brillinger, David R. (2002a) John W. Tukey: His life and professional contributions. *Annals of Statistics* (30) 1535–75.

———. (2002b) John Wilder Tukey (1915–2000). *Notices of the American Mathematical Society* (49:2) 193–201.

———. (2002c) John W. Tukey's work on time series and spectrum analysis. *Annals of Statistics* (30) 1595–1618.

Casella G et al. (2003) Tribute to John W. Tukey. *Statistical Science* (18) 283–84.

Computer History Museum. "Selling the Computer Revolution." www.computerhistory.org/brochures/companies. Acc. March 7, 2007.

Dempster, Arthur P. (2002) John W. Tukey as "philosopher." *Annals of Statistics* (30) 1619–28.

Gnanadesikan R, Hoaglin DC. (1993) *A Discussion with Elizabeth and John Tukey, Parts I and II*. DVD. American Statistical Association.

Jones, Lyle V., ed. (1986) *The Collected Works of John W. Tukey*. Vol. III: *Philosophy and Principles of Data Analysis: 1949–1964*. Wadsworth and Brooks/Cole.

Leonhardt, David. (2000) John Tukey: Statistician who coined 2 crucial words. www.imstat.org/Bulletin/Sept2000/node18.html. Acc. April 7, 2007.

Link, Richard F. (1989) Election night on television. In *Statistics: A Guide to the Unknown*, eds., Judith M. Tanur et al. Wadsworth and Brooks/Cole. 104–12.

McCullagh, Peter. (2003) John Wilder Tukey. *Biographical Memoirs of the Fellows of the Royal Society London* (49) 537–55.

Mosteller F, Tukey JW. (1954) Data analysis, including statistics. In *The Collected Works of John W. Tukey*, vol. IV, ed. Lyle V. Jones.

Robinson, Daniel J. (1999) *The Measure of Democracy: Polling, Market Research, and Public Life, 1930–1945*. University of Toronto Press.

Tedesco, John. Huntley, Chet. Museum of Broadcast Communications. www.museum .tv/archives/etv/H/htmlH/huntleychet/huntleychet.htm. Acc. July 17, 2007.

Tukey, John W. (1962) The future of data analysis. *Annals of Mathematical Statistics* (33) 1–67.

———. (1984) *The Collected Works of John W. Tukey: Time Series: 1949–1964, 1965–1984*. Vols. 1, 2. Ed. David R. Brillinger. Wadsworth Advanced Books and Software.

———. (1984) *The Collected Works of John W. Tukey: Philosophy and Principles of Data Analysis, 1949–1953; 1965–1986*. Vols. 3, 4. Ed. Lyle V. Jones.Wadsworth Advanced Books and Software.

Waite CH, Brinkley, David. Museum of Broadcast Communications. www.museum.tv/ archives/etv/B/htmlB/brinkleydav/brinkleydav.html. Acc. July 16, 2007.

Chapter 14. Three Mile Island

Anonymous. (1982) Using experience to calculate nuclear risks. *Science* (217) 338.

Apostolakis, George E. (1988) Editorial: The interpretation of probability in probabilistic safety assessments. *Reliability Engineering and System Safety* (23) 247–52.

———. (1990) The concept of probability in safety assessments of technological systems. *Science* (250) 1359(6).

———. (2004) How useful is quantitative risk assessment? *Risk Analysis* (24) 515–20.

Barnett, Vic. (1973, 1982, 1999) Comparative Statistical Inference. John Wiley and Sons.

Bather, John. (1996) A conversation with Herman Chernoff. *Statistical Science* (11) 335–50.

Bier, Vicki M. (1999) Challenges to the acceptance of probabilistic risk analysis. *Society for Risk Analysis* (19) 703–10.

Cooke, Roger M. (1991) *Experts in Uncertainty: Opinion and Subjective Probability in Science*. Oxford University Press.

Feynman, Richard P. (1986) Appendix F: Personal observations on the reliability of the Shuttle. http://www.Science.ksc.nasa.gov/shuttle/missions/51-l/docs/rogers-commission/Appendix-F.txt.

———. (1987) Mr. Feynman Goes to Washington. *Engineering and Science*. California Institute of Technology. (vol. 50) 6–22.

Harris, Bernard. (2004) Mathematical methods in combating terrorism. *Risk Analysis* (24:4) 985–88.

Martz HF, Zimmer WJ. (1992) The risk of catastrophic failure of the solid rocket boosters on the space shuttle. *American Statistician* (46) 42–47.

Russell, Cristine. (1974) Study gives good odds on nuclear reactor safety. *BioScience* (24) 605–06.

Selvidge, Judith. (1973) A three-step procedure for assigning probabilities to rare events. In *Utility, Probability, and Human Decision Making*, ed. Dirk Wendt and Charles Vlek. D. Reidel Publishing.

U.S. Atomic Energy Commission. (1974) *Reactor Safety Study: An Assessment of Accident Risks in U.S. Commercial Nuclear Power Plants*. WASH-1400. NUREG-75/014. National Technical Information Service.

Webb, Richard E. (1976) *The Accident Hazards of Nuclear Power Plants*. University of Massachusetts Press.

Wilson, Richard. (2002) Resource letter: RA-1: Risk analysis. *American Journal of Physics* (70) 475–81.

Chapter 15. The Navy Searches

Andrews, Capt. Frank A., ed. (1966) *Aircraft Salvops Med: Sea Search and Recovery of an Unarmed Nuclear Weapon by Task Force 65, Interim Report*. (1966) Chief of Naval Operations, U.S. Navy.

Arkin, William, and Handler, Joshua. (1989) *Naval Accidents 1945–1988, Neptune Paper No. 3*. Greenpeace.

Associated Press (August 31, 1976) Soviet sub and U.S. frigate damaged in crash. *New York Times* 5; U.S. frigate and Soviet submarine collide. *The Times of London* 5.

Belkin, Barry. (1974) Appendix A: An alternative measure of search effectiveness for a clearance operation. Wagner, Associates. Unpublished.

Centers for Disease Control and Prevention (CDC). (April 20, 2005) Plutonium. Department of Health and Human Services Agency, CDC Radiation Emergencies Radio-isotope Brief. www.bt.cdc.gov/radiation Accessed July 29, 2006.

Church BW et al. (2000) *Comparative Plutonium-239 Dose Assessment for Three Desert Sites: Maralinga, Australia; Palomares, Spain; and the Nevada Test Site, USA Before and After Remedial Action*. Lawrence Livermore National Laboratory.

Craven, John Piña. (2002) *The Silent War: The Cold War Battle Beneath the Sea*. Simon and Schuster.

Feynman, Richard P. (1985) *Surely You're Joking, Mr. Feynman!* W. W. Norton.

Gonzalez II, Ruiz SS. (1983) *Doses from Potential Inhalation by People Living Near Plutonium Contaminated Areas*. Oak Ridge National Laboratory.

Handler J, Wickenheiser A, Arkin WM. *Naval Safety 1989: The Year of the Accident, Neptune Paper No. 4*. Greenpeace.

http/www.destroyersonline.com/usndd/ff1047/f1047pho.htm. DestroyersOnLine web page. *Voge* collision picture.

Lewis, Flora. (1967) *One of Our H-Bombs is Missing*. McGraw-Hill. A Pulitzer Prize winner.

Moody, Dewitt H. (2006) 40th anniversary of Palomares. *Faceplate* (10:2) 15–19.

Nicholson, John H. (1999) Foreword in *Under Ice: Waldo Lyon and the Development of the Arctic Submarine* by William Leary. Texas A&M University Press.

Otto, M. (1998) Course 'Filling Station.' Foreign Technology Division, Wright-Patterson AFB.

Place WM, Cobb FC, Defferding CG. (1975) *Palomares Summary Report*. Defense Nuclear Agency, Kirtland Air Force Base.

Reuters (October 11, 2006) Radioactive snails lead to Spain–U.S. atomic probe.

Richardson HR, Stone LD. (1971) Operations analysis during the underwater search for Scorpion. *Naval Research Logistics Quarterly* (18) 141–57.

———. (1984) Advances in search theory with application to petroleum exploration, Report to National Science Foundation. Daniel H. Wagner, Associates.

Richardson HR, Discenza JH. (1980) The United States Coast Guard Computer-Assisted Search Planning System (CASP). *Naval Research Logistics Quarterly* (27:4) 659–80.

Richardson HR, Weisinger JR. (1984) The search for lost satellites. *Proceedings of the 7th MIT-ONR Workshop on C3 Systems*, eds., M. Athans, A. Levis.

Richardson, Henry R. (1986) Search theory. Center for Naval Analyses, Alexandria, Va.

Richardson HR, Stone LD, Monarch WR, Discenza JH. (2003) *Proceedings of the SPIE Conference on Optics and Photonics, San Diego*. SPIE.

Sontag S, Drew C, with Drew AL. (1998) *Blind Man's Bluff: The Untold Story of American Submarine Espionage*. Public Affairs.

Stone, Lawrence D. (1975) *Theory of Optimal Search*. Academic Press.

———. (1983) The process of search planning: Current approaches and continuing problems. *Operations Research* (31) 207–33.

———. (1989) What's happened in search theory since the 1975 Lanchester Prize? *Operations Research* (37) 501–06.

———. (1990) Bayesian estimation of undiscovered pool sizes using the discovery record. *Mathematical Geology* (22) 309–32.

Stone LD, Barlow CA, Corwin TL. (1999) *Bayesian Multiple Target Tracking*. Artech House.

Taff LG. (1984) Optimal searches for asteroids. *Icarus* (57) 259–66.

U.S. Air Force Medical Service. (July 29, 2006) Air Force Releases Reports on Palomares, Spain and Thule Airbase, Greenland Nuclear Weapons Accidents. AFMS.mil/latestnews/palomares.htm. Acc. July 29, 2006.

U.S. Department of Energy. (Palomares, Spain medical surveillance and environmental monitoring. www.eh.doe.gov/health/ihp/indalo/spain/html. Acc. July 29, 2006.

Wagner, Daniel H. (1988) *History of Daniel H. Wagner, Associates 1963–1986*. Daniel H. Wagner, Associates.

Part V. Victory

Chapter 16. Eureka!

Alder, Berni J. (1990) Transcript of interview with Berni J. Alder conducted June 18, 1990. American Institute of Physics, Center for the History of Physics.

Bayarri MJ, Berger JO. (2004) The interplay of Bayesian and frequentist analysis. *Statistical Science* (19) 58–80.

Berger, James O. (2000) Bayesian analysis: A look at today and thoughts of tomorrow. *JASA* (95) 1269.

———. (2006) The case for objective Bayesian analysis. *Bayesian Analysis* (1:3) 385–402.

Besag, Julian. (1974) Spatial interaction in the statistical analysis of lattice systems. *JRSS B* (36) 192–236.

Britton JL, ed. (1992) *Collected Works of A. M. Turing: Pure Mathematics*. North-Holland.

Cappé O, Robert CP. (2000) Markov chain Monte Carlo: 10 years and still running! *JASA* (95) 1282–86.

Cooke, Roger M. (1991) *Experts in Uncertainty: Opinion and Subjective Probability in Science*. Oxford University Press.

Couzin, Jennifer. (2004) The new math of clinical trials. *Science* (303) 784–86.

DeGroot, Morris H. (1986b) A conversation with Persi Diaconis. *Statistical Science* (1:3) 319–34.

Diaconis P, Efron B. (1983) Computer-intensive methods in statistics. *Scientific American* (248) 116–30.

Diaconis, Persi. (1985) Bayesian statistics as honest work. *Proceedings of the Berkeley Conference in Honor of Jerzy Neyman and Jack Kiefer* (1), eds., Lucien M. Le Cam and Richard A. Olshen. Wadsworth.

Diaconis P, Holmes S. (1996) Are there still things to do in Bayesian statistics? *Erkenntnis* (45) 145–58.

Diaconis P. (1998) A place for philosophy? The rise of modeling in statistical science. *Quarterly of Applied Mathematics* (56:4) 797–805.

DuMouchel WH, Harris JE. (1983) Bayes methods for combining the results of cancer studies in humans and other species. *JASA* (78) 293–308.

Efron, Bradley. (1986) "Why isn't everyone a Bayesian?" *American Statistician* (40) 1.

Gelfand, Alan E. (2006) Looking back on 15 years of MCMC: Its impact on the statistical (and broader) research community. Transcript of Gelfand's speech when awarded Parzen Prize for Statistical Innovation.

Gelfand AE, Smith AFM. (June 1990) Sampling-based approaches to calculating marginal densities. *JASA* (85:410) 398–409.

Gelfand AE et al. (December 1990) Illustration of Bayesian inference in normal data models using Gibbs Sampling. *JASA* (85:412) 972–85.

Geman S, Geman D. (1984) Stochastic relaxation, Gibbs distributions and the Bayesian restoration of images. *IEEE Trans. Pattern Anal. Mach. Intell.* (6) 721–40.

Gill, Jeff. (2002) *Bayesian Methods: A Social and Behavioral Sciences Approach*. Chapman and Hall.

Hanson KM. (1993) Introduction to Bayesian image analysis. In *Medical Imaging: Image Processing*, ed. MH Loew. Proc. SPIE (1898) 716–31.

Hastings WK. (1970) Monte Carlo sampling methods using Markov chains and their applications. *Biometrika* (57:1) 97–109.

Hively, Will. (1996) The mathematics of making up your mind. *Discover* (17:5) 90(8). Early popular-level description of Bayes.

Householder, Alston S. (1951) *Monte Carlo Method: Proceedings of a Symposium Held June 29, 30, and July 1, 1949*. National Bureau of Standards Applied Mathematics Series 12. v.

Hubert, Peter J. (1985) Data analysis: In search of an identity. In *Proceedings of the Berkeley Conference in Honor of Jerzy Neyman and Jack Kiefer*, eds., Lucien M. Le Cam and Richard A. Olshen. Wadsworth.

Hunt BR. (1977) Bayesian methods in nonlinear digital image restoration. *IEEE Transactions on Computers* (C-26:3) 219–29.

Kay, John. (2003) What is the chance of your being guilty? *Financial Times* (London). June 29, 21.

Kuhn, Thomas S. (1962). *The Structure of Scientific Revolutions*. University of Chicago Press.

Leonhardt, David. (2001) Adding art to the rigor of statistical science. *New York Times,* April 28. B 9.

Lindley, DV. (1965) *Introduction to Probability and Statistics from a Bayesian Viewpoint.* Cambridge University Press.

Luce, R. Duncan. (2003) Whatever happened to information theory in psychology? *Review of General Psychology* (7:2) 183–88.

Malakoff, David. (1999) "Bayes offers a 'new' way to make sense of numbers," "A brief guide to Bayes theorem," "The Reverend Bayes goes to court," and "An improbable statistician." *Science* (286:5444) 1460ff.

Markoff, John. (2000) Microsoft sees software 'agent' as way to avoid distractions. *New York Times,* July 17 C1.

Metropolis N, Ulam S. (1949) The Monte Carlo method. *JASA* (44:247) 335–41.

Metropolis, Nicholas. (1987) The beginning of the Monte Carlo method. *Los Alamos Science* (15) 125–30.

Neiman, Fraser D. (2000) Coincidence or causal connection? The relationship between Thomas Jefferson's visits to Monticello and Sally Heming's conceptions. *William and Mary Quarterly,* 3d ser. (57:1) 198–210.

Press, S. James (1986) [Why isn't everyone a Bayesian?]: Comment. *American Statistician* (40) 9–10.

Raftery, Adrian E. (1986) Choosing models for cross-classifications. *American Sociological Review* (51:1) 145–46.

Raftery AE, Zeh JE. (1998) Estimating bowhead whale population size and rate of increase from the 1993 census. *JASA* (93:442) 451–63.

Robert C, Casella G. (2008) A history of Markov chain Monte Carlo—subjective recollections from incomplete data. Unpublished draft kindly provided by C. Robert.

Royal Statistical Society. "News Release: Royal Statistical Society concerned by issues raised in Sally Clark case." October 23, 2001. http://www.rss.org.uk/archive/reports/sclark.html Acc. February 13, 2004.

Salsburg, David. (2001) *The Lady Tasting Tea: How Statistics Revolutionized Science in the Twentieth Century.* W. H. Freeman.

Sivia DS. (1996) *Data Analysis: A Bayesian Tutorial.* Clarendon Press.

Smith, Adrian F.M. (1983) Comment. *JASA* (78:382) 310–11.

———. (1984) Present position and potential developments—some personal views; Bayesian statistics. *JRSS A* (147) 245–59.

Spiegelhalter DJ et al. (1999) An Introduction to Bayesian methods in health technology assessment. *British Medical Journal* (319) 508–12.

Spiegelhalter DJ, Abrams KR, Myles JP. (2004) *Bayesian Approaches to Clinical Trials and Health-Care Evaluation.* John Wiley.

Spiegelhalter, David J. (2004) Incorporating Bayesian ideas into health-care evaluation. *Statistical Science* (19) 156–74.

Taylor BL, Gerrodette T. (1993) The uses of statistical power in conservation biology: the vaquita and Northern Spotted Owl. *Conservation Biology* (7) 489–787.

Weinberger, Steven E. (2008) Diagnostic evaluation and initial management of the solitary pulmonary nodule. Online in *UpToDate,* ed. Basow, DS. Waltham, Mass.

Chapter 17. Rosetta Stones

Abazov VM et al. (2007) Search for production of single top quarks via t c g and tug flavor-changing-neutral-current couplings. *Physical Review Letters* (99) 191802.

Anderson, Philip W. (1992) The Reverend Thomas Bayes, needles in haystacks, and the fifth force. *Physics Today* (45:1) 9, 11.

Aoki, Masanao. (1967) *The Optimization of Stochastic Systems.* Academic Press.

Berger JO. (2003) Could Fisher, Jeffreys and Neyman have agreed on testing? *Statistical Science* (18:1) 1–12.

Brockwell AE, Rojas AL, Kass RE. (2004) Recursive Bayesian decoding of motor cortical signals by particle filtering. *Journal of Neurophysiology* (91) 1899–1907.

Broemeling, Lyle D. (2007) *Bayesian Biostatistics and Diagnostic Medicine.* Chapman and Hall.

Brown, Emery N, et al. (1998) A statistical paradigm for neural spike train decoding applied to position prediction from ensemble firing patterns of rat hippocampal place cells. *Journal of Neuroscience* (18) 7411–25.

Campbell, Gregory. (2009) Bayesian statistics at the FDA: The trailblazing experience with medical devices. Emerging Issues in Clinical Trials, Rutgers Biostatistics Day, April 3, 2009. http://www.stat.rutgers.edu?iob/bioconf09/slides/campbell.pdf. Acc. October 8, 2009.

Committee on Fish Stock Assessment Methods, National Research Council. (1998) *Improving Fish Stock Assessments.* National Academy of Sciences.

Dawid, AP. (2002) Bayes's theorem and weighing evidence by juries. In *Bayes's Theorem,* ed. Richard Swinburne. 71–90.

Doya, Kenji, et al, eds. (2007) *Bayesian Brain: Probabilistic Approaches to Neural Coding.* MIT Press.

Efron, Bradley. (2005). Bayesians, frequentists, and scientists. *JASA* (469) 1–11.

———. (2005) Modern science and the Bayesian-frequentist controversy. www-stat.Stanford.edu/~brad/papers/NEW-Mod-Sci_2005. Acc. June 13, 2007.

———. (2006) Microarrays, Empirical Bayes, and the two-groups model. www-stat.Stanford.edu/brad/papers/twogroups.pdf Acc. June 13, 2007.

Frith, Chris. (2007) *Making Up the Mind: How the Brain Creates Our Mental World.* Blackwell.

Gastwirth JL, Johnson WO, Reneau DM. (1993) Bayesian analysis of screening data: Application to AIDS in blood donors. *Canadian Journal of Statistics* (19) 135–50.

Geisler WS, Kersten D. (2002) Illusions, perception and Bayes. *Nature Neuroscience* (5:6) 508–10.

Goodman J, Heckerman D. (2004) Fighting spam with statistics. *Significance* (1) 69–72.

Goodman, Steven N. (1999) Toward evidence-based medical statistics, Parts 1 and 2. *Annals of Internal Medicine* (130:12) 995–1013.

———. (2005) Introduction to Bayesian methods I: measuring the strength of evidence. *Clinical Trials* (2:4) 282–90.

Greenspan, Alan. (2004) Risk and uncertainty in monetary policy. With panel discussion by Martin Feldstein, Mervyn King, Janet L. Yellen. *American Economic Review* (94:2) 33–48.

Helm, Leslie. (Oct. 28, 1996) Improbable Inspiration. *Los Angeles Times* B1. http://www.LATimes.com.

Heuer, Richards J. Jr., ed. (1978) *Quantitative Approaches to Political Intelligence: The CIA Experience.* Westview.

Hillborn R, Mangel M. (1997) *The Ecological Detective: Confronting Models with Data.* Princeton University Press.

Hively, Will. (1996) The mathematics of making up your mind. *Discovery* (17) 98(8).

Kass, Robert E. (2006) Kinds of Bayesians (Comment on articles by Berger and by Goldstein). *Bayesian Analysis* (1) 437–40.

Kaye, David H. (in press) *The Double Helix and the Law of Evidence.* Harvard University Press.

Kaye DH, Bernstein D, Mnookin J. (2004) *The New Wigmore, A Treatise on Evidence: Expert Evidence.* Aspen Publishers.

Kersten D, Mamassian P, Yuille A. (2004) Object perception as Bayesian inference. *Annual Review of Psychology* (55) 271–305.

Kiani R, Shadlen MN. (2009) Representation of confidence associated with a decision by neurons in the parietal cortex. *Science* (324) 759–64.

Knill DC, Pouget A. (2004) The Bayesian brain: The role of uncertainty in neural coding and computation. *Trends in Neurosciences* (27) 712–19.

Körding KP, Wolpert DM. (2004) Bayesian integration in sensorimotor learning. *Nature* (427) 244–47.

Leamer, Edward E. (1983) Let's take the con out of econometrics. *American Economic Review* (73) 31–43.

Lebiere, Christian. (1999) The dynamics of cognition: An ACT-% model of cognitive arithmetic. *Kognitionswissenschaft* (8) 5–19.

Linden G, Smith B, York J. (2003) Amazon.com recommendations. *IEEE Internet Computing* (7:1) 76–80.

Ludlum, Robert. (2005) *The Ambler Warning.* St. Martin's.

O'Hagan A, Luce BR. (2003) *A Primer on Bayesian Statistics in Health Economics and Outcomes Research.* MEDTAP International.

Pearl, Judea. (1988) *Probabilistic Reasoning in Intelligence Systems: Networks of Plausible Inference.* Morgan Kaufman Publishers.

Pouget A et al. (2009) Neural Computations as Laplacian (or is it Bayesian?) probabilistic inference. In draft.

Quatse JT, Najmi A. (2007) Empirical Bayesian targeting. Proceedings, 2007 World Congress in Computer Science, Computer Engineering, and Applied Computing, June 25–28, 2007.

Schafer JB, Konstan J, Riedl J. (1999) Recommender systems in E-commerce. In *ACM Conference on Electronic Commerce* (EC-99) 158–66.

Schafer JB, Konstan J, Riedl J. (2001) Recommender systems in E-commerce. *Data Mining and Knowledge Discovery* (5) 115–53.

Schneider, Stephen H. (2005) *The Patient from Hell.* Perseus Books.

Spolsky, Joel. (2005) (http://www.joelonsoftware.com/items/2005/10/17.html).

Swinburne, Richard, ed. (2002) *Bayes's Theorem.* Oxford University Press.

Taylor BL et al. (2000) Incorporating uncertainty into management models for marine mammals. *Conservation Biology* (14) 1243–52.

Unwin, Stephen D. (2003) *The Probability of God: A Simple Calculation that Proves the Ultimate Proof.* Random House.

Wade, Paul R. (1999) A comparison of statistical methods for fitting population models to data. In *Marine Mammal Survey and Assessment Methods*, eds., Garner et al. Rotterdam: AA Balkema.

———. (2000) Bayesian methods in conservation biology. *Conservation Biology* (14) 1308–16.

———. (2001) Conservation of exploited species in an uncertain world: Novel methods and the failure of traditional techniques. In *Conservation of Exploited Species*, eds., JD Reynolds et al. Cambridge University Press. 110–44.

Weaver, Warren. (1955) Translation. In *Machine Translation of Languages: Fourteen Essays*, eds., WN Locke, AD Booth. MIT Technology Press and John Wiley.

———. (1963) *Lady Luck: The Theory of Probability.* Dover Publications.

Westerfield, H. Bradford, ed. (1995) *Inside CIA's Private World: Declassified Articles from the Agency's Internal Journal, 1955–1992.* Yale University Press.

Wolpert DM, Ghahramani Z. (2005) Bayes' rule in perception, action, and cognition. In *The Oxford Companion to the Mind*, ed., Gregory RL. Oxford Reference OnLine.

Wolpert DM. (December 8, 2005) The puppet master: How the brain controls the body. Francis Crick Lecture, Royal Society. Online.

Zellner, Arnold. (2006) Bayesian econometrics: Past, present and future. HGB Alexander Research Foundation, University of Chicago. Paper 0607.

Epilogue

Ferrante, Olivier. (2011) Interviews with author, telephone June 16, 2011, and Le Bourget, France, Sept. 23, 2011.

Ferrante O, Kutzleb M, Purcell M. (2011) AF447 Underwater search and recovery operations: A shared government-industry process. ISASI 2011 Conference.

Metron. (2011) "Search Analysis for the Location of the AF447 Underwater Wreckage," published on the BEA website Jan. 20, 2011, at http://www.bea.aero/fr/enquetes/vol.af.447/metron.search.analysis.pdf.

Stone, Lawrence D. (Aug. 2011) In search of Air France Flight 447. INFORMS-Today (38:4). http://www.informs.org/ORMS-Today/Public-Articles/August-Volume-38-Number-4/In-search-of-air-france-flight-447.

Appendixes

Campbell, Michael J. (2008) The doctor sees the light. *Significance* (5:4) 172.

Frith, Chris. (2007) *Making Up The Mind: How the Brain Creates our Mental World.* Blackwell.

Elmore Joann G et al. (April 16, 1998) Ten-year risk of false positive screening mammograms and clinical breast examinations. *New England Journal of Medicine* (338:16) 1089–96.

Kerlikowske Karla et al. (November 24, 1993) Positive predictive value of screening mammography by age and family history of breast cancer. *JAMA* (270: 20) 2444–50.

Kolata, Gina (November 23, 2009) Behind cancer guidelines, quest for data. *New York Times*.

National Cancer Institute (2009) Breast cancer screening: harms of screening. Accessed October 16 2009. pp. 1–4.

Weaver DL et al. (2006) Pathologic findings from the Breast Cancer Surveillance Consortium: population-based outcomes in women undergoing biopsy after screening mammography. *Cancer* (106) 732. Cited in Fletcher, Suzanne W. (2010) Screening for breast cancer. www.uptodate.com

reading list

For those who want to start learning to use Bayesian methods, here are some recommendations for introductory how-to-do-Bayes books at different levels of difficulty.

Most of these suggestions come from an annotated reading list published online by Jon Starkweather at the University of North Texas Computing and Information Technology Center. For details, consult Jon Starkweather (2011) *Go Forth and Propagate: Book Recommendations for Learning and Teaching Bayesian Statistics.* As published in Benchmarks RSS Matters, September 2011, http://web3.unt.edu/benchmarks/issues/2011/09/rss-matters. http://www.unt.edu/rss/class/Jon/Benchmarks/BayesBooks_L_JDS_Sep2011.pdf.

Here are some introductions to Bayesian methods, in alphabetical order:

BM Bolker. *Ecological Models and Data in R.* Starkweather considers this "one of the most down-to-Earth introductions to Bayesianism."

Peter D Hoff. *A First Course in Bayesian Statistical Methods.* For those with some familiarity with probability.

John K Kruschke. *Doing Bayesian Data Analysis.* For the advanced undergraduate in social or biological sciences who—in Kruschke's words—grew up on Lake Wobegon. Or for first-year graduate students who didn't grow up there. Prerequisite: algebra and basic calculus.

Peter M Lee. *Bayesian Statistics: An Introduction*. For those with some upper-level undergraduate math. Uses set theory. Few real-world data examples.

WA Link and RJ Baker. *Bayesian Inference with Ecological Applications* assumes a limited knowledge of statistics.

SM Lynch. *Introduction to Applied Bayesian Statistics and Estimation for Social Scientists*. Starkweather again: "Very accessible to the newcomer to Bayesianism."

GG Woodworth. *Biostatistics: A Bayesian Introduction*. For bio-oriented researchers who do not necessarily have much prior knowledge of statistics.

Conflict of interest must be noted here: Starkweather considers *The Theory That Would Not Die* "essential for anyone interested in Bayesian methods."

Have fun!

index